ALLE ZEIT WACH 1842

T. G. Cooper

The Epididymis, Sperm Maturation and Fertilisation

With 8 Figures and 21 Tables

Springer-Verlag
Berlin Heidelberg New York
London Paris Tokyo

Dr. Trevor G. Cooper
Max-Planck-Gesellschaft zur Förderung der Wissenschaften e.V.
Klinische Forschungsgruppe für Reproduktionsmedizin
an der Frauenklinik der Universität Münster
Steinfurter Straße 107, 4400 Münster, FRG

The cover photograph is of a decapsulated, partially unravelled length of tubule from the rat epididymis from which spermatozoa have been removed through the cannulae inserted into each end. For the photograph it has been filled with air, but when continuously perfused with physiological solutions and immersed in an organ bath it can provide information on the permeability of the epididymal epithelium.

ISBN-13:978-3-642-71473-3 e-ISBN-13:978-3-642-71471-9
DOI: 10.1007/978-3-642-71471-9

Library of Congress Cataloging-in-Publication Data. Cooper, T. G. (Trevor G.), 1949– . The epididymis, sperm maturation, and fertilisation. Includes bibliographies and index. 1. Epididymis. 2. Spermatozoa. 3. Fertilization. I. Title. [DNLM: 1. Epididymis. 2. Fertilization. 3. Sperm Maturation. WJ 800 C778e] QP255.C66 1986 612'.61 86-20354
ISBN-13:978-3-642-71473-3 (U.S.)

Softcover reprint of the hardcover 1st edition 1986

2127/3130-543210

This work is dedicated to Marie-Claire Orgebin-Crist, who first showed me the way to epididymal enlightenment, and my Chinese scientist colleagues to whom I hope I have done the same.

Preface

The epididymis has great significance in the reproductive biology of the male and it is gaining recognition as an organ worthy of study in its own right – where the secretory and absorptive activities of one tissue (the epithelium) profoundly modify the function of another (the spermatozoon). Apart from cases of epididymal agenesis or physical blockage, however, it is not yet known to what extent malfunctions of the epididymis contribute to "unexplained" male infertility, but its importance as a target for antifertility agents in the male is now widely appreciated.

This monograph evolved from two lectures on epididymal function given at the National Research Institute for Family Planning, Beijing and the Sichuan Provincial Family Planning Research Institute, Chengdu, China in 1983. In order to stress for this audience the central importance of the epididymis in fertilisation, and so highlight its potential for attack by antifertility agents, the talks attempted to put the epididymis in reproductive perspective by stressing the functional development of the sperm cells during their sojourn in the epididymis, rather than merely listing the changes observed in them. This required a discussion of the mechanisms of fertilisation, necessarily obtained from studies on mature spermatozoa, and the contrast between the success of these sperm and the limitations suffered by immature sperm in performing certain steps in fertilisation was used to reveal which functions of the developing sperm were influenced by the epididymis. While more is becoming known about epididymal secretions, especially specific proteins, there is as yet no full appreciation of the significance of these secretory products for sperm function, although in the last five years headway has been made by the recognition of specific secretory products of the epididymis that initiate and modulate sperm motility and influence sperm-egg binding.

Even now the most frequently cited review of epididymal function is one published in 1975 (Hamilton and Greep, 1975). Since then, reviews of epididymal function have ranged from short synopses (Hacker, 1984) or slightly longer works (Turner, 1979; Hinton, 1980; Orgebin-Crist, 1981, 1985; Orgebin-Crist and Olson, 1984) to those devoted to only one aspect of maturation (Dacheux and Paquignon, 1981). Longer tracts (Orgebin-Crist and Hochereau-de Reviers, 1980; Orgebin-Crist et al., 1981) have appeared in books nominally about testicular function and the one book devoted to epididymal function (Bollack and Clavert, 1981) was not comprehensive in its coverage.

Because of this, the wider aspects of the subject covered in these lectures required access to numerous journals, many of which were unavailable to the Chinese Institutes.

This work then, is an attempt to bring together the information relevant to the approach cited above, including a tabulation of the composition of luminal fluid in various regions of the epididymis, the "milieu exterior" of utmost importance to the maturing spermatozoa. The topics covered are necessarily many and overviews of the mechanisms under study at present have been included at the expense of detailed criticism of methodology. Despite these shortcomings, it is hoped that the information contained therein will prove useful as a backdrop to, and provide a framework from which to see, future studies on the epididymis.

The impetus to originate these ideas was the opportunity to present lectures in China, and for this I am grateful to Dr. F. T. G. WEBB of the WHO Special Programme of Research in Human Reproduction, Geneva and Dr. CHANG CHAI-FEN, at that time the Director of the National Research Institute for Family Planning, Beijing. I was encouraged by my Chinese colleagues there and Professor E. NIESCHLAG here to expand these thoughts and I thank them and Springer Verlag for providing the opportunity to publish them. Material was gathered during the tenure of a Wellcome Trust Senior Fellowship and a Project Grant from the Medical Research Council at Bristol University, England, where the resources of the University library proved invaluable. I am indebted to Dr. M. REA for introducing me so painlessly to the benefits of the computer, without which the work would not have been completed. Finally, I am most grateful to my wife for her gentle encouragement and patient toleration at times of exasperation experienced during the compilation of the work.

T. G. COOPER

References

Bollack C and Clavert A (1981) (Eds) Epididymis and Fertility: Biology and Pathology. Progress in Reproductive Biology, Volume 8

Dacheux JL and Paquignon M (1981) Physiologie épididymaire et maturation des spermatozoides. Actualités gynécologiques Ser 12, Masson Press, Paris. pp 201–213

Hacker PJ (1984) Un aperçú sur les connaissances actuelles de la physiologie de l'épididyme. J Urol (Paris) 90:57–62

Hamilton DW and Greep RO (1975) (Eds) Male Reproductive System, Volume 5, Section VII, Endocrinology, Handbook of Physiology. American Physiological Society, Washington DC

Hinton BT (1980) The epididymal microenvironment. A site of attack for a male contraceptive? Invest Urol 18:1–10

Orgebin-Crist MC (1981) Epididymal physiology and sperm maturation. In: Epididymis and Fertility: Biology and Pathology. Eds C Bollack and A Clavert, S Karger, Basel. pp 80–89

Orgebin-Crist MC (1985) Epididymal maturation of spermatozoa. In: Recent Progress in Cellular Endocrinology of the Testis. Eds JM Saez, MG Forest, A Dazord and J Bertrand, INSERM, Paris. pp 85–90

Orgebin-Crist MC and Hochereau-de Reviers DE (1980) Sperm formation and maturation – role of testicular and epididymal somatic cells. In: 9th Int Congr Anim Reprod and AI, Madrid, Spain. Editorial Garsi, Madrid. pp 59–82

Orgebin-Crist MC and Olson GE (1984) Epididymal Sperm Maturation. In: The Male in Farm Animal Reproduction. Ed M Courot, Martinus-Nijhoff, Boston. pp 80–102

Orgebin-Crist MC, Olson GE and Danzo BJ (1981) Factors influencing maturation of spermatozoa in the epididymis. In: Intragonadal Regulators of Reproduction. Eds P Franchimont and CP Channing. Academic Press, New York. pp 393–417

Turner TT (1979) On the epididymis and its function. Invest Urol 16:311–321

Contents

Part I. Maturation of Spermatozoa in the Epididymis

1. Acquisition in the Epididymis of Sperm Fertilising Ability

During their time in the epididymis, spermatozoa develop the ability to fertilise eggs both in-vivo and in-vitro (see Tables 1 and 2). Populations of spermatozoa capable of penetrating more than 50% of eggs are found in the distal corpus epididymidis of rabbits, rams and possibly boars, and in the proximal cauda epididymidis of rats, hamsters and possibly guinea pigs. Once fertilising ability is aquired there is no further significant increase distally in the epididymis (Overstreet and Bedford, 1974, 1976) and a decrease in the vas deferens (Lubicz-Nawrocki and Chang, 1978).

Comparisons between species is made difficult by differences in the gross morphology of the epididymis and generalisation to humans is complicated by the different form of the primate epididymis (see Section III.A.2). Sperm taken from the human caput epididymidis may fertilise eggs in-vivo (Young, 1951; Hanley, 1956) but the success in achieving pregnancies varies (Hanley, 1955). Evidence for sperm maturation in the human epididymis has come from the results of surgery in cases of infertility due to epididymal blockage. Sperm appear in human ejaculates if the lumen of the vas deferens is anastomosed to that of the epididymis and the more distally the connection is made, the greater is the chance of pregnancies resulting (Table 1). Using these criteria, sperm fertilising capacity would appear to be acquired in rather more proximal regions of the human epididymis than in other species; however, the results of surgery are not strictly comparable to experimental studies in animals because spermatozoa trapped in an occluded duct may have come into contact with epididymal secretions for a considerable time before surgery, and these may influence sperm fertilising ability (see Section I.2). Sperm taken from the corpus epididymidis of fertile men fail to affect normal fertilisation in-vitro (Mahadevan and Trounson, 1985).

a. The Nature of the Infertility of Immature Spermatozoa

Immature spermatozoa either fail to fertilise eggs at all, or as a population demonstrate a reduced ability to do so compared with mature cells. Fertilisation by "young" epididymal spermatozoa is associated with smaller litters in many species and demonstrated areas of incompetence include polyspermy in rabbits but not hamsters; delayed fertilisation in rabbits, sheep and man and embryonic mortality in rabbits, sheep and pigs (Orgebin-Crist, 1967, 1968; Fournier-Delpech et al., 1977, 1979, 1981; Horan and Bedford, 1972; Holtz and Smidt, 1976; Mahadevan and Trounson, 1985). Nevertheless, it is not clear whether such observations reflect delayed fertilisation brought about by differences in the transport of spermatozoa to the eggs, the time-course of pre-fertilisation changes (see Part II) or inadequate activation of the eggs (see Orgebin-Crist and Jahad, 1977).

b. The Time Required for Sperm Maturation

Recent compilations of the times taken for immature testicular spermatozoa to reach the region of the epididymis where fertilising ability is conferred (Amann et al., 1976; Courot, 1981; Orgebin-Crist and Olson, 1984) show that maturation takes from 2.5 to 5 days depending on species. In the rabbit spermatozoa pass through the corpus epididymidis in one day (Orgebin-Crist, 1965) so fertility is conferred on the majority of the spermatozoa from the upper corpus in less than 24 h.

2. Involvement of the Epididymis in the Development of Sperm Fertilising Ability

Mature spermatozoa are older than immature cells and this raises the important question whether sperm maturation depends on the age of the sperm cells or a specific epididymal contribution. Despite early claims by Young and colleagues (Young 1929a,b, 1931; Young and Simeone, 1939; Simeone and Young, 1931; Simeone, 1933; Glover, 1969) that spermatozoa mature uninfluenced by epididymal secretions, the weight of evidence now suggests that the epididymis has a central role in influencing its contained spermatozoa (see Orgebin-Crist et al., 1975, 1976).

a. Ageing Spermatozoa Artificially

Allowing spermatozoa to age in-vivo by ligating the excurrent ducts does not promote the fertilising capacity of rabbit spermatozoa trapped within the testis (Cooper and Orgebin-Crist, 1975, 1977). Likewise, the fertilising capacity of spermatozoa from the initial segments or proximal caput regions of the epididymis is not induced by ageing them in these regions, whereas fertilising ability can be acquired by spermatozoa from the proximal corpus epididymidis (Bedford, 1967; Orgebin-Crist, 1967b).

A report of pregnancies resulting from insemination of macerated human testicular tissue (Adler and Makris, 1951) is inconclusive since the patients suffered from epididymitis, and the spermatozoa may have come into contact with accumulated epididymal secretions, which are now considered important for maturation of spermatozoa.

Ageing of spermatozoa alone is insufficient for fertilising ability of immature sperm to be acquired, but it should be remembered that the composition of trapped testicular fluid (Cooper et al., 1976; Guerrero et al., 1976; Free and Jaffe, 1979) and the structure of the epididymis (Bedford, 1967; Orgebin-Crist, 1969; Glover, 1969) are altered by ligation. Where in-vitro ageing of testicular spermatozoa failed to promote fertilising ability (Setchell et al., 1969; Voglmayr et al., 1978) conditions of incubation may have been inadequate for maintaining the fertilising ability of spermatozoa. Nevertheless, where the fertilising capacity of mature spermatozoa can be maintained for long periods in-vitro (undiluted within epididymal tubules maintained in culture) there is no acquisition of fertilising capacity by spermatozoa in the caput epididymidis (Orgebin-Crist and Hoffmann, 1976). Where an improvement in the motility of caput spermatozoa was observed after ageing within epididymal tubules maintained in-vitro (Klinefelter and Hamilton, 1984), an epididymal secretion product known to initiate motility of immature sperm was present in the medium.

b. Steroid Support for Sperm Maturation

The fertilising capacity of spermatozoa within the upper corpus epididymidis can be promoted by ligating the epididymis at the mid-corpus regions, but this ability is lost on castration (Orgebin-Crist, 1973) and hypophysectomy (Orgebin-Crist and Davies, 1974) and maintained in these animals by androgens (see Orgebin-Crist et al., 1972, 1973, 1975; Orgebin-Crist and Hoffmann, 1976) and pregnenolone (Vreeburg et al., 1976; Anthony et al., 1984). In-vitro experiments have confirmed that androgens are required for the maturation of rabbit spermatozoa in proximal corpus tubules (see Orgebin-Crist and Tichenor, 1972, 1973; Orgebin-Crist and Hoffmann, 1976; Orgebin-Crist et al., 1976) and sperm maturation is also androgen-dependent in hamsters (Lubicz-Nawrocki and Glover, 1974; Lubicz-Nawrocki, 1976) and rats (Nelson, 1941; Dyson and Orgebin-Crist, 1973).

c. Luminal Steroids and Sperm Maturation

The support of sperm maturation by systemic administration of steroids or extra-tubular androgens in-vitro suggests that luminal steroids are unimportant for the process. Indeed, ligating the efferent ducts, which should limit the access of luminal steroids from the testis (see Section III.B.2.a) does not affect sperm maturation in rats (Dyson and Orgebin-Crist, 1973) or rabbits (Orgebin-Crist and Davies, 1974); however, sperm in the caut epididymidis at the time of the ligation would already have been exposed to luminal androgens. Frankel and Eik-Nes (1970) argued that luminal androgens were important, postulating that inherently fertile spermatozoa migrate away from an inhibitory influence of steroids as they mature. Although steroid concentrations are high in testicular fluid (Waites, 1977) and in luminal fluid (Table 11) this theory has not received experimental support: testicular spermatozoa do not fertilise eggs when removed from testicular fluid, and this fluid is not inhibitory to the expression of fertilising capacity of mature cells (Cooper and Orgebin-Crist, 1975). Recent experiments, however, have suggested that luminal androgens may support sperm maturation when circulating levels of androgens are low (Anthony et al., 1984). There is no evidence that androgens directly promote the fertilising ability of testicular spermatozoa in-vitro (Voglmayr and Gandhi, 1978).

d. Extra-tubular Steroids and Sperm Maturation

The presence of androgen-receptors along the length of the epididymis (see Section III.B.3.b) supports the view that steroids act only indirectly on the maturation of spermatozoa, mediated by the epididymal epithelium. Administration of anti-androgens in-vivo does not prevent sperm maturation in intact rabbits (Cummins and Orgebin-Crist, 1974) but does block androgen-supported maturation in castrated rats (Dyson and Orgebin-Crist, 1973) and in-vitro (Orgebin-Crist et al., 1976). In the hamster the androgenic support of sperm maturation differs from that of the maintenance of sperm fertilising capacity, and is not prevented by administration of progesterone (Lubicz-Nawrocki, 1973b), thought to block the conversion of testosterone to 5a-dihydrotestosterone (see Section III.B.5.d).

Synthesis of RNA and protein appear to mediate the steroid action, because blocking these processes prevents the androgen-dependent acquisition of sperm fertilising ability (Orgebin-Crist and Jahad, 1978). Androgens promote incorporation of precursors into RNA and proteins of epididymal tissue (see Section III.B.5.a).

3. The Nature of the Androgen-Dependent Epididymal Secretions Involved

An active "principal" from sperm- and steroid-free epididymal tubules can influence the fertilising capacity of spermatozoa in adjacent co-cultured epididymal tubules (Orgebin-Crist and Jahad, 1979). This factor must be of low molecular weight if entry into the lumen *via* the severed ends of the tubules were minimal and passage between the epithelial cells is blocked by tight junctions (see Section III.A.8). Indeed, under conditions that promote the fertilising capacity of upper corpus spermatozoa in-vitro there is an accumulation of carnitine by these cells (Casillas and Chaiyungpan, 1982). However, there is evidence that proteins can cross the epididymal epithelium (see Section III.A.8) and thus epididymal proteins cannot be totally excluded as mediators of the effect seen above.

4. Response of Immature Spermatozoa to Epididymal Secretions

If sperm maturation were simply a matter of immature spermatozoa gaining access to secretion products of the epididymis found more distally, it should be feasible to mimic maturation in the epididymis in-vitro, but attempts have largely proved negative (Table 3).

5. Scope of this Review

This review summarises what is known about the infertility of immature spermatozoa in the light of current knowledge of fertilisation, which is based on work with mature sperm cells. Much of this has been done in-vitro and this has enabled the effect of changes in the environment of the fertilising spermatozoa to be monitored; for example, extracellular calcium and potassium ions and intracellular cyclic AMP (cAMP) can promote fertilisation in-vitro (see Yanagimachi, 1981; Fraser, 1984).

In succeeding parts of this review the effects of these agents will be seen to relate to particular processes of fertilisation, namely sperm motility (Section II.A), capacitation (Section II.B) and its attendant changes (Sections II.C,D,E), sperm-egg fusion (Section II.F) and post-fusion events (Section II.G). Figure 1 depicts the main sequence of events of the fertilising spermatozoon (see Hinrichsen et al., 1984): understanding the underlying mechanisms of these events is important, for failure of sperm to undergo these events may be the cause of human infertility; conversely, blocking any step in the sequence should prevent fertilisation occurring. In Part II of the review each step of the fertilisation process is discussed in relation to the ability of immature sperm to perform it and in this way the changes that occur on sperm maturation are revealed and the role of the epididymis in influencing them can be inferred. The secretions of the epididymis and their interaction with spermatozoa are subsequently discussed in Part III with references being made to epididymal function in the human. Interfering with epididymal function is a promising area for the development of antifertility agents for men: spermatozoa would continue to be ejaculated and there would be none of the disadvantages associated with interference with testicular function (see Part IV).

6. Summary

The epididymis provides both the secretions and the time that are necessary for immature spermatozoa leaving the testis to develop into mature cells capable of fertilising eggs. The resorptive properties of the epididymis may also modify the sperm cells so that they can interact with the secretions.

7. References

Adler L and Makris A (1951) Successful artificial insemination with macerated testicular tissue. Fert Steril 2:459–461

Amann RP, Johnson L, Thompson DL and Pickett BW (1976) Daily spermatozoal production, epididymal spermatozoal reserves and transit times of spermatozoa through the epididymis of the rhesusmonkey. Biol Reprod 15:586–592

Anthony CT, Danzo BJ and Orgebin-Crist MC (1984a) Investigations on the relationship between sperm fertilizing ability and androgen-binding protein in the restricted rat. Endocrinology 114:1413–1418

Anthony CT, Danzo BJ and Orgebin-Crist MC (1984b) Investigations of the relationship between sperm fertilizing ability and androgen-binding protein in the hypophysectomised, pregnenolnone injected rat. Endocrinology 114:1419–1425

Bedford JM (1967) Effects of duct ligation on the fertilizing ability of spermatozoa from different regions of the rabbit epididymis. J exp Zool 166:271–282

Casillas ER and Chaiyungpan S (1982) Carnitine content of rabbit epididymal spermatozoa in organ culture. J Reprod Fert 65:247–251

Cooper TG and Orgebin-Crist MC (1975) The effect of epididymal and testicular fluids on the fertilising capacity of testicular and epididymal spermatozoa. Andrologia 7:85–93

Cooper TG and Orgebin-Crist MC (1977) Effect of aging on the fertilizing capacity of testicular spermatozoa from the rabbit. Biol Reprod 16:258–266

Cooper TG, Danzo BJ, DiPietro DL, McKenna TJ and Orgebin-Crist MC (1976) Some characteristics of rete testis fluid from rabbits. Andrologia 8:87–94

Courot M (1981) Transport and maturation of spermatozoa in the epididymis of mammals. In: Epididymis and Fertility: Biology and Pathology. Eds. C Bollack and A Clavert, Karger, Munich. pp 67–79

Cummins JM and Orgebin-Crist MC (1974) Effects of the anti-androgen SK and F 7690 on the fertility of epididymal spermatozoa in the rabbit. Biol Reprod 11:56–63

Dyson ALMB and Orgebin-Crist MC (1973) Effect of hypophysectomy, castration and androgen replacement upon the fertilizing ability of rat epididymal spermatozoa. Endocrinology 93:391–402

Fournier-Delpech S, Colas G, Courot M and Ortavant R (1977) Observations on the motility and fertilizing ability of ram epididymal spermatozoa. Ann biol anim Biochem Biophys 17:987–990

Fournier-Delpech S, Colas G, Courot M, Ortavant R and Brice G (1979) Epididymal sperm maturation in the ram: motility, fertilizing ability and embryonic survival after uterine insemination in the ewe. Ann biol anim Biochem Biophys 19:597–605

Fournier-Delpech S, Colas G and Courot M (1981) Observations sur les premiers clivages des oeufs intratubulaires de brebis apres fecondation des spermatozoides epididymaires ou ejacules. Comptes Rendus Acad Sci (Paris) 292:515–517

Frankel AI and Eik-Nes KB (1970) Testosterone and dehydroepiandrosterone in the epididymis of the rabbbit. J Reprod Fert 23:441–445

Fraser LR (1984) Mechanisms controlling mammalian fertilization. In: Oxford Reviews of Reproductive Biology, Vol. 6. Ed JR Clarke, Clarendon Press Oxford UK. pp 174–225

Free MJ and Jaffe RA (1979) Collection of rete testis fluid from rats without previous efferent duct ligation. Biol Reprod 20:269–278

Glover TD (1969) Some aspects of function in the epididymis. Experimental occlusion of the epididymis in the rabbit. Int J Fert 14:2145–221

Guerrero R, Ritzen EM, Purvis K, Hansson V and French FS (1975) Concentration of steroid hormones and androgen binding protein (ABP) in rabbit efferent duct fluid. In: Hormonal Regulation of Spermatogenesis. Eds. FS French, EM Ritzen and V Hansson, Plenum Press, New York. pp 213–221

Hanley HG (1955) The Surgery of Male Infertility. Ann Rep Royal Coll Surg, England. pp. 159–183

Hanley HG (1956) Pregnancy following artificial insemination from epididymal cyst. Proc Soc Study Fertil 8:20–21

Hinrichsen AC, Hinrichsen MJ and Schill WB (1984) Molecular events leading to fertilization. A review. Andrologia 16:321–341

Holtz W and Smidt D (1976) The fertilizing capacity of epididymal spermatozoa in the pig. J Reprod Fert 46:227–229

Horan AH and Bedford JM (1972) Development of the fertilizing ability of spermatozoa in the epididymis of the Syrian hamster. J Reprod Fert 30:417–423

Klinefelter GR and Hamilton DW (1984) Organ culture of rat epididymal tubules in a perifusion chamber. J Androl 5:243–258

Lubicz-Nawrocki CM (1973) Anomalous effects of progesterone on the maturation and survival of spermatozoa in the epididymis of the golden hamster. J Endocr 58:199–205

Lubicz-Nawrocki CM (1976) The effect of metabolites of testosterone on the development of fertilizing ability by spermatozoa in the epididymis of castrated hamsters. J exp Zool 197:89–96

Lubicz-Nawrocki CM and Chang MC (1978) The influence of the testis on the fertilizing life of spermatozoa in the ligated vas deferens of the golden hamster. J Reprod Fert 53:147–153

Lubicz-Nawrocki CM and Glover TD (1974) The effect of castration and testosterone replacement on sperm maturation in the hamster. J Reprod Fert 37:251–235

Mahadevan MM and Trounson AO (1985) Removal of the cumulus oophorus from the human oocyte for in vitro fertilization. Fert Steril 43:263–267

Nelson WO (1940) Maintenance of spermatogensis in hypophysectomised rats. Am J Physiol 129:430

Orgebin-Crist MC (1965) The passage of spermatozoa labelled with thymidine-3H through the ductus epididymidis of the rabbit. J Reprod Fert 10:241–251
Orgebin-Crist MC (1967a) Sperm maturation in rabbit epididymis. Nature (Lond.) 216:816–818
Orgebin-Crist MC (1967b) Maturation of spermatozoa in the rabbit epididymis: fertilizing ability and embryonic mortality in does inseminated with epididymal spermatozoa. Ann biol Anim Biochem Biophys 7:373–389
Orgebin-Crist MC (1968) Maturation of spermatozoa in the rabbit epididymis: Delayed fertilization in does inseminated with epididymal spermatozoa. J Reprod Fert 16:29–33
Orgebin-Crist MC (1969) Studies on the function of the epididymis. Biol Reprod Suppl 1:155–175
Orgebin-Crist MC (1973) Maturation of spermatozoa in the rabbit epididymis: effect of castration and testosterone replacement. J expl Zool 185:301–310
Orgebin-Crist MC and Davies J (1974) Functional and morphological effects of hypophsectomy and androgen replacement in the rabbit epididymis. Cell Tissue Res 148:188–201
Orgebin-Crist MC and Hoffmann LH (1976) The effect of testosterone and testosterone metabolites on epididymal function. In: Regulatory mechanisms of male reproductive physiology. Eds. CH Spilman, TJ Lobl and KT Korton, Exerpta Medica, Amsterdam, Elsevier. pp 141–142
Orgebin-Crist MC and Jahad N (1977) Delayed cleavage of rabbit ova after fertilization by young epididymal spermatozoa. Biol Reprod 16:358–363
Orgebin-Crist MC and Jahad N (1978) The maturation of rabbit epididymal spermatozoa in organ culture: inhibition by antiandrogens and inhibitors of ribonucleic acid and protein synthesis. Endocrinology 103:46–53
Orgebin-Crist MC and Jahad N (1979) The maturation of rabbit spermatozoa in organ culture: stimulation by epididymal cytoplasmic extracts. Biol Reprod 21:511–515
Orgebin-Crist MC and Olson GE (1984) Epididymal Sperm Maturation. In: The Male in Farm Animal Reproduction. Ed M Courot, Martinus-Nijhoff, Boston, US. pp 80–102
Orgebin-Crist MC and Tichenor P (1972) Technique for studying sperm maturation in vitro. Nature (Lond.) 239:227–228
Orgebin-Crist MC and Tichenor P (1973) Effect of testosterone on sperm maturation in vitro. Nature (Lond.) 245:328–329
Orgebin-Crist MC, Dyson ALMB and Davies J (1972) Hormonal regulation of epididymal sperm maturation. Int Congress Series No 273, Exerpta Medica pp 18–24
Orgebin-Crist MC, Davies J and Tichenor P (1973) Maturation of spermatozoa in the rabbit epididymis: effect of hypophsectomy and castration. In: Regulation of mammalian reproduction, Eds and CC Thomas, Springfield, Illinois. pp 189–203
Orgebin-Crist MC, Danzo BJ and Davies J (1975) Endocrine control of the development and maintenance of sperm fertilizing ability in the epididymis. In: Handbook of Physiology, Section VII Endocrinology, vol V, Male Reproductive System, Eds. RO Greep and DW Hamilton, Am Physiol Soc, Washington DC. pp 319–338.
Orgebin-Crist MC, Danzo BJ and Cooper TG (1976a) Re-examination of the dependence of epididymal sperm viability on the epididymal environment. J Reprod Fert Suppl 24:115–128
Orgebin-Crist MC, Jahad P and Hoffmann LH (1976b) The effects of testosterone, 5α-dihydrotestosterone, 3α-androstanediol, and 3β-androstenediol on the maturation of rabbit epididymal spermatozoa in organ culture. Cell Tiss Res. 167:515–525
Overstreet JW and Bedford JM (1974) Transport, capacitation and fertilizing ability of epididymal spermatozoa. J exp Zool 189:203–214
Overstreet JW and Bedford JM (1976) Embryonic mortality in the rabbit is not increased after fertilization by young epididymal spermatozoa. Biol Reprod 15:54–57
Setchell BP, Scott TW, Voglmayr JK and Waites GMH (1969) Characteristics of testicular spermatozoa and the fluids which transport them into the epididymis. Biol Reprod Suppl 1:40–66
Simeone FA (1933) A neuromuscular mechanism in the ductus epididymis and its inhibition by sympathetic denervation. Am J Physiol 103:582–591
Simeone FA and Young WC (1931) A study of the function of the epididymis. IV. The fate of non-ejaculated spermatozoa in the genital tract of the male guinea pig. J expl Biol 8:163–175

Voglmayr JK and Gandhi JS (1978) Survival of ram testicular spermatozoa in vitro: effects of glucose, glucose metabolites, rete testis fluid proteins, selected androgens and phospholipids. Theriogeneology 9:463–478

Voglmayr JK, White IG and Parks RP (1978) The fertilizing capacity of ram testicualr spermatozoa, freshly collected and after storage in cauda epididymal fluid. Theriogenology 10:313–321

Vreeburg JTM, Bielske and Ooms M (1976) Maturation and survival of spermatozoa in the epididymis of pregnenolone treated hypophysectomised rats. Endocrinology 99:824–830

Waites GMH (1977) Fluid secretion. In: The Testis, vol 4. Eds AD Johnson and WR Gomes, Acad. Press, New York. pp 91–123

Yanagimachi R (1981) Mechanisms of Fertilization in Mammals. In: Fertilization and Embryonic Development In Vitro. Eds L Mastroianni and JD Biggers, Plenum Press London. pp 81–182

Young D (1951) Surgical problems of the vas deferens. Proc Soc Study Fert 3:40–45

Young WC (1929a) A study of the function of the epididymis. I. Is the attainment of full spermatozoon maturity attributable to some specific action of the epididymal secretion? J Morphol Physiol 47:479–495

Young WC (1929b) A study of the function of the epididymis. II. The importance of an ageing process in sperm for the length of the period during which fertilizing capacity is retained by sperm isolated in the epididymis of the guinea pig. J Morphol Physiol 48:475–491

Young WC (1931) A study of the function of the epididymis. III. Functional changes undergone by spermatozoa during their passage through the epididymis and vas deferens in the guinea pig. J exp Biol 8:151–162

Young WC and Simeone FA (1930) Development and fate of spermatozoa in the epididymis and vas deferens in the guinea pig. Proc Soc exp Biol Med 27:838–841

Part II. Fertilisation

A. Sperm Motility

1. The Need for Flagellar Activity of Sperm for Them to Reach the Egg

The rapidity of sperm ascent to the oviducts in domestic species, where the distances covered are large, and the velocity of sperm determined in-vitro is known to be too low to account for the time taken for the ascent, sustains the belief that the female tract promotes bulk transfer of sperm cells. Movements of the female tract do occur, but whether these alone are sufficient for sperm to reach the egg is still debated (see Blandau, 1969; Hunter et al., 1983). Many experiments have been poorly designed (see Overstreet and Katz, 1977); e.g. the appearance of spermatozoa in the oviducts after insemination of immotile cells is poor evidence for their passive transport, especially where motility could be reinitiated in the female tract (*e.g.* Howe and Black, 1963). Conversely, where immotile spermatozoa fail to pass the cervix (Noyes et al., 1958; Baker and Degen, 1972), the treatments rendering the cells immotile (temperature shock or detergents) were not considered. The damage may have altered their surfaces and their susceptibility to removal by invading leucocytes (Bedford, 1965). In addition to the use of damaged sperm to assess normal sperm transport, most studies have involved recovery of sperm *post-mortem*, with all the complications arising from relaxation of normally constricting muscle tone.

Despite these reservations it seems likely that the two anatomical barriers in the female tract, the cervix and the uterotubal junction, do serve as obstacles to sperm transport, although their relative significance may vary between species. In those with copious cervical mucus, sperm flagellar activity may be essential; for although the "phalanges" of seminal plasma that form upon its contact with cervical mucus in-vitro reflect the biophysical properties of the fluids (Perloff and Steinberger, 1963; Moghissi et al., 1964), the caudal movement of the mucus, carrying with it seminal plasma, dictates that the fertilising spermatozoon escape by its own motility through the mucus. Certainly, infertility in humans may result from the inability of antibody-coated spermatozoa to move out of semen (Kremer and Jager, 1976; Kremer et al., 1978).

In rodents the uterus is fluid-filled at oestrus, the cervix is contracted and passage of semen through the cervix is achieved at ejaculation (Blandau, 1945) when a cervical (vaginal) plug is deposited. In these species passage of spermatozoa through the uterotubal junction may be limiting (see Gaddum-Rosse, 1981). Although motile spermatozoa may be recovered from uterine fluid during their sojourn there, this may be an expression of their viability, rather than evidence that their own motility is needed for them to pass through the uterine fluids, since a thorough mixing of uterine contents follows mating in rodents (Rossman, 1937; Warren, 1938; Yamanaka and Soderwall, 1960).

By 1–2 h of mating rat uterine fluid has gelated and sperm cells in it are coated with seminal proteins (Joshi et al., 1972), but whether these proteins promote, preserve or compromise the motility of the cells is unclear. Motile spermatozoa recovered from the vaginal plug of guinea pigs are non-fertilising after intravaginal insemination (Martan and Shepherd, 1976), although vaginal insemination of fresh epididymal spermatozoa is successful in this species (Blandau and Young, 1939). The timing of gelation relative to sperm movement to the oviducts may be critical. Intermittant opening at a fixed time *post-coitum* in mice has been postulated (Zamboni, 1972), but no such opening was observed when the rat uterotubal junction of rats was observed in-vitro (Gaddum-Rosse, 1981).

In the latter study the uterus and oviduct were examined in-vitro under conditions where viability of the tissue was reflected in normal muscular contractions. Spermatozoa were inseminated at normal mating or by insemination into the uterus in-vitro, and the uterotubal junction was examined from the oviductal side for the appearance of sperm cells by high-power magnification. Their arrival was not related to contractions of the uterus, and only motile sperm cells were observed to swim through the junctional folds and continue moving with progressive movement. Since immobilised spermatozoa never passed beyond the junction these results argue convincingly that sperm motility in this species is needed for the male gamete to negotiate the junction.

In most species sperm transport in the female has not been examined in critical detail, and one recent review (Overstreet and Katz, 1977) placed less emphasis than hitherto on the role of the cervix and uterotubal junction as barriers to transport in the rabbit. When detailed observations on sperm movement have been made in this species (Overstreet and Cooper, 1978; Cooper et al., 1979) two waves of transport were observed, with the ascent of vanguard spermatozoa being rapid but of no relevance to fertilisation. This rapid ascent has been shown to occur only when sperm are inseminated in seminal plasma (Overstreet and Tom, 1982) and this observation is relevant to studies on epididymal spermatozoa, which are not usually inseminated in seminal plasma.

It has been known for some time that the ascent of sperm cells to the site of fertilisation in the oviducts is faster around the time of ovulation in many species. It is of interest to know if this accelerated transport is due to the sperm cell's own motility, but because of the association with the characteristic motility of capacitated spermatozoa (Hyperactivated motility) the discussion is pursued further in Section II.E.9.a.

2. The Ability of Immature Spermatozoa to Reach the Site of Fertilisation

Testicular spermatozoa from rams (Setchell et al., 1969; Voglmayr et al., 1978) and rabbits (Lambiase and Amann, 1973; Cooper and Orgebin-Crist, 1975) can pass the uterotubal junction, although they cannot fertilise eggs, but bovine testicular spermatozoa do not enter the oviducts from the uterus (Amann and

Griel, 1974). Rabbit spermatozoa from the caput and lower corpus epididymidis can be found in flushings from the oviducts, although there are fewer than more mature cells, after insemination of the same sperm numbers into the uterus (Orgebin-Crist, 1967a, 1968). While this suggests that immature sperm suffer from reduced transport in the female tract, no differences were found in transport across the cervix or the uterotubal junction of mixed populations of marked distal corpus and ejaculated spermatozoa (Overstreet and Bedford, 1974, 1976).

After uterine insemination, spermatozoa from the rat caput epididymidis are not observed in the oviduct (Blandau and Rumery, 1964) and they do not pass the uterotubal junction in-vitro (Gaddum-Rosse, 1981). Live spermatozoa from the hamster caput and corpus epididymidis can enter the oviducts in this species (Cummins, 1976), as can sperm cells from the lower corpus epididymidis of mice (Pavlok, 1974).

3. Development in the Epididymis of Sperm Motility In-Vitro

a. Terminology

Assessing sperm motility is fraught with difficulty because descriptions of motility are not standardised and at least five parameters can be used as a basis for assessment (see Morton and Sagadraca, 1981); subjective estimates of "partial, weak" or "complete, strong" flagellation, for example, will not be reproducible between laboratories. In addition, the use of terminology is inconsistent. Some workers refer to the burst of motility on dilution of previously immobile, but mature, spermatozoa as "Activation" (Morton et al., 1974; Cascieri et al., 1976), but others used this term to describe the vigorous motion of capacitated sperm cells (now termed "Hyperactivation"; see Section III.E). Hyperactivation was previously a term used by Tabor and Rosenthal (1956) to describe a very different, high frequency, low amplitude but non-progressive motion of mature spermatozoa stimulated by spermine in-vitro. Another source of confusion is the application of the term "pre-motile" to describe mature, but immobile, spermatozoa within the the cauda epididymidis (Morton et al., 1979), a term that is also used to describe immature spermatozoa in the caput epididymidis which have not yet developed their ability to move in-vitro (Hinton et al., 1979). The term "circling" also has different meanings to different workers (see Sections II.A.3.b and III.E.1). References made in this review are solely to objective measurements of motility.

b. Maturation of Motility

Many investigators have commented upon the increasing "vigour" of motility of spermatozoa taken from more distal regions of the epididymis (see Bedford, 1975), but few objective measurements have been made (Table 4). These confirm that increased motility occurs distally, generally in parallel with increasing fertilising ability in-vivo, suggesting some relationship between the two (Dacheux and Paguignon, 1980b).

Microscopical observations reveal differences in the type of beating pattern in maturing sperm cells. Spermatozoa from the seminiferous tubules display weak vibratory motion, if any, although they fail to register in an objective assay (Hinton et al., 1979). Motility of sperm from the rete testis is low whether they are collected from conscious or anaesthetised animals. In contrast, immature spermatozoa tend to move in circles after dilution in-vitro whether taken from the caput epididymidis of rats (Fray et al., 1972; Wyker and Howards, 1977; Hinton et al., 1979; Turner and Giles, 1981), rabbits (Gaddum, 1968; Gaddum et al., 1968; Orgebin-Crist, 1967a; Acott et al., 1979; Pholpramool and Chaturapanich, 1979), hamsters (Kann and Serres, 1980), cats, dogs, guinea pigs, boars or monkeys (Acott et al., 1979). Human epididymal spermatozoa differ, in that no circling is evident by spermatozoa taken from proximal sites (Mooney et al., 1972; Bedford et al., 1973; Jouannet, 1981). Caudal spermatozoa from all species, like normal ejaculated spermatozoa, progress forwards in-vitro and the changes in motility between segments are not abrupt but the proportion of cells displaying different patterns of motility varies from region to region (Gaddum, 1968).

The motility patterns of immature sperm observed in-vitro may be artefacts arising from dilution of spermatozoa in hypo-osmotic media, secondary to a retroflexion of the sperm tail that occurs under these conditions to caput sperm from rats (Blandau and Rumery, 1964; Fray et al., 1972; Wyker and Howards, 1977; Cooper, 1985) and hamsters (Serres and Kann, 1984). Epididymal fluid from the caput is hypertonic to blood plasma (Table 5) and retroflexion is not observed if caput cells are diluted in medium isotonic to their normal environment, although it was not ascertained if these cells still swam in circles (Serres and Kann, 1984).

In the rat, hamster and mouse, circling of immature cells is associated with a stiff, curved mid-piece lacking rotation (Fray et al., 1972; Kann and Serres, 1980; Saling, 1982) whereas bovine caput epididymal spermatozoa have a flexible mid-piece which can rotate (Acott et al., 1983; Acott and Hoskins, 1983), although at a lower rate than that of mature cells. Conversion to forward motility on maturation is thus associated with a straightening of the mid-piece and the initiation of rotation in rodents, but a reduced flexibility and increased rotation in the bull.

Quantitative studies have revealed that in the hamster, beat frequency of testicular spermatozoa and caput sperm (Mohri and Yanagimachi, 1980) and progressive velocity of caput sperm (Kann and Serres, 1980) are lower than in mature spermatozoa. Likewise, in the bull, beat frequency, progressive velocity and rotation frequency are lower, whereas amplitude of beat is greater in immature spermatozoa (Acott et al., 1983; Acott and Hoskins, 1983).

4. The Nature of the Reduced Motility of Immature Spermatozoa

a. Sperm Structure and Function

i. Mature Spermatozoa

Sperm swim because of active sliding between adjacent axonemal doublets situated longitudinally throughout the flagellum (see Satir, 1979; Amelar et al., 1980).

The mechanochemical coupler is the ATPase dynein which projects from one tubule and is thought to form cross-bridges during beating with tubulin subunits of its neighbour (see Linck, 1979). Cytochemical demonstrations of flagellar Ca^{2+} and Mg^{2+} ATPases have localised the enzymes mainly in or around the outer dense fibres (see Gordon et al., 1983) but few reports have detected activity within the axonemal filaments of the microtubules (Nagano, 1965; Gordon and Barrnett, 1967; Yasazumi et al., 1975).

The proteins responsible for the tail movement (termed "flactin" (for flagellar actin) and "spermosin" (for sperm myosin) by Young and Nelson (1968, 1969) are different from the proteins that mediate length changes in muscle. While actin has been identified in spermatozoa from a variety of species (see Clarke and Yanagimachi, 1978 Tamblyn, 1980; Clarke et al., 1982) and non-muscle myosin is present in bovine spermatozoa (Tamblyn, 1981), they are not present in the sperm tail and thus cannot mediate sperm motility. A variety of cytoskeletal proteins have been identified in human sperm cells, of which tubulin occurs along the length of the tail (Virtanen et al., 1984). Chemically dissecting the sperm and determining the ATPase activity in the fractions (Young and Smithwick, 1975a,b, 1976) threw little light on the ATPase directly involved in flagellation until contamination with mitochondrial ATPase was eliminated (Smithwick and Young, 1977, 1978) but recently characterisation of both flagellar ATPase (Young and Smithwick, 1983) and tubulin (Little and Rohrrich, 1983) has been achieved.

In sperm cells to which nucleotides are made accessible to the axoneme (impaled or demembranated models), flagellar beating can be re-initiated (the models "reactivated") by ATP (Bishop and Hoffman-Berling, 1959) and the frequency of beat is affected by ATP concentration in hamsters (Morton, 1973; Mohri and Yanagimachi, 1980) and bulls (Rikmenspoel et al., 1978). That inhibition of the dynein ATPase prevents reactivation of motility (Kobayashi et al., 1978; Bouchard et al., 1981) suggests that motility is related to the utilisation of ATP as substrate. ATP also decreases the stiffness of the flagellum (Lindemann et al., 1973; Rikmenspoel et al., 1981). Sliding of adjacent filaments is normally restricted by their anchorage at the neck and end-piece of the sperm tail and the presence of accessory structures – circumferential nexin links and radial spokes – linking doublets to each other or to the central pair of microtubules, respectively (see Linck, 1979). When these accessory structures are removed by proteolytic enzymes or reduction of disulphide bonds with dithiothreitol (DTT), bends that are generated are not propagated distally and the reactivated axonemes are extruded from the sperm tail (see Summers, 1974; Lindemann and Gibbons, 1975; Olson and Linck, 1977; Mohri and Yano, 1980, 1982). The mitochondrial sheath proximally and the fibrous sheath distally impose further constraints on movement in as yet unidentified ways, but the outer dense fibres associated with the axonemal doublets do not seem to provide mechanical power (Rikmenspoel, 1984).

Initiation and propagation of bends in the sperm tail are distinct events, for bends created by physical deformation of the axonemes are propagated in axonemes whose wave initiation has been blocked by impalement (Lindemann and Rikmenspoel, 1972a) or Ni^{2+} treatment (Lindemann, 1980; Lindemann et al., 1980). Human spermatozoa are immotile if they lack dynein arms (Afzelius et al.,

1975; Pederson and Rebbe, 1975) or the central pair of microtubules (Baccetti et al., 1979).

ii. Immature Spermatozoa

The structure of the sperm tail changes little on passage through the epididymis. Striations on the outer dense fibres of rat caput sperm (Phillips and Olson, 1975) have the same periodicity on caudal cells (Olson and Sammons, 1980); the outer mitochondrial membrane is thicker than the inner membrane in both caput (Elfvin, 1968) and cauda sperm (Friend and Heuser, 1981); and flagellar ATPase activity can be localised histochemically in testicular (Chakraborty and Nelson, 1974), caput and cauda sperm (Nagano, 1965). There are increases in disulphide bonds in a variety of structures (see Bedford, 1975) and some of these proteins have been characterised from mature and immature spermatozoa (Sattasayai and Panyim, 1982).

All testicular spermatozoa have cytoplasmic droplets at the neck position and relatively straight mid-pieces, but when the droplet has migrated to the end of the mid-piece the flagellum is curved in rats (Fray et al., 1972) and rabbits (Gaddum, 1968) and it straightens out on further maturation. It is not known to what extent this reflects artefacts (see Section II.3.b) or increases in total disulphydryl bonds in the tail (Calvin and Bedford, 1971; Calvin et al., 1973; Bedford and Calvin, 1974). The decline in free sulphydryl (SH) groups parallels the decrease in bound flagellar zinc (Calvin et al., 1973; Calvin and Bleau, 1974) and a role for SH groups in zinc binding has been proposed. As zinc binds to tubulin (Chapvil, 1973) this ion could modulate flagellar movement.

The extrusion of axonemal doublets from demembranated, DTT-treated cells from the rat rete testis (T.G. Cooper, unpublished observations) and caput epididymidis (Chulavatnatol et al., 1978), and the similar velocity of axonemal sliding from DTT-treated caput and cauda hamster spermatozoa (Mohri and Yano, 1982) suggest that the reduced motility of intact immature spermatozoa reflects inadequate supplies of ATP or a defect in the control of dynein-tubulin interactions, rather than an inadequacy in the flagellar components themselves.

b. Sperm Energy Metabolism

i. Mature Spermatozoa

Objective measurements of sperm motility and metabolism indicate that motility is supported by ATP derived from both glycolysis and respiration (Rikmenspoel, 1965a; Rikmenspoel et al., 1969), although the extent to which these processes support motility varies between species. The motility of human spermatozoa, for example, can be supported solely from glycolysis (Hong et al., 1983), whereas less than 5% of the ATP requirement of bovine spermatozoa comes from glucolysis (Inskeep and Hammerstedt, 1985). In the hamster respiration supports the amplitude of the beat better than glycolysis (Fisher-Fischbein et al., 1985).

The compartmentation of mitochondrial enzymes in the mid-piece (Mohri et al., 1965) and the retention of glycolytic ezymes by hypotonically demembranated spermatozoa (Storey and Kayne, 1975, 1980b) has implications for the access to, and utilisation of, ATP in the flagellum: respiratory-derived ATP has to diffuse along the length of the flagellum, but glycolytic ATP can be produced close to its site of action. Gonse (1962) postulated that glycolysis provided energy for immediate movement, whereas respiration produced energy to be stored, but it is not known if there is strict association of glycolysis and respiration with, *e.g.*, sliding of filaments and ion pumping, as exists in smooth muscle (Gluck and Paul, 1977; Paul et al., 1979).

Utilisation of ATP by flagellar ATPase is commensurate with both its production from flagellar pyruvate kinase (Storey and Kayne, 1980) and mitochondria (Halangk et al., 1985). Despite one calculation to the contrary (Nelson, 1967), theoretical considerations have indicated that diffusion is adequate to supply sufficient mitochondrial ATP to the sperm tail (Raff and Blum, 1968; Rikmenspoel et al., 1969; Nevo and Rikmenspoel, 1970; Adam and Wei, 1975). In support of this view, ATP entering the head of an impaled, immotile bull sperm does initiate flagellar motility (Lindemann and Rikmenspoel, 1971, 1972a,b) and there is no decrease in amplitude of the wave as it passes along the tail in respiratory-supported motility, as would be expected if ATP concentrations within the flagellum were limiting (Rikmenspoel et al., 1969).

ii. Immature Spermatozoa

Immature spermatozoa generally have a greater oxidative and synthetic capacity than more mature cells whether they are taken from the testis (see Voglmayr, 1975) or from the epididymis of guinea pigs (Frenkel et al., 1973a; Shilon et al. 1978), rats ((Paz)Frenkel et al., 1978) or rams (Inskeep and Hammerstedt, 1982), but histochemistry of dehydrogenases (Abou-Haila and Fain-Maurel, 1983) and oligomycin-inhibition of mitochondrial function in rat testicular (Ford, 1981) or epididymal spermatozoa (Table 6) suggest that few changes occur to mitochondria on maturation. There is a marked decrease in the total LDH activity of porcine spermatozoa as they pass into the epididymis as well as a shift from the aerobic isoenzyme 1 to number 5 which mediates anaerobic pathways (Egbunike et al., 1986), but in rats the fall in total LDH activity as sperm mature in the epididymis is the same for total and LDH-X activity (Vermouth et al., 1986).

The entry of glucose into testicular spermatozoa may exceed that of ejaculated cells (Hiipaka and Hammerstedt, 1979) but there is no general trend in the maturational activities of glycolytic enzymes (Hammerstedt, 1975; Terner et al., 1975; Mongkolsirikieat and Boonsaeng, 1980; Ford, 1982; Ford and Harrison, 1983; Panse et al., 1983).

Total ATPase activity is lower in mature sperm cells from the ram (Voglmayr et al., 1969) and rat (Chulavatnatol and Yindepit, 1976; Chulavatnatol et al., 1978; Majumder, 1981) but this is not necessarily true for axonemal ATPase. There is no consistent trend on maturation in the total cellular content of ATP, which is higher in immature spermatozoa from rams (Voglmayr et al., 1967), and guinea pigs (Frenkel et al., 1973b), lower in immature sperm from rats (Chulavatnatol et al., 1977) and not established beyond doubt in bulls (Hammerstedt, cited Hoskins, 1973; Hoskins et al., 1975a; Cascieri et al., 1976). An important consideration than total concentration of nucleotide is its disposition within the cell, but unfortunately, this is not known.

Immature spermatozoa are unlikely to be immotile in-vitro because their metabolism is reduced by deprivation of intracellular water (Salisbury, 1956) or inhibitory concentrations of ATP (Hoskins, 1973) as these factors would be expected to change on removal from the male tract.

c. Control of Sperm Motility by Cyclic Nucleotides

i. Mature Spermatozoa

The adenylate cyclase of spermatozoa differs from that of somatic cells (Hildebrandt et al., 1985) and there are no known hormonal activators of it (see Hoskins and Casillas, 1975a,b; Casillas et al., 1980). The enzyme was initially thought to lack the guanine nucleotide regulatory subunit which could be provided by erythrocyte membranes (Stengel and Hanoune, 1981; Stengel et al., 1982), but further study showed that the increased enzyme activity in such reconstitution experiments was due to a stimulation of the red cell enzyme by a sperm-derived factor (Jakobs et al., 1983; Johnson et al., 1983; Hanski and Garty, 1983) subsequently identified as a protease (Johnson et al., 1985).

Although cyclic AMP (cAMP) by itself is unable to promote flagellation of demembranated sperm models, it does alter the beat frequency (Lindemann, 1978, 1980; Mohri and Yanagimachi, 1980) and amplitude (Tash and Means, 1982; Yeung, 1984) of ATP-induced beating and it promotes the forward progression of previously motile, but non-progressive, cells (Fentie and Lindemann, 1978; Yeung, 1984) reactivated by ATP. In the intact cell increased sperm motility associated with increased intracellular cAMP is not always mediated by protein kinase (Babcock et al., 1981; Babcock and Lardy, 1982) and some protein kinases in sperm are activated by GTP and polyamines (Casillas et al., 1980; Atmar et al., 1981). Furthermore, caffeine may also inhibit cholinesterase (Nachmanssohn and Schneeman, 1945) and so act *via* the cholinergic system (see Section II.A.4.f), or displace membrane-bound calcium (Peterson et al., 1979) and so act *via* divalent metal ions (see Section II.A.4.d).

However, there is much evidence to support an action of cAMP mediated by cAMP-dependent protein kinases which are present in spermatozoa from bulls, rams and rats (see Tash and Means, 1983). Many sperm proteins can be phosphorylated by labelled phosphate injected in-vivo (Chulavatnatol et al., 1982) but the physiological substrates for protein kinases are not established (see Brandt and Hoskins, 1980; Lindemann et al., 1983), possibly because of the high phosphoprotein phosphatase activity of sperm cells (Tang and Hoskins, 1975). Phosphorylation of the α-subunit of tubulin (Tongkao and Chulavatnatol, 1979; Tash and Means, 1982) would be anticipated to affect flagellar microtubule sliding, but cAMP does not influence the rate of microtubule sliding from the DTT-treated naked axoneme (Mohri and Yano, 1982). A flagellar protein, axokinin, has been found to be phosphorylated in dog sperm undergoing reactivated motility and addition of the phosphorylated protein promotes demembranated sperm motility (Tash et al., 1984).

ii. Immature Spermatozoa

In the bull activities of both cAMP phosphodiesterase (PDE) and adenylate cyclase (AC) decrease as sperm mature (Cascieri et al., 1976; Stephens et al., 1979; Casillas et al., 1980) yet cyclic AMP concentrations are higher in mature sperm than immature ones (Hoskins et al., 1974). The sensitivity of bovine sperm adenylate cyclase to forskolin, a stimulator of the catalytic subunit, also decreases on maturation (Vijayaraghavan and Hoskins, 1985). The failure of increased intracellular concentrations of cAMP to promote forward motility in testicular spermatozoa from sheep (Hammerstedt and Hay, 1980) and bulls (Cascieri et al., 1976) indicates that the ability of sperm to respond to cAMP is also underdeveloped in immature spermatozoa, and in the bull, caput sperm contain lower protein kinase activity than cauda cells (Hoskins et al., 1974), *i.e.* both the "trigger" (cAMP) and the ability to respond to it (protein kinase) are underdeveloped in immature spermatozoa. This may explain why much higher concentrations of caffeine (a PDE inhibitor) are required to stimulate kinetic activity of immature spermatozoa than mature ones in bulls (Garbers et al., 1971; Hoskins, 1973) and boars (Dacheux and Paquignon, 1980a). In the ram, cAMP concentrations in spermatozoa also increase as they mature but a threshold concentration has been invoked to explain the relationship between sperm motility and cAMP content by Amann et al. (1982). Although these changes in cAMP were not confirmed by Pariset et al. (1985) a good correlation between the content of cAMP-dependent protein kinase and the aquisition of flagellation by maturing spermatozoa was observed. It is not known if cAMP levels were raised in rat sperm from the caput epididymidis whose motility failed to respond to caffeine (Paz(Frenkel) et al., 1978) but in this species immature sperm contain more cAMP than mature cells (del Rio and Raisman, 1978).

In demembranated models there is no difference in the initiation of motility by caput and cauda spermatozoa in response to ATP and cAMP in rats when performed at alkaline pH (Treetipsatit and Chulavatnatol, 1982; Chulavatnatol and Treetipsatit, 1983), so the intracellular pH may be an important regulator of motility of intact cells. There is evidence that intracellular pH in immature bovine spermatozoa is lower than that in mature cells (Vijayaraghavan et al., 1985). In contrast, some changes in the axoneme were found in the hamster sperm model where fewer testicular sperm responded, they required a lag period before responding and displayed lower beat frequencies than caudal cells. In addition, testicular spermatozoa never showed the great flagellar bending assumed by more mature cells (Mohri and Yanagimachi, 1980; Mohri et al., 1983) and cAMP never promoted early forward motility in cauda cells as it did in caput cells (Yeung, 1984). The mid-piece of caput cells was also found to be more rigid and never developed the flexibility assumed by more mature cells on treatment with cAMP (Yeung, 1984). Similar observations have been made in the ram with testicular spermatozoa requiring a lag period before responding to conditions that maximally activate mature epididymal spermatozoa (White and Voglmayr, 1986).

Thus there appear to be differences in the ability of immature axonemes to respond to cAMP that may explain, in part, the maturation of motility. In the bull the difference in intracellular pH between caput and cauda sperm could explain the phenomenon (Vijayaraghavan et al., 1985), and in the rat there are differences in the size of the regulatory subunit of cAMP-dependent protein kinase obtained from caput and cauda sperm (Atherton et al., 1985). Other pertinent differences may include the accessibility of ATP to proteins and the greater number of free SH groups on phosphorylated proteins from immature cells (Chulavatnatol et al., 1982).

d. Control of Sperm Motility by Ions

i. Mature Spermatozoa

Decreasing Na^+ concentrations in-vitro reduces bovine sperm motility, but whether sperm are rendered totally immotile depends on the nature of the ion used to replace sodium (McGradey and Nelson, 1972; McGradey et al., 1974). By manipulating external K^+ concentrations not only is the percentage of motile ejaculated bovine cells reduced, but the velocity and amplitude of beat are reduced and the beat frequency is increased (McGradey and Nelson, 1972). High K^+ concentrations also depress motility of epididymal spermatozoa from rats (Wong and Lee, 1983, 1985), mice (Alvarez and Storey, 1984) and hamsters (Mrsny and Meizel, 1985).

The plasma membrane of intact spermatozoa, which maintains the optimum concentration of these ions around the axoneme, makes the effects of extracellular ions on sperm motility difficult to interpret. Whereas the effects of K^+ on beat amplitude (Lindemann and Rikmenspoel, 1972a,b) and Mg^{2+} on beat frequency (Rikmenspoel et al., 1978; O'Day and Rikmenspoel, 1979) shown in impaled cells are explicable by the ion requirements and Mg^{2+}-dependence of dynein ATPase (Mohri and Ogawa, 1975), the maintenance of forward progression (Wong et al., 1981a,b) and the decline in motility (Reed and Lardy, 1972) induced by K^+, NH_4^+, Na^+ and monovalent cation ionophores could be explained by subsequent changes in intracellular pH. Sodium ion-induced forward motility of rat sperm is associated with acidification of the medium (reflecting a presumed alkalinisation of the cell interior: Wong et al., 1981a,b) and the optimum pH for reactivation of demembranated cells is alkaline (see Tash and Means, 1983). On the other hand, the inhibitory action of high K^+ may be related to promotion of lipid peroxidation (Alvarez and Storey, 1983b, 1984).

The action of Ca^{2+} in stimulating sperm motility is less clear, but could be *via* activation of adenylate cyclase (Hyne and Garbers, 1979a,b; Garbers et al., 1982). Outwardly directed Ca^{2+} ATPases (see Bradley and Forrester, 1985) and Ca^{2+}-Na^+ antiporters (Bradley and Forrester, 1980a; Rufo et al., 1984) demonstrated on sperm plasma membranes would act in concert with mitochondrial uptake of calcium (see Babcock and Lardy, 1982; Breitbart et al., 1985) to reduce intracellular Ca^{2+}.

The free cytosolic Ca^{2+} concentration may be critical in the control of sperm motility and agents expected to modulate intracellular Ca^{2+} would be anticipated to modify motility. Quercetin, which inhibits the Ca^{2+} transport ATPase of sperm (Bradley and Forrester, 1980c) and would be anticipated to increase intracellular calcium, reduces ovine sperm motility (Breitbart et al., 1985). It was also found to reduce calcium uptake in spermatozoa, possibly *via* an action at calcium channels (Breitbart et al., 1985). The calmodulin inhibitor trifluoperazine, at concentrations that promote uptake of Ca^{2+} into porcine sperm mitochondria (Peterson et al., 1983) also reduces sperm motility. In contrast, when displacing calcium from membrane-bound sites this agent increase in bovine sperm motility (Singh et al., 1983). Motility of bovine epididymal spermatozoa associated with increased extra-mitochondrial Ca^{2+} can also be achieved by suitable treatment of cells with ionophores (Babcock et al., 1976, 1978, 1980); however, Ca^{2+} inhibits reactivation of naked axonemes (Mohri and Yano, 1980, 1982; Mohri and Yanagimachi, 1980; see Tash and Means, 1983).

Calmodulin may mediate the action of calcium on sperm motility, since it is the major acidic glycoprotein of sperm (Brooks and Siegel, 1973) and is present along the length of the flagellum (Jones et al., 1980a,b; Feinberg et al., 1981; see Means et al., 1982; Gordon et al., 1983; Pariset et al., 1985). Calmodulin may act to store Ca^{2+} in the outer dense fibres (Gordon et al., 1983) or reduce the concentration of cAMP by stimulating phosphodiesterase activity of sperm (Wasco et al., 1984). On the other hand, Ca^{2+}-calmodulin antagonists that inhibit sperm motility (Levin et al., 1981) and induce circular activity in bovine and canine spermatozoa (Tash and Means, 1982; Singh et al., 1983) may be acting to displace Ca^{2+} from membrane-bound sites (Singh et al., 1983) rather than inhibit calmodulin-dependent events.

Ion fluxes across sperm membranes generate membrane potentials as in other non-excitable cells (Williams, 1970). Although the magnitude of the potential in spermatozoa varies with their degree of motility, especially amplitude (McGradey and Nelson, 1972, 1973) and forward progression (Wong and Lee, 1983), it is not thought to control motility *per se*, but rather reflect the partition across the membrane of the ions which primarily control motility (Wong and Lee, 1983). Inhibiting ion pumps in spermatozoa by ouabain decreases forward motility by reducing frequency, amplitude and membrane potential (McGradey, 1979). Using ^{3}H-ouabain to monitor ion transport (Hoffmann, 1969), glycolysis was better able than respiration to support ion pumping in spermatozoa from sheep, bulls and boars (O'Donnell and Ellory, 1970a,b).

ii. Immature Spermatozoa

Intracellular concentrations of K^+, Ca^{2+} and Mg^{2+} are higher than, and Na^+ lower than their respective concentrations in surrounding fluids (Table 12). There is a trend towards higher concentrations of intracellular Na^+ and Ca^{2+} in immature cells, but no definite trend in K^+ and Mg^{2+} concentrations. The values quoted were obtained by metal ion determination, but different values have recently been obtained for mature bovine spermatozoa (14mM-Na^+, 120mM-K^+) using a non-invasive fluorescent technique (Babcock, 1983) which has yet to be applied to immature sperm cells. Non-invasive (fluorescent dye or nuclear magnetic resonance) techniques have confirmed that intracellular pH is alkaline (Babcock, 1983; Vijayaraghavan et al., 1985; Smith et al., 1985). There is a decrease in intracellular calmodulin concentration in maturing ovine spermatozoa (Pariset et al., 1985), which could have repercussions for Ca^{2+}-dependent events.

Few studies have been made of the permeability of sperm cells' plasma membranes. The permeability of bovine spermatozoa to water is higher than other cell types when examined by osmotic entry of water (see Drevius, 1975) and immature spermatozoa from bulls, rams and boars have different osmotic responses from mature cells (O'Donnell, 1969; Jones, 1971). Immature sperm cells suffer different osmotic insults from mature cells on removal from the tract (see Section II.3.b) and also different restraints to osmotic swelling (Cooper, 1985).

Indirect evidence has been taken to indicate that immature sperm cell membranes are more permeable to a range of compounds: the greater production of lactate from fructose 6-phosphate (bovine spermatozoa: Stephens et al., 1979) or incorporation of dUTP into RNA at 4 °C (rat sperm: McLaughlin and Terner, 1973) than in mature sperm cells implied a greater entry of substrate rather than more extensive metabolism, and the greater uptake of Ca^{2+} into bovine caput sperm (Hoskins et al., 1983) and Na^+ and K^+ into porcine corpus sperm (Crabo and Hunter, 1975; Zimmermann et al., 1979) than into caudal cells has been demonstrated by tracer techniques. This may reflect more permeable membranes or lower activities of ion pumps in immature cell membranes.

The activity of one "ecto" ATPase on mature rat spermatozoa decreased on maturation (Majumder, 1981) but the enzyme was not inhibited by ouabain (Majumder and Biswas, 1979). In contrast, a surface ATPase on mature rat sperm was found by Chulavatnatol and Yindepit (1976) to be inhibited by ouabain to a greater extent than that of caput sperm. Using labelled ouabain to determine ion pump activity, O'Donnell and Ellory (1969) found caput sperm from the boar epididymis to have equal activity to that of ejaculated cells. In contrast to cauda sperm, immature ovine caput spermatozoa have no detectable Ca^{2+}-ATPase activity on their flagellar membranes (Bradley and Forrester, 1985).

If immature sperm cells are more permeable than mature cells to ions present in epididymal fluid in the caput (see Section III.C.1) then supra-optimal concentrations of ions or pH may explain their poor motility, since no differences in the pH optimum for reactivation of mature and immature sperm has been observed (Treetipsattit and Chulavatnatol, 1982; Chulavatnatol and Treetipsatit, 1983). Demembranated testicular spermatozoa from the ram, however, are more sensitive to inhibition by calcium ions that mature cells (White and Voglmayr, 1986). The ions reaching the cell interior are those available in epididymal fluid and this reflects activity of the epididymal epithelium (see Section III.C.2).

e. Control of Sperm Motility by Protein Methylation

i. Mature Spermatozoa

The carboxymethylation of proteins (Gagnon and Heisler, 1979) and phospholipids (Hirato and Axelrod, 1980) may control fluidity of membranes, but the consequences for ion movements into, and the motility of, sperm is not well established. Protein carboxymethylase (PCM) activity is absent from immotile human sperm cells (Gagnon et al., 1980a, 1982), but the immotility could result from deficiency of other enzymes. Similarly, the ability of a PCM inhibitor to prevent reactivation of demembranated rat sperm cells by ATP could be explained by its inhibition of dynein ATPase activity (Bouchard et al., 1981). Sperm cells are able to utilise methionine to methylate proteins (Castaneda et al., 1983; Goh and Hoskins, 1985; Fisher-Fischbein et al., 1985), but transmethylase inhibitors have no consistent effect on the motility of hamster sperm (Meizel, 1981; Llanos and Meizel, 1983). Metabolic inhibitors that decrease the production of ATP decrease motility and protein methylation in the hamster (Fisher-Fischbein et al., 1985) and agents that increase the intracellular concentration of S-adenosylhomocysteine (an inhibitor of S-adenosylmethionine-mediated protein methylation) reduce the motility of bovine spermatozoa (Goh and Hoskins, 1985). The nature and role of the methylated proteins are unknown; PCM can utilise calmodulin as substrate (Gagnon et al., 1980b) so Ca^{2+}-dependent systems could be involved.

ii. Immature Spermatozoa

Both PCM and methyl acceptor proteins (MAP) are present in increasing amounts in maturing rat and rabbit spermatozoa (Bouchard et al., 1980; Purvis et al., 1982) and protein methyl esterase activity, which hydrolyses the methyl esters formed by PCM, decreases as sperm mature in bulls and rats (Gagnon et al., 1984). In the hamster PCM activity towards endogenous substrates in caput and cauda sperm is similar but the amount of methyl acceptor proteins is reduced in the more mature spermatozoa (Castaneda et al., 1983).

f. Control of Sperm Motility by Cholinergic Agents

i. Mature Spermatozoa

The role of cholinergic agents in regulating sperm motility has been reviewed by its main proponent (Nelson, 1975, 1978, 1979). Although both acetyl- and butyryl-choline are detectable in spermatozoa from bulls, rats, rabbits and man (Bishop et al., 1976, 1977; Harbison et al., 1976a), acetylcholine is not thought to be stored to a great extent by spermatozoa, so that the regulation of its synthesis *via* cholineacetyltransferase (ChAT) and its breakdown *via* acetylcholine-esterase (AChE) is considered important for regulating its intracellular concentration. Cholineacetyltransferase has been determined in spermatozoa from rams (Stewart and Forrester, 1976, 1978a), bulls (Bishop et al., 1976), rats, rabbits and man (Harbison et al., 1976a), although there is now concern that this enzyme has been grossly overestimated by confusion with the activity of carnitine acetyltransferase (Goodman and Harbison, 1981; Goodman et al., 1984). Cholinesterase is especially active in the tails of spermatozoa from pigs (Sekine, 1951), bulls (Nelson, 1964, 1966) and rams (Stewart and Forrester, 1976, 1978a).

Acetylcholine and AChE inhibitors modulate sperm motility in intact sperm cells (see Ramasastry and Sadavangvivad, 1978), with inhibitors of ChAT reducing the number and vigour of human sperm tail beating (Ramasastry et al., 1981) and altering beat amplitude and frequency of bovine spermatozoa without altering their velocity (McGradey and Nelson, 1976). Acetylcholine itself stimulates the motility human sperm in an atropine-sensitive and eserine-augmented fashion

(Sanyal and Khanna, 1971). This suggests that muscarinic sites are involved, but the only characterised muscarinic receptors from sperm cells (in the mouse) differ from those of other tissues, are found only in the sperm head, and are not thought to be involved in motility, since motility was only affected at much higher concentrations of the agonist 3-quinuclidinyl benzilate that those found to block sperm-egg binding (Florman and Storey, 1981, 1982b). It may be the nicotinic receptors on the plasma membranes of ovine (Stewart and Forrester, 1978b) and bovine (Sastry et al., 1979) sperm tails that are involved in flagellar movement.

An action *via* Ca^{2+} movements within the cells has been proposed (see Nelson et al., 1980a,b,c; Stewart and Forrester, 1978c); certainly acetylcholine has no effect on the exposed axoneme of the ATP-reactivated, impaled spermatozoa (Lindemann and Rikmenspoel, 1972a). Acetylcholine can modulate the flux of calcium ions in and out of ram sperm mitochondria (Stewart and Forrester, 1979). However, it is not clear why administration of ChAT inhibitors in-vivo (Harbison et al., 1976a) or AChE inhibitors in-vitro (McGradey and Nelson, 1976) lead to the same pattern of movement (circling) in mature spermatozoa when opposite effects on intracellular concentrations of acetylcholine would be anticipated.

ii. Immature Spermatozoa

ChAT activity of spermatozoa increases upon maturation in the epididymis of rats and rabbits (Harbison et al., 1976a,b) whereas AChE activity decreases upon maturation in boars, bulls and rats (Egbunike, 1980, 1982) but is little changed in the ram (Wright et al., 1981). Acetylcholine may accumulate more in mature cells, but it does not initiate motility in pre-motile sperm from the rat caput epididymidis (Hinton et al., 1981; Hinton and Setchell, 1980b).

g. Control of Sperm Motility by Adrenergic Agents

i. Mature Spermatozoa

The decline in motility that occurs on sperm dilution can be prevented by "Sperm Motility Factors" (SMF) from a variety of sources (see Bavister, 1979). They are now thought to comprise catecholamines (see Bavister et al., 1979) and/or the sulphur amino acids taurine and hypotaurine (see Meizel et al., 1980; Gwatkin, 1983). SMF is present in the epididymis (Bavister et al., 1978) and both the sulphur amino acids have been identified in epididymal fluid (Jones, 1978). SMF can also be extracted from spermatozoa (Bavister and Yanagimachi, 1977) so they may act at intracellular sites.

Catecholamines do not interact with normal adrenergic receptors in the hamster sperm (Cornett and Meizel, 1980), are surface-active in the bull (Lindahl, 1978) and may chelate ions, thus perturbing membranes (Chaix et al., 1950; Chenoweth, 1956; Maas and Colburn, 1965). Taurine and hypotaurine may inhibit lipid peroxidation of sperm and thus reduce the damaging action of peroxides on sperm motility (Alvarez and Storey, 1983a) and inhibit the

Na^+/K^+-ATPase (Mrsny and Meizel, 1985), thus reducing the inhibitory effect of potassium ions on sperm motility (see Section II.A.4.d.i).

ii. Immature Spermatozoa

The interaction of these agents with immature spermatozoa has not been examined.

5. Involvement of the Epididymis in the Maturation of Motility

Trapping spermatozoa in the testes by ligating the excurrent ducts increases the rate of flagellar beating of spermatozoa subsequently taken from seminiferous tubules (Glover, 1962; Orgebin-Crist, 1967b) but no forward motility is induced in these sperm or others trapped within the rete testis of rabbits (Cooper and Orgebin-Crist, 1975, 1977) or rams (O'Shea and Voglmayr, 1970). Sperm trapped in the caput epididymidis become more active, but rotate faster without developing forward forward progression in rats (Burgos and Tovar, 1974) and rabbits (Bedford, 1967; Orgebin-Crist, 1967b). Similarly, after vasectomy (Mooney et al., 1972) or epididymal obstruction (Lindholmer, 1974) sperm in the proximal human epididymis become more active but do not display forward progression. In the rabbit only sperm from the upper corpus epididymidis develop forward motility after a period of ageing in that region and these develop fertilising ability (Orgebin-Crist, 1967b; Bedford, 1967). Increased motility and fertilising ability is also observed after ligation of the distal guinea pig epididymis (Young, 1931), but in the hamster increased motility in the ligated corpus epididymidis is divorced from increased in-vivo fertilising capacity (Horan and Bedford, 1972; Cummins, 1976).

In-vitro ageing of testicular spermatozoa in rete testis fluid does not promote their forward motility in rams (Voglmayr et al., 1967; Murdoch and White, 1968) or bulls (Voglmayr et al., 1970) and incubation in epididymal fluid (Voglmayr et al., 1977) or substrates (Voglmayr and Gandhi, 1978) is similarly ineffective. These studies would suggest that ageing alone for short periods does not permit the expression of latent motility. An epididymal influence on the acquisition of sperm motility is suggested by castration and androgen-replacement studies which prevent and maintain, respectively, forward motility induction of sperm retained in the upper corpus epididymidis (Orgebin-Crist, 1973). Some credence is given to this by the observation that sperm motility is initially acquired adjacent to the ligatures and then proceeds proximally (Gaddum and Glover, 1965) where epididymal secretions would be expected to accumulate.

6. Induction of Motility in Immature Spermatozoa In-Vitro

a. Carnitine

i. The Initiation of Flagellar Motility

The parallel rise in carnitine concentration in epididymal luminal fluid (Table 16) and sperm motility (Table 4) suggests that the two phenomena are related (see Hinton, 1980) and rat caput sperm retained in epididymal tubules cultured with carnitine do become motile (Klinefelter and Hamilton, 1984). "Pre-motile" spermatozoa from the proximal caput epididymidis become motile, although non-progressive, on incubation for short periods in concentrations of carnitine found more distally in the epididymis (Hinton and Setchell, 1980; Setchell and Hinton, 1981; Hinton et al., 1981). After longer incubations or with higher concentrations of carnitine, motility is depressed.

ii. Action of Carnitine

The depression of motility of caput sperm in substrate-free media containing carnitine may be explained by a rapid depletion of endogenous substrates; the motility of mature spermatozoa in substrate-free media is also reduced by high concentrations of carnitine in parallel with a decline in pyruvate oxidation, but both are restored on addition of substrate (Turner and Giles, 1981). This suggests that the action of carnitine is to provide substrate as acetylcarnitine (see Carter et al., 1980; Section III.E.5.b) and acetylcarnitine can also initiate motility in premotile cells (Hinton et al., 1981). Carnitine does not stimulate endogenous respiration of immature bovine spermatozoa (Casillas, 1973) or improve the efficiency of ATP production by immature ovine spermatozoa (Inskeep and Hammerstedt, 1982) as it does in mature spermatozoa, so there is also some change on maturation in the response of the sperm cell to carnitine.

b. Acidic Epididymal Glycoprotein (AEG)

i. The Initiation of Flagellar Motility

Sperm motility can be induced in pre-motile proximal caput sperm cells by incubation in acidic epididymal glycoprotein (AEG: see Lea et al., 1978; Lea and French, 1981), but because other proteins were equally effective the action of AEG was considered non-specific (Pholpramool et al., 1983). However, as AEG is a major component of epididymal fluid (see Table 20) it may have some physiological importance.

ii. Action of AEG

The action of AEG in promoting motility is not known. Other epididymal proteins alter ion fluxes across the membranes of maturing spermatozoa (Zimmermann et al., 1979; Hoskins et al., 1983) and these may have consequences for motility (see Section III.G.12.b.i).

c. Forward Motility Protein (FMP)

i. The Initiation of Forward Motility

A protein in seminal plasma, together with a PDE inhibitor, promotes the forward motility of previously circling, immature caput spermatozoa in-vitro (Hoskins et al., 1978, 1979). The effect cannot be duplicated by albumin (Hoskins et al., 1975b) and is not species-specific (Acott et al., 1979), although seminal plasma from the bull is incapable of converting the circling movement of ovine caput spermatozoa to forward progression (Amann et al., 1982). The glycoprotein responsible for the change in motility (FMP) has been characterised (Acott and Hoskins, 1978; Table 20).

The reduced FMP activity of seminal plasma from vasectomised monkeys suggested an epididymal origin of the protein, and FMP activity has been found in cauda epididymal plasma from bulls (Brandt et al., 1978; Acott et al., 1983) and hamsters (Kann and Serres, 1980). However, this does not prove the epididymal secretion of the protein, since the concentration of FMP in bovine epididymal fluid is less than that anticipated to arise from the concentration of FMP in rete testis fluid, judging from the relative concentrations of sperm cells in the two fluids (Sexton et al., 1971). On the other hand, the lower than expected concentrations of FMP in sperm-free fluid could be explained by its uptake by caput cells during maturation: binding of FMP to caput cells has been demonstrated in-vitro (Acott and Hoskins, 1981; see Section III.G.8.f).

ii. Action of FMP

At least some of the response of caput sperm cells to FMP in-vitro can be expained by a reduction of sticking of the cells to the observation chamber (Stephens et al., 1981), but another action is to decrease the flexibility of the mid-piece of bovine spermatozoa (Acott and Hoskins, 1983; Acott et al., 1983) and prevent the retroflexion of the neck of hamster sperm that occurs in hypo-osmotic media (Serres and Kann, 1984). Clues to the mechanism of action may be found from studying causes of the reverse action: the induction of circular motility in progressing mature spermatozoa. This can be caused by (a) high extracellular Ca^{2+} (Nelson et al., 1980a); (b) addition of AChE inhibitors (McGradey and Nelson, 1976); (c) local anaesthetics and Ca^{2+}-transport inhibitors (Singh et al., 1983) and (d) anti-calmodulin agents (Tash and Means, 1982). These observations point to changes in intracellular Ca^{2+} mediating the response because: (a) intracellular Ca^{2+} may reflect the extra-cellular concentration (McGradey and Nelson, 1974); (b) the AChE inhibitor eserine (Rama Sastry and Sadavangvivad, 1978) and acetylcholine itself increase sperm membrane permeability towards calcium ions (Nelson et al., 1982) and alter calcium fluxes across sperm mitochondria (Stewart and Forrester, 1979); (c) Ca^{2+}-calmodulin antagonists structurally similar to local anaesthetics may act like them to displace Ca^{2+} from membrane-bound sites (Singh et al., 1983).

Conversion of circling to forward motility on maturation would therefore be consistent with a reduction of intracellular Ca^{2+}, and there is a reduction in the uptake of extracellular calcium by maturing bovine spermatozoa (Hoskins et al., 1983). In support of this, a Ca^{2+}-calmodulin inhibitor, thought to reduce intracellular calcium, promotes forward motility in some caput sperm from the hamster epididymis (Serres and Kann, 1984).

FMP may act to reduce intracellular calcium by acting on sperm cell surface membrane pumps. An anology may be with a human seminal protein (calsemin) that promotes initiation of motility in quiescent, but mature, epididymal spermatozoa in parallel with stimulation of the outwardly-directed Ca^{2+}-ATPase (Bradley and Forrester, 1981).

The epididymis contains an androgen-dependent PDE (Djoseland et al., 1980) and the gradual movement of immature spermatozoa into epididymal fluid that contains PDE (Tash, 1976), an androgen-dependent PDE inhibitor (Morton et al., 1974; Terner et al., 1980) and FMP could explain the development of forward motility. The inability of PDE inhibitors together with egg-yolk or FMP from seminal plasma or epididymal fluid to initiate forward motility of testicular spermatozoa from rabbits (T.G. Cooper, M.C. Orgebin-Crist and D.D. Hoskins, unpublished data; Orgebin-Crist et al., 1981) and boars (Dacheux and Paquignon, 1980b), caput sperm from rats (Wyker and Howards, 1977; Turner and Giles, 1981, 1982b) and caput or corpus sperm from rams (Amann et al., 1982) suggests that in these species yet other changes occur to spermatozoa in the proximal epididymis (*e.g.* intracellular pH) which allow them to respond to these factors. If Na^+/H^+ exchange mediates the alkalinisation of the cell interior at the induction of motility (Wong et al., 1981a; Babcock, 1983) then the epididymis would influence quiescence of spermatozoa by depriving them of the sodium ions necessary for this exchange. Immature spermatozoa in more proximal regions of the epididymis exposed to higher concentrations of Na^+ (Table 13) may not suffer a raising of pH if the cation-exchanger is not fully developed, or may not be affected by any raising of intracellular pH if cAMP concentrations are too low.

d. Other Factors Influencing Forward Motility

Caffeine alone cannot promote forward motility of testicular spermatozoa from bulls or rams (Cascieri et al., 1976; Hammerstedt and Hay, 1980) but it can promote forward motility in some testicular spermatozoa from rabbits and boars and caput sperm from boars (Brackett et al., 1978; Dacheux and Paquignon, 1983a). In the additional presence of egg-yolk, caffeine stimulates forward motility in testicular spermatozoaa from boars, rams, rabbits and goats (Dacheux and Paquignon, 1980a) and acts on corpus rather than caput spermatozoa from the rat (Turner and Giles, 1981). It is not clear if the egg-yolk supplies energy substrates of modifies intracellular pH.

Steroids can also promote forward migration of mature spermatozoa (see Section III.B.7.d). A peptide present in cauda but not corpus epididymal fluid of the rat and secreted into the lumen of the perfused cauda can promote the Na^+- and Ca^{2+}-dependent initiation of forward motility of rat sperm in Ca^{2+}-free solution (Wong et al., 1980, 1982; see Section III.G.13), possibly by displacing bound Ca^{2+}. Sulphydryl blocking agents that inhibit sperm motility (Dawson et al., 1979; Majumder and Chaudhury, 1984) may interfere with membrane transport.

i. Immature Spermatozoa

Incubation of testicular spermatozoa with a variety of steroids does not promote their forward motility (Voglmayr and Gandhi, 1978). Decreases in membrane sulphydryl groups that occur on maturation (Reyes et al., 1976; Reyes-Fuentes and Chavarnia-Dante, 1980; Reyes and Chavarnia, 1981) would render immature sperm cells more susceptible than mature ones to the action of sulphydryl agents. A scheme displaying various possible changes in immature spermatozoa that may contribute to their increased motility on maturation in the epididymis is given in Figure 2.

7. Relationship of Induced Forward Motility to Fertilising Ability

Caffeine- and egg-yolk-treated, forward progressing testicular or caput spermatozoa are unable to fertilise eggs when placed in the oviducts (Dacheux and Paquignon, 1980b, 1983a), but caffeine- and FMP-treated hamster caput spermatozoa achieve fertilisation after uterine insemination (Kann and Raynaud, 1982). In the latter case, the percentage of eggs fertilised was only about 1/4 of those fertilised by the same number of mature (cauda) cells, however, demonstrating that forward motility by itself is not enough to promote complete fertilisation. Where sperm do not need to move at all to reach the eggs (in-vitro fertilisation) immature sperm still fail to fertilise eggs (Table 2), emphasising the importance of other factors in the fertilisation process.

8. Summary

The induction within the epididymal canal of the capacity of spermatozoa to undergo forward progressive motility when released from the tract is of importance for them to swim out of seminal plasma or negotiate the cervix or uterotubal junction. Changes in the intracellular composition of the sperm appear to be directly responsible for the improvement in motility but the epididymis is involved in facilitating these changes by providing or removing ions or cofactors that act on or in the flagellum. The ability to perform hyperactiated motility in the vicinity of the egg must also be developed during epididymal passage. It is important for forward motility and hyperactivation to be separeted temporally and the epididymis may be involved in this by secreting decapacitation factors.

9. References

Abou-Haila A and Fain-Maurel MA (1983) Enzyme activities in Cynomolgus monkey and mouse spermatozoa evaluated histochemically during epididymal maturation. Arch Androl 11:101–113

Acott TS and Hoskins DD (1978) Bovine forward motility protein. Partial purification and characterization. J Biol Chem 253:6744–6750

Acott TS and Hoskins DD (1981) Bovine sperm forward motility protein: binding to epididymal spermatozoa. Biol Reprod 24:234–240

Acott TS and Hoskins DD (1983) Cinematographic analysis of bovine epididymal sperm motility: epididymal maturation and forward motility protein. J Submicroscop Cytol 15:77–82
Acott TS, Johnson DJ, Brandt H and Hoskins DD (1979) Sperm forward motility protein: tissue distribution and species cross reactivity. Biol Reprod 20:247–252
Acott TS, Katz DF and Hoskins DD (1983) Movement characteristics of bovine epididymal spermatozoa. Effects of forward motility protein and epididymal maturation. Biol Reprod 29:389–399
Adam DE and Wei J (1975) Mass transport of ATP within the motile sperm. J Theoret Biol 49:125–145
Afzelius BA, Eliasson R, Johnson O and Lindholmer C (1975) Lack of dynein arms in immotile human spermatozoa. J Cell Biol 66:225–232
Alvarez JG and Storey BT (1983a) Taurine, hypotaurine, epinephrine and albumin inhibit lipid peroxidation in rabbit spermatozoa and protect against loss of motility. Biol Reprod 29:548–555
Alvarez JG and Storey BT (1983b) Role of superoxide dismutase in protecting rabbit spermatozoa from O_2 toxicity due to lipid peroxidation. Biol Reprod 28:1129–1136
Alvarez JG and Storey BT (1984) Assessment of cell damage caused by spontaneous lipid peroxidation in rabbit spermatozoa. Biol Reprod 30:323–331
Amann RP and Griel LC (1974) Fertility of bovine spermatozoa from rete testis, cauda epididymidis and ejaculated semen. J Dairy Sci 57:212–219
Amann RP, Hay SR and Hammerstedt RH (1982) Yield, characteristics, motility and cAMP content of sperm isolated from seven regions of ram epididymis. Biol Reprod 27:723–733
Amelar RD, Dubin L and Schoenfeld C (1980) Sperm Motility. Fert Steril 34:197–215
Atherton RW, Khatoon S, Schoff PK and Hadley BE (1985) A study of rat epididymal sperm adenosine 3′,5′-monophosphate-dependent protein kinase: maturation differences and cellular location. Biol Reprod 32:155–171
Atmar VJ, Kuehn GD and Casillas ER (1981) A polyamine-dependent protein kinase from bovine epididymal spermatozoa. J Biol Chem 256:8275–8278
Babcock DF (1983) Examination of the intracellular ionic environment and of ionophore action by nullpoint measurements employing the fluorescent chromophore. J Biol Chem 258:6380–6389
Babcock DF and Lardy HA (1982) Alterations in membrane permeability to Ca^{2+} and their consequences during maturation of mammalian spermatozoa. In: Membranes and Transport, Vol 1. Ed AN Martonisi, Plenum Press, London. pp 671–676
Babcock DF, First NL and Lardy HA (1976) Action of ionophore A23187 at the cellular level. Separation of effects at the plasma and mitochondrial membranes. J Biol Chem 251:3881–3886
Babcock DF, Stamerjohn DS and Hutchison T (1978) Calcium redistribution in individual cells correlated with ionophore activity on motility. J exp Zool 204:391–400
Babcock DF, Singh JP and Lardy HA (1981) Changes in internal calcium concentrations modulate mammalian sperm motility without mediation of cyclic AMP. In: Calcium-binding proteins: Structure and Function. Eds FL Siegel, E Carafoli, RH Kretsinger, DH MacLennon and RH Wasserman, Elsevier North Holland. pp 479–481
Baccetti B, Burrini AG, Maver A, Pallini V and Renieri T (1979) "9+0" immotile spermatozoa in an infertile man. Andrologia 11:437–443
Baker RD and Degen AA (1972) Transport of live and dead boar spermatozoa within the reproductive tract of gilts. J Reprod Fert 28:369–377
Bavister BD (1979) Elimination of the adverse effect of dilution on hamster sperm motility in vitro. In: The Spermatozoon. Eds DW Fawcett and JM Bedford, Urban and Schwarzenberg, Baltimore. pp 169–172
Bavister BD and Yanagimachi R (1977) The effects of sperm extracts and energy sources on the motility and acrosome reaction of hamster spermatozoa in vitro. Biol Reprod 16:228–237
Bavister BD, Rogers BJ and Yanagimachi R (1978) The effects of cauda epididymal plasma on the motility and acrosome reaction of hamster and guinea pig spermatozoa in vitro. Biol Reprod 19:358–363

Bavister BD, Chen AF and Fu PC (1979) Catecholamine requirement for hamster sperm motility in vitro. J Reprod Fert 56:507–513

Bedford JM (1965) Effect of environment on phagocytosis of rabbit spermatozoa. J Reprod Fert 9:249–256

Bedford JM (1967) Effects of duct ligation on the fertilizing ability of spermatozoa from different regions of the rabbit epididymis. J exp Zool 166:271–282

Bedford JM (1975) Maturation, transport, and fate of spermatozoa in the epididymis. In: Handbook of Physiology, Section VII Endocrinology, Volume V, Male reproductive system. Eds RO Greep and DW Hamilton, American Physiol Soc, Washington DC. pp 303–317

Bedford JM and Calvin HI (1974) Changes in the -S-S- linked structures of the sperm tail during epididymal maturation with comparative observations in submammalian species. J exp Zool 187:181–204

Bedford JM, Calvin H and Cooper GW (1973) The maturation of spermatozoa in the human epididymis. J Reprod Fert Suppl 18:199–213

Bishop DW and Hoffmann-Berling H (1959) Extracted mammalian sperm models. I. Preparation and reactivation with adenosine triphosphate. J Cell Comp Physiol 53:445–466

Bishop MR, Ramasastry BV, Schmidt DE and Harbison RD (1976) Occurrence of choline acetyltransferase and acetylcholine and other quarternary ammonium compounds in mammalian spermatozoa. Biochem Pharmacol 25:1617–1622

Bishop MR, Ramasastry BV and Stavinoha WB (1977) Identification of acetylcholine and propionylcholine in bull spermatozoa by integrated pyrolysis, gas chromatography and mass spectrometry. Biochem Biophys Acta 500:440–444

Blandau RJ (1945) On the factors involved in sperm transport through the cervix uteri of the albino rat. Am J Anat 77:253–272

Blandau RJ (1969) Gamete transport – Comparative Aspects. In: The Mammalian Oviduct. Composition, Biology and Methodology. Eds ESE Hafez and RJ Blandau, University of Chicago Press, Chicago. pp 129–162

Blandau RJ and Rumery RE (1964) The relationship of swimming movements of epididymal spermatozoa to their fertilizing capacity. Fert Steril 15:571–579

Blandau RJ and Young WC (1939) The effects of delayed fertilization on the development of the guinea pig ovum. Am J Anat 64:303–329

Bouchard P, Gagnon C, Phillips DM and Bardin CW (1980) The localization of protein carboxymethylase in sperm tails. J Cell Biol 86:417–423

Bouchard P, Penningroth SM, Cheung A, Gagnon C and Bardin CW (1981) Erythro-9-(3-)2-hydroxynonyl))adenine is an inhibitor of sperm motility that blocks dynein ATPase and protein carboxymethylase activities. Proc Nat Acad Sci US 78:1033–1036

Brackett BG, Hall JL and Oh YK (1978) In-vitro fertilizing ability of testicular, epididymal, and ejaculated rabbit spermatozoa. Fert Steril 29:571–582

Bradley MP and Forrester IT (1980a) A sodium-calcium exchange mechanism in plasma membrane vesicles isolated from ram sperm flagella. FEBS Letts 121:15–18

Bradley MP and Forrester IT (1980b) The inhibition of sperm motility and flagellar plasma membrane (Ca^{2+} + Mg^{2+}) ATPase activity by quercetin. Proc U Otago Med Sch 58:69–70

Bradley MP and Forrester IT (1981) Stimulation of ram caudal sperm motility by a seminal plasma factor calsemin. Proc U Otago Med Sch 59:6–8

Bradley and MP and Forrester IT (1985) Sperm calcium homeostasis during maturation. In: Male Fertility and its Regulation. Eds TJ Lobl and ESE Hafez, MTP Press, Lancaster. pp 437–449

Brandt H and Hoskins DD (1980) A cAMP-dependent phosphorylated motility protein in bovine epididymal sperm. J Biol Chem 255:982–987

Brandt H, Acott TS, Johnson DJ and Hoskins DD (1978) Evidence for an epididymal origin of bovine sperm forward motility protein. Biol Reprod 19:830–835

Breitbart H, Rubinstein S and Nass-Arden L (1985) The role of Ca^{2+} and Ca^{2+}-ATPase in maintaining motility in ram spermatozoa. J Biol Chem 260:11548–11553

Brooks JC and Siegel FL (1973) Calcium-binding phosphoprotein: the principal acidic protein of mammalian sperm. Biochem Biophys Res Commun 55:710–716

Burgos MH and Tovar ES (1974) Sperm motility in the rat epididymis. Fert Steril 25:985–991

Calvin HI and Bedford JM (1971) Formation of disulphide bonds in the nucleus and accessory structures of mammalian spermatozoa during maturation in the epididymis. J Reprod Fert Suppl 13:65–75

Calvin HI and Bleau G (1974) Zinc-thiol complexes in keratinoid structures of rat spermatozoa. Exp Cell Res 86:280–285

Calvin HI, Yu CC and Bedford JM (1973) Effects of epididymal maturation, zinc (II) and copper (II) on the reactive sulphydryl content of structural elements in rat spermatozoa. Exp Cell Res 81:333–341

Carter AL, Stratman FW, Hutson SM and Lardy HA (1980) The role of carnitine and its esters in sperm metabolism. In: Carnitine Biosynthesis, Metabolism and Functions. Eds RA Frenkel and JD McGarry, Acad Press, New York. pp 251–264

Cascieri M, Amann RP and Hammerstedt RH (1976) Adenine nucleotide changes at initiation of bull sperm motility. J Biol Chem 251:787–793

Casillas ER (1973) Accumulation of carnitine by bovine spermatozoa during maturation in the epididymis. J Biol Chem 248:8227–8232

Casillas ER, Elder CM and Hoskins DD (1980) Adenylate cyclase activity of bovine spermatozoa during maturation in the epididymis and the activation of sperm particulate adenylate cyclase by GTP and polyamines. J Reprod Fert 59:297–302

Castaneda E, Bouchard P, Saling P, Phillips D, Gagnon C and Bardin CW (1983) Endogenous protein carboxyl methylation in hamster spermatozoa: changes associated with capacitation in vitro. Int J Androl 6:482–496

Chaix P, Morin GA and Jezequel J (1950) Sur la participation des phosphates aux reactions d'oxydation de l'adrenaline. Biochem Biophys Acta 5:472–476

Chakraborty J and Nelson L (1974) Organization and redistribution of adenosine-triphosphatase during spermiogenesis in the mouse. Biol Reprod 10:85–97

Chapvil M (1973) New aspects in the biological role of zinc: a stabilizer of macromolecules and biological membranes. Life Sci 13:1041–1049

Chenoweth MB (1956) Chelation as a mechanism of pharmacological action. Pharm Rev 8:57–87

Chulavatnatol M and Treetipsatit N (1983) Initiation of sperm flagellar movement using rat demembranated sperm model: nucleotide specificities. In: The Sperm Cell. Ed J Andre, Martinus-Nijhoff, London. pp 364–367

Chulavatnatol M and Yindepit S (1976) Changes in surface ATPase of rat spermatozoa in transit from the caput to the cauda epididymidis. J Reprod Fert 48:91–97

Chulavatnatol M, Hasibuan I, Yindepit S and Eksiittikul T (1977) Lack of effect of a-chlorohydrin on the ATP content of rat, mouse and human spermatozoa. J Reprod Fert 50: 137–139

Chulavatnatol M, Eksittikul T and Toowicharanont P (1978) Control of epididymal sperm motility; an approach to male fertility regulation. Int J Androl Suppl 2:629–638

Chulavatnatol M, Panyin S and Wittisuwannakul D (1982) Comparison of phosphorylated proteins in intact rat spermatozoa from caput and cauda epididymidis. Biol Reprod 26:197–207

Clarke GN and Yanagimachi R (1978) Actin in mammalian sperm heads. J exp Zool 205:125–132

Clarke GN, Clarke FM and Wilson S (1982) Actin in human spermatozoa. Biol Reprod 26:319–327

Cooper GW, Overstreet JW and Katz DF (1979) The motility of rabbit spermatozoa recovered from the female reproductive tract. Gamete Research 2:35–42

Cooper TG (1985) Prevention of hypo-osmotic swelling by detergents provides clues to the membrane structure of rat sperm. Int J Androl 8:159–167

Cooper TG and Orgebin-Crist MC (1975) The effect of epididymal and testicular fluids on the fertilising capacity of testicular and epididymal spermatozoa. Andrologia 7:85–93

Cooper TG and Orgebin-Crist MC (1977) Effect of aging on the fertilizing capacity of testicular spermatozoa from the rabbit. Biol Reprod 16:258–266

Cornett E and Meizel S (1980) 9-AAP, a fluorescent β-adrenergic antagonist, enters the hamster sperm acrosome in a manner inconsistent with binding to β-adrenergic receptors. J Histo Cyto 28:462–464

Crabo BG and Hunter AG (1975) Sperm maturation and epididymal function. In: Control of

Male Fertility. Eds JJ Sciarra, C Markland and JJ Speidel, Harper Row, Hagerstown. pp 2–23

Cummins JM (1976) Effects of epididymal occlusion on sperm maturation in the hamster. J exp Zool 197:187–190

Dacheux JL and Paquignon M (1980a) Effects of caffeine on ram and boar spermatozoa; influence of their stage of maturation and the medium; initiation of progressive motility of testicular spermatozoa. In: Testicular Development, Structure, and Function. Eds A Steinberger and E Steinberger, Raven Press, New York. pp 513–522

Dacheux JL and Paquignon M (1980b) Relations between the fertilizing ability, motility and metabolism of epididymal spermatozoa. Reprod Nutr Develop 20:1085–1099

Dacheux JL and Paquignon M (1983) Influence of initiation of forward motility on the fertilizing ability of immature boar spermatozoa in in vivo homologous and in vitro heterologous systems of insemination In: The Sperm Cell. Ed J Andre, Martinus-Nijhoff, The Hague. pp 99–102

Dawson MI, Goerver RTN and Tait A (1979) The effect of N-(methoxy polyethylene glycol)p-hydroxymercuribenzamide, a polymeric sulphydyl group reagent, on spermatozoan motility. Contraception 20:159–165

del Rio AG and Raisman R (1978) cAMP in spermatozoa taken from different segments of the rat epididymis. Experientia 34:670–671

Djøseland O, Gordeladze JO, Høglo C, Halse JI and Haugen HN (1980) Evidence for androgen-dependent phosphodiesterasse activity in rat seminal vesicle and epididymis. Int J Androl 3:363–366

Drevius LO (1975) Permeability of the bull-sperm membrane. In: The Functional Anatomy of the Spermatozoon. Ed BA Afzelius, Pergamon Press, Oxford. pp. 373–383

Egbunike GN (1980) Changes in the acetylcholinesterase activity of mammalian spermatozoa during maturation. Int J Androl 3:459–468

Egbunike GN (1982) Effect of chloroquine on the motility and acetylcholinesterase activity of porcine spermatozoa during epididymal maturation. Andrologia 14:503–508

Egbunike GN, Branscheid W, Pfisterer J and Holtz W (1986) Changes in porcine sperm lactate dehydrogenase isoenzymes during sperm matruation. Andrologia 18:108–113

Elfvin LG (1968) An ultrastructural difference between the outer and inner membrane of the middle piece mitochondria in rat spermatozoa. J Ultrastruct Res 24:259–268

Feinberg J, Weinman J, Weinman S, Walsh MP, Harricane MC, Gabaion J and Demaille JG (1981) Immunocytochemical and biochemical evidence for the presence of calmodulin in bull sperm flagellum. Isolation and characterization of sperm calmodulin. Biochem Biophys Acta 673:303–311

Fentie IH and Lindemann CB (1978) Simultaneous comparisons of motility and ATPase activity in Triton X-100 extracted rat sperm. J Cell Biol 79:290a (abstr)

Fisher-Fischbein J, Gagnon C and Bardin CW (1985) The relationship between glycolysis, mitochondrial respiration, protein-carboxylmethylation and motility in hamster epididymal spermatozoa. Int J Androl 8:403–416

Florman HM and Storey BT (1981) Inhibition of in vitro fertilization of mouse eggs: 3-quinuclidinyl benzilate specifically blocks penetration of zonae pellucidae by mouse spermatozoa. J exp Zool 216:159–167

Florman HM and Storey BT (1982) Characterization of cholinomimetic agents that inhibit in vitro fertilization in the mouse. Evidence for a sperm-specific binding site. J Androl 3:157–164

Ford WCL (1981) The oxidation of glycerol 3-phosphate by testicular and epididymal spermatozoa. Comp Biochem Physiol 68B:289–293

Ford WCL (1982) The mode of action of 6-chloro-6-deoxy sugars as antifertility agents in the male. In: Progress Towards a Male contraceptive. Eds SL Jeffcoate and M Sandler, John Wiley and Sons Ltd., Chichester. pp 159–184

Ford WCL and Harrison A (1983) The activity of glyceraldehyde 3-phosphate dehydrogenase in spermatozoa from different regions of the epididymis in laboratory rodents treated with a-chlorohydrin or 6-chloro-deoxyglucose. J Reprod Fert 69:147–156

Fray CS, Hoffer AP and Fawcett DW (1972) A reexamination of motility patterns of rat epididymal spermatozoa. Anat Rec 173:301–308

Frenkel G, Peterson RN and Freund M (1973a) Changes in the metabolism of guinea pig sperm from different segments of the epididymis. Proc Soc exp Biol Med 143:1231–1236

Frenkel G, Peterson RN and Freund M (1973b) The role of adenine nucleotides and the effect of caffeine and dibutyryl cyclic AMP on the metabolism of guinea pig epididymal spermatozoa. Proc Soc exp Biol Med 144:420–425

Friend DS and Heuser JE (1981) Orderly particle arrays on the mitochondrial outer membrane in rapidly frozen sperm. Anat Rec 199:159–175

Gaddum P (1968) Sperm maturation in the male reproductive tract. Development of motility. Anat Rec 161:471–482

Gaddum P and Glover TD (1965) Some reactions of rabbit spermatozoa to ligation of the epididymis. J Reprod Fert 9:119–130

Gaddum P, Blandau RJ and Hayashi R (1968) Sperm maturation in the male reproductive tract. Film. University of Washington Press

Gaddum-Rosse P (1981) Some observations on sperm transport through the uterotubal junction of the rat. Am J Anat 160:333–341

Gagnon C and Heisler S (1979) Minireview. Protein carboxy-methylation: role in exocytosis and chemotaxis. Life Sci 25:993–1000

Gagnon C, Sherins RJ, Mann T, Bardin CW, Amelar RD and Dubin L (1980a) Deficiency of protein carboxyl-methylase in spermatozoa of necrospermic patients. In: Testicular Development, Structure, and Function. Eds A Steinberger and E Steinberger, Raven Press, New York US. pp 491–495

Gagnon C, Kelly S, Manganiello V, Vaughn M, Strittmatter W, Hoffman A and Hirata F (1980b) Protein carboxyl-methylase modifies calmodulin function. Ann NY Acad Sci 356:385–386

Gagnon C, Sherins RJ, Phillips DM and Bardin CW (1982) Deficiency of protein-carboxyl methylase in immotile spermatozoa of infertile men. New Engl Med J 306:821–825

Gagnon C, Harbour D, de Lamirande E, Bardin CW and Dacheux JL (1984) Sensitive assay detects protein methylesterase in spermatozoa: decrease in enzyme activity during epididymal maturation. Biol Reprod 30:953–958

Garbers DL, Lust WD, First NL and Lardy HA (1971) Effects of phosphodiesterase inhibitors and cyclic nucleotides on sperm respiration and motility. Biochemistry 10:1825–1831

Garbers DL, Tubb DJ and Hyne RV (1982) A requirement of bicarbonate for Ca^{2+}-induced elevations of cyclic AMP in guinea pig spermatozoa. J Biol Chem 257:8980–8984.

Glover TD (1962) The response of rabbit spermatozoa to artificial cryptorchidism and ligation of the epididymis. J Endocr 23:317–328

Gwatkin RBL (1983) Effect of compounds structurally related to taurine and of taurine uptake inhibitors on the motility of hamster sperm in vitro. Gamete Res 4:347–350

Glück E and Paul RJ (1977) The aerobic metabolism of porcine carotid artery and its relationship to isometric force. Energy cost of isometric contraction. Pflügers Arch Ges Physiol 370:9–18

Goh P and Hoskins DD (1985) The involvement of methyl transfer reactions and S-adenosylhomocysteine in the regulation of bovine sperm motility. Gamete Research 12:399–409

Gonse PH (1962) Respiration and oxidative phosphorylation in relation to sperm motility. In: Spermatozoon Motility. Ed DW Bishop, Amer Assoc Adv Sci, Washington DC. pp 99–132

Goodman DR and Harbison RD (1981) Characterization of enzymatic acetylcholine synthesis by mouse brain, rat sperm, and purified carnitine acetyltransferase. Biochem Pharmacol 30:1521–1528

Goodman DR, Adatsi FK and Harbison RD (1984) Evidence for the extreme overestimation of choline acetyltransferase in human sperm, human seminal plasma and rat heart: a case of mistaking carnitine acetyltransferase for choline acetyltransferase. Chem Biol Interact 49:39–53

Gordon M and Barrnett RJ (1967) Fine structural cytochemical localizations of phosphatase activities of rat and guinea pig. Exp Cell Res 48:395–412

Gordon M, Morris EG and Young RJ (1983) The localization of Ca^{2+}-ATPase and Ca^{2+} binding proteins in the flagellum of the guinea pig. Gamete Res 8:49–55

Halangk W, Bohnensack R and Kunz W (1985) Interdependence of mitochondrial ATP production and extramitochondrial ATP utilization in intact spermatozoa. Biochem Biophys Acta 808:316–322

Hammerstedt RH (1975) Tritium release from (2-^{3}H)D-glucose as a monitor of glucose consumption by bovine sperm. Biol Reprod 12:545–551

Hammerstedt RH and Hay SR (1980) Effect of incubation temperature on motility and cAMP content of bovine sperm. Arch Biochem Biophys 199:427–437

Hanksi E and Garty GNB (1983) Activation of adenylate cyclase by sperm membranes. The role of guanine nucleotide binding proteins. FEBS Letts 162:447–452

Harbison RD, Dwivedi C and Evans MA (1976a) A proposed mechanism for trimethylphosphate induced sterility. Toxicol Appl Pharmacol 35:481–490

Harbison RD, Orgebin-Crist MC, Dwivedi C and Rama Sastry BV (1976b) Cholineacetyltransferase activity in sperm of rabbit and man: possible involvement in TMP-induced sterility. In: Recent Advances in Contraceptive Technology. Proc Int Symp. pp 125–130

Hiipaka RT and Hammerstedt RH (1979) Changes in 2-deoxyglucose transport during epididymal maturation of ram sperm. Biol Reprod 19:1030–1035

Hildebrandt JD, Codina J, Tash JS, Kirchick HJ, Lipschultz L, Sekura (1985) The membrane-bound spermatozoa adenylyl cyclase system does not share coupling characteristics with somatic cell adenylyl cyclases. Endocrinology 116:1357–1366

Hinton BT (1980) The epididymal microenvironment. A site of attack for a male contraceptive? Invest Urol 18:1–10

Hinton BT and Setchell BP (1980) Concentration and uptake of carnitine in the rat epididymis. A micropuncture study. In: Carnitine Biosynthesis, Metabolism, and Function. Eds RA Frenkel and JD McGarry, Acad Press, New York. pp 237–250

Hinton BT, Dott HM and Setchell BP (1979) Measurement of the motility of rat spermatozoa collected by micropuncture from the testis and from different regions along the epididymis. J Reprod Fert 55:167–172

Hinton BT, Brooks DE, Dott HM and Setchell BP (1981) Effects of carnitine and some related compounds on the motility of rat spermatozoa from the caput epididymidis. J Reprod Fert 61:59–64

Hirata F and Axelrod J (1980) Phospholipid methylation and biological signal transmission. Science 209:1082–1090

Hoffmann JF (1969) The interaction between tritiated ouabain and the Na-K pump in red blood cells. J Gen Physiol 54:343S–350S

Hong CY, Chiang BN and Wei YH (1983) Mitochondrtial respiration inhibitors and human sperm motility: implications in the development of spermicides. Brit J Clin Pharm 16:487–490

Horan AH and Bedford JM (1972) Development of the fertilizing ability of spermatozoa in the epididymis of the Syrian hamster. J Reprod Fert 30:417–423

Hoskins DD (1973) Adenine nucleotide mediation of fructolysis and motility in bovine epididymal spermatozoa. J Biol Chem 248:1135–1140

Hoskins DD and Casillas ER (1975a) Function of cyclic nucleotides in mammalian spermatozoa. In: Handbook of Physiology, Section and Endocrinology, Volume 5 Male Reproductive System. Eds DW Hamilton and RO Greep, American Physiol. Soc., Washington DC. pp 453–460

Hoskins DD and Casillas ER (1975b) Hormones, second messengers, and the mammalian spermatozoon. In: Molecular mechanisms of gonadal hormone action. Vol 1. Eds JA Thomas and RL Singhal, HM and M Medical Scientific Publishers, Aylesbury. pp 283–324

Hoskins DD, Stephens DT and Hall ML (1974) Cyclic adenosine 3′:5′-monophosphate and protein kinase levels in developing bovine spermatozoa. J Reprod Fert 37:131–133

Hoskins DD, Munsterman D and Hall ML (1975a) The control of bovine sperm glycolysis during epididymal transit. Biol Reprod 12:566–572

Hoskins DD, Hall ML and Munsterman D (1975b) Induction of motility in immature bovine spermatozoa by cyclic AMP phosphodiesterase inhibitors and seminal plasma. Biol Reprod 13:168–176

Hoskins DD, Brandt H and Acott TS (1978) Initiation of sperm motility in the mammalian epididymis. Fed Proc 37:2534–2542

Hoskins DD, Johnson D, Brandt H and Acott TS (1979) Evidence for a role for a forward motility protein in the epididymal development of sperm motility. In: The Spermatozoon. Eds DW Fawcett and JM Bedford, Urban and Schwarzenberg, Baltimore. pp 43–53

Hoskins DD, Acott TS, Critzchlow L and Vijayaraghavan S (1983) Studies on the roles of cyclic AMP and calcium in the development of bovine sperm motility. J Submicroscop Cytol 15:21–27

Howe GR and Black DL (1963) Migration of rat and foreign spermatozoa through the utero-tubal junction of the oestrous rat. J Reprod Fert 5:95–100

Hunter RHF, Cook B and Poyser NL (1983) Regulation of oviduct function in pigs by local transfer of ovarian steroids and prostaglandins: a mechanism to influence sperm ascent. Europ J Obstet Gynaecol reprod Biol 14:225–232

Hyne RV and Garbers DL (1979a) Calcium-dependent increase in adenosine-3′,5′-monophosphate and induction of the acrosome reaction in guinea pig spermatozoa. Proc Nat Acad Sci US 76:5699–5703

Hyne RV and Garbers DL (1979b) Regulation of guinea pig sperm adenylate cyclase by calcium. Biol Reprod 21:1135–1142

Inskeep PB and Hammerstedt RH (1982) Changes in metabolism of ram sperm associated with epididymal transit or induced by endogenous carnitine. Biol Reprod 27:735–743

Inskeep PB and Hammerstedt RH (1985) Endogenous metabolism of sperm in response to altered cellular ATP requirements. J Cell Physiol 123:180–190

Jakobs KH, Johnson RA and Schultz G (1983) Activation of human platelet adenylate cyclase by a bovine sperm component. Biochem Biophys Acta G756:369–375

Johnson RA, Awad JA, Jakobs KS and Schultz G (1983) Activation of brain adenylate cyclase by a factor derived from bovine sperm. FEBS Lett 152:11–16

Johnson RA, Jakobs KH and Schultz G (1985) Extraction of the adenylate cyclase-activating factor of bovine sperm and its identification as a trypsin-like protease. J Biol Chem 260:114–121

Jones RC (1971) Studies of the structure of the head of boar spermatozoa from the epididymis. J Reprod Fert Suppl 13:51–64

Jones R (1978) Comparative biochemistry of mammalian epididymal plasma. Comp Biochem Biophys 61B:365–370

Jones HP, Lenz RW, Palevitz BA and Cormier MJ (1980a) Calmodulin localization in mammalian spermatozoa. Proc Nat Acad Sci US 77:2772–2776

Jones HP, Lenz RW, Palevitz BA and Cormier MJ (1980b) Localization of calmodulin in mammalian spermatozoa by immunofluorescence methods. Annals NY Acad Sci US 356:393–394

Joshi MS, Yaron A and Lindner HR (1972) Intrauterine gelation of seminal plasma components in the rat after coitus. J Reprod Fert 30:27–37

Jouannet P (1981) Movement of human spermatozoa from caput epididymis. In: Epididymis and Fertility: Biology and Pathology. Eds C Bollack and A Clavert, Karger, Munich. pp 100–101

Kann ML and Serres C (1980) Development and initiation of sperm motility in the hamster epididymis. Reprod Nutr Develop 20:1739–1749

Kann ML and Raynaud F (1982) In vivo fertilization after initiation of sperm motility in the hamster epididymis. Reprod Nutr Develop 22:455–463

Klinefelter GR and Hamilton DW (1984) Organ culture of rat caput epididymal tubules in a perifusion chamber. J Androl 5:243–258

Kobayashi T, Martensen T, Nath J and Flavin M (1978) Inhibition of dynein ATPase by vanadate, and its possible use as a probe for the role of dynein in cytoplasmic motility. Biochem Biophys Res Commun 81:1313–1318

Kremer J and Jager S (1976) The sperm-cervical mucus contact test: a preliminary report. Fert Steril 27:335–340

Kremer J, Jager W, Kuiken J and van Slochteren-Draaisma (1978) Recent advances in diagnosis and treatment of infertility due to antisperm antibodies. In: Spermatozoa, Antibodies and Infertility. Eds J Cohen and WF Hendry, Blackwell, Oxford. pp 117–127

Lambiase JT and Amann RP (1973) Infertility of rabbit testicular spermatozoa collected in their native environment. Fert Steril 24:65–67

Lea OA and French FS (1981) Characterization of an acidic glycoprotein secreted by principal cells of the rat epididymis. Biochem Biophys Acta 668:370–376

Lea OA, Petrusz P and French FS (1978) Purification and localization of acidic epididymal glycoprotein (AEG): a sperm coating protein secreted by the rat epididymis. Int J Androl Suppl 2:592–607

Lee WM, Tsang AYF and Wong PYD (1981) Effects of divalent and lanthanide ions on motility initiation in rat caudal epididymal spermatozoa. Brit J Pharmac 73:633–638

Levin RM, Amsterdam JD, Winokur A and Wein AJ (1981) Effects of psychotropic drugs on human sperm motility. Fert Steril 36:503–506

Linck RW (1979) Advances in the ultrastructural analysis of the sperm flagellar axoneme. In: The Spermatozoon. Eds DW Fawcett and JM Bedford, Urban and Schwarzenberg, Baltimore. pp 99–115

Lindahl PE (1978) Head-to-head association in bovine spermatozoa induced by catecholamines. Exp Cell Res 113:421–433

Lindemann CB (1978) A cAMP-induced increase in the motility of demembranated bull sperm models. Cell 13:9–18

Lindemann CB (1980) Requirements for motility in mammalian sperm. In: Testis structure, Development, and Function. Eds A. Steinberger and E. Steinberger, Raven Press, New York. pp 473–479

Lindemann CB and Gibbons IR (1975) Adenosine triphosphate-induced motility and sliding of filaments in mammalian sperm extracted with Triton X-100. J Cell Biol 65:147–162

Lindemann CB and Rikmenspoel R (1971) Intracellular potentials in bull spermatozoa. J Physiol 219:127–138

Lindemann CB and Rikmenspoel R (1972a) Sperm flagella: autonomous oscillations of the contractile system. Science 175:337–338

Lindemann CB and Rikmenspoel R (1972b) Sperm flagellar motion maintained by ADP. Exp Cell Res 73:255–259

Lindemann CB, Rudd WG and Rikmenspoel R (1973) The stiffness of the flagella of impaled bull sperm. Biophys J 13:437–448

Lindemann CB, Fentie I and Rikmenspoel R (1980) A selective effect of Ni^{2+} on wave initiation in bull sperm flagella. J Cell Biol 87:420–426

Lindemann CB, Lipton M and Shlafer R (1983) The interaction of cAMP with modeled bull sperm. Cell Motility 3:199–210

Lindholmer C (1974) The importance of seminal plasma for human sperm motility. Biol Reprod 10:533–542

Little M and Rohricht C (1983) Pig sperm tail tubulin. Its extraction and characterization. Exp Cell Res 147:15–22

Llanos MN and Meizel S (1983) Phospholipid methylation increases during capacitation of golden hamster sperm in vitro. Biol Reprod 28:1043–1051

Maas JW and Colburn RW (1965) Co-ordination chemistry and membrane function with particular reference to the synapse and catecholamine transport. Nature 208:41–46

Majumder GC (1981) Enzymic characteristics of ecto-adenosine triphosphatase in rat epididymal intact spermatozoa. Biochem J 195:103–110

Majumder GC and Biswas R (1979) Evidence for the occurence of an ecto-(adenosine triphosphatase) in rat epididymal spermatozoa. Biochem J 183:737–743

Majumder GC and Chaudhuri DP (1984) Occurence of -SH containing molecules on the goat sperm external surface that are essential for flagellar motility. Andrologia 16:219–223

Martan J and Shepherd BA (1976) The role of the copulatory plug in reproduction of the guinea pig. J exp Zool 196:79–84

McGrady A (1979) The effect of ouabain on membrane potential and flagellar wave in ejaculated bull spermatozoa. J Reprod Fert 56:549–553

McGrady and Nelson L (1972) Cationic influences on sperm biopotentials. Exp Cell Res 73:192–196

McGrady AV and Nelson L (1973) Electrophysiology of bull spermatozoa. Correlations with motility. Exp Cell Res 76:349–352

McGrady AV and Nelson L (1974) Effect of calcium on bull spermatozoan membrane potential. Exp Cell Res 87:398–400

McGrady AV and Nelson L (1976) Cholinergic effects on bull and chimpanzee sperm motility. Biol Reprod 15:248–253
McGrady AV, Nelson L and Ireland M (1974) Ionic effects on the motility of bull and chimpanzee spermatozoa. J Reprod Fert 40:71–76
McLaughlin J and Terner C (1973) Ribonucleic acid synthesis by spermatozoa from the rat and hamster. Biochem J 133:635–639
Means AR, Tash JS and Chaouleas JG (1982) Physiological implications of the presence, distribution, and regulation of calmodulin in Eukaryotic cells. Physiol Rev 62:1–39
Meizel S (1981) Inhibition of the hamster sperm acrosome reaction by transmethylase inhibitors. J exp Zool 217:443–446
Meizel S, Lui CW, Working PK and Mrsny RJ (1980) Taurine and hypotaurine: their effects on motility, capacitation and the acrosome reaction of hamster sperm in vitro and their presence in sperm and reproductive tract fluids of several mammals. Develop Growth Diff 22:483–494
Moghissi KS, Dabich D, Levine J and Neuhaus OW (1964) Mechanism of sperm migration. Fert Steril 15:15–23
Mohri H and Ogawa K (1975) Tubulin and dynein in spermatozoon motility. In: The Functional Anatomy of the Spermatozoon. Ed BA Afzelius, Pergamon Press, Oxford. pp 161–167
Mohri H and Yanagimachi R (1980) Characteristics of motor apparatus in testicular, epididymal and ejaculated spermatozoa. A study using demembranated sperm models. Exp Cell Res 127:191–196
Mohri H and Yano Y (1980) Analysis of mechanism of flagellar movement with golden hamster spermatozoa. Biomed Res 1:552–555
Mohri H and Yano Y (1982) Reactivation and microtubule sliding in rodent spermatozoa. Cell Motility Suppl 1:143–147
Mohri H, Mohri T and Ernster L (1965) Isolation and enzymic properties of the midpiece of bull spermatozoa. Exp Cell Res 38:217–246
Mohri H, Ishijima S and Hiramoto Y (1983) Movement characteristics of hamster spermatozoa. In: The Sperm Cell. Ed J Andre, Martinus-Nijhoff, London. pp 345–348
Mongkolsirikieat S and Boonsaeng V (1980) Increase in sperm type hexokinase activity of rat spermatozoa during maturation. Int J Androl 3:671–678
Mooney JK, Horan AH and Lattimer JK (1972) Motility of spermatozoa in the human epididymis. J Urol 108:443–445
Morton B (1973) Reactivation of progressive motility in hamster sperm modified by Triton X-100. Exp Cell Res 79:106–110
Morton BE and Sagadraca R (1981) Quantitation of sperm population migration: capillary scanning assay. Archs Androl 7:219–227
Morton B, Harrigan-Lum J, Albagli L and Joos T (1974) The activation of motility in quiescent hamster sperm from the epididymis by calcium and cyclic nucleotides. Biochem Biophys Res Commun 56:372–379
Morton BE, Fraser CF and Albagli L (1979) Studies on factors in hamster caudal epididmal plasma and other sources which inhibit sperm dilution damage. Fert Steril 32:99–106
Mrsny RJ and Meizel S (1985) Inhibition of hamster sperm Na^+, K^+-ATPase activity by taurine and hypotaurine. Life Sci 36:271–275
Murdoch RN and White IG (1968) Metabolic studies of testicular, epididymal, and ejaculated spermatozoa of the ram. Aust J Biol Sci 21:111–121
Nachmansohn D and Schneemann H (1945) The effect of drugs on cholinesterase. J Biol Chem 159:239–240
Nagano T (1965) Localization of adenosine triphosphatase activity in the rat sperm tail as revealed by electron microscopy. J Cell Biol 25:101–112
Nelson L (1962) Cytochemical aspects of spermatozoan motility. In: Spermatozoan Motility, Ed DW Bishop, Amer Assoc Adv Sci, Washington DC. pp 171–187
Nelson L (1966) Enzyme distribution in "naturally-decapacitated" bull spermatozoa: acetylcholinesterase, adenylpyrophosphatase and adenosinetriphosphatase. J exp Physiol 68: 113–116
Nelson L (1967) Sperm Motility. In: Fertilization. Comparative Morphology, Biochemistry and Immunology. Eds CB Metz and A Monroy, Vol I, Acad Press, London. pp 27–97

Nelson L (1975) Control of sperm motility: a neurochemical approach. In: The Functional Anatomy of the Spermatozoon. Ed BA Afzelius, Pergamon Press, Oxford. pp 169–176

Nelson L (1978) Chemistry and neurochemistry of sperm motility control. Fed Proc 37:2543–2547

Nelson L (1979) Second messenger control of sperm motility. In: Motility in Cell Function. Eds FA Pope, JW Sanger and VT Nachmias, Acad Press, London. pp 453–455

Nelson L, Chakraborty J, Young M, Goodwin A, Kock E and Gardner ME (1980a) Control of sperm cell motility: neurochemical regulation of calcium transport. In: Testicular Development, Structure, and Function. Eds A Steinberger and E. Steinberger, Raven Press, New York. pp 503–511

Nelson L, Young MJ and Gardner ME (1980b) Sperm motility and calcium transport: a neurochemically controlled process. Life Sci 26:1739–1749

Nelson L, Gardner ME and Young MJ (1980c) Calcium binding in relation to sperm motility control. In: Calcium Binding Proteins: structure and function. Eds FL Siegel, E Carafoli, DH MacLennon and RH Wasserman, Elsevier North Holland. pp 495–498

Nelson L, Gardner ME and Young MJ (1982) Regulation of calcium distribution in bovine sperm cells: cytochemical evidence for motility control mechanisms. Cell Motility 2:225–242

Nevo AC and Rikmenspoel R (1970) Diffusion of ATP in sperm flagella. J Theoret Biol 26:11–18

Noyes RW, Adam CE and Walton A (1958) Transport of spermatozoa into the uterus of the rabbit. Fert Steril 9:288–299

O'Day PM and Rikmenspoel R (1979) Electrical control of flagellar activity in impaled bull spermatozoa. J Cell Sci 35:123–138

O'Donnell JM (1969) Electrical counting and sizing of mammalian spermatozoa and cytoplasmic droplets. J Reprod Fert 19:263–272

O'Donnell JM and Ellory JC (1969) Cardiac glycoside binding and alkali cation transport in mature and epididymal bull spermatozoa. J Reprod Fert 20:359–360 Abs

O'Donnell JM and Ellory JC (1970a) The binding of cardiac glycosides to bull spermatozoa. Experientia 26:20–21

O'Donnell JM and Ellory JC (1970b) The binding of ouabain to spermatozoa of boar and ram. J Reprod Fert 23:181–184

Olson GE and Linck RW (1977) Observations on the structural components of flagellar axonemes and central pair of microtubules from rat sperm. J Ultrastruct Res 61:21–43

Olson GE and Sammons DW (1980) Structural chemistry of outer dense fibres of rat sperm. Biol Reprod 22:319–332

Orgebin-Crist MC (1967a) Maturation of spermatozoa in the rabbit epididymis: fertilizing ability and embryonic mortality in does inseminated with epididymal spermatozoa. Ann biol Anim Biochem Biophys 7:373–389

Orgebin-Crist MC (1967b) Sperm maturation in rabbit epididymis. Nature (Lond.) 216:816–818

Orgebin-Crist MC (1968) Maturation of spermatozoa in the rabbit epididymis: Delayed fertilization in does inseminated with epididymal spermatozoa. J Reprod Fert 16:29–33

Orgebin-Crist MC (1973) Maturation of spermatozoa in the rabbit epididymis: effect of castration and testosterone replacement. J exp Zool 185:301–310

Orgebin-Crist MC, Olson GE and Danzo BJ (1981) Factors influencing maturation of spermatozoa in the epididymis. In: Intragonadal Regulators of Reproduction, Eds P Franchimont and CP Channing, Academic Press, New York. pp 393–417

O'Shea T and Voglmayr JK (1970) Metabolism of glucose, lactate, and acetate by testicular and ejaculated spermatozoa of the ram. Biol Reprod 2:326–332

Overstreet JW and Bedford JM (1974) Transport, capacitation and fertilizing ability of epididymal spermatozoa. J exp Zool 189:203–214

Overstreet JW and Bedford JM (1976) Embryonic mortality in the rabbit is not increased after fertilization by young epididymal spermatozoa. Biol Reprod 15:54–57

Overstreet JW and Cooper GW (1978) Sperm transport in the reproductive tract of the female rabbit: I. The rapid transit phase of transport. Biol Reprod 19:101–114

Overstreet JW and Katz DF (1977) Sperm transport and selection in the female genital tract. In: Development in Mammals. Ed MH Johnson, North Holland Pub Co, Oxford. Vol 2:31–65

Overstreet JW and Tom RA (1982) Experimental studies of rapid sperm transport in rabbits. J Reprod Fert 66:601–606

Panse GT, Jayarraman S and Sheth AR (1983) Shift of glycolysis as a marker of sperm maturation. Archs Androl 11:137–140

Pariset CC, Feinberg JMF, Dacheux JL and Weinman SJ (1985) Changes in calmodulin level and cAMP-dependent protein kinase activity during epididymal maturation of ram spermatozoa. J Reprod Fert 74:105–112

Paul RJ, Bauer M and Pease W (1979) Vascular smooth muscle: aerobic glycolysis linked to sodium and potassium transport processes. Science 206:1414–1416

Pavlok A (1974) Development of the penetrating activity of mouse spermatozoa in vivo and in vitro. J Reprod Fert 36:203–205

Paz(Frenkel) G, Kaplan R, Yedwab G, Homonnia ZT and Kraicer PF (1978) The effect of caffeine on rat epididymal spermatazoa: motility, metabolism and fertilizing capacity. Int J Androl 1:145–152

Pederson H and Rebbe H (1975) Absence of arms in the axoneme of immoble human spermatozoa. Biol Reprod 12:541–544

Perloff WH and Steinberger E (1963) In vitro penetration of cervical mucus by spermatozoa. Fert Steril 14:231–236

Peterson RN, Seyler D, Bundman D and Freund M (1979) The effect of theophylline and dibutyryl cyclic AMP on the uptake of radioactive calcium and phosphate ions by boar and human spermatozoa. J Reprod Fert 55:385–390

Peterson RN, Ashraf M and Russell LD (1983) Effect of calmodulin antagonists on Ca^{2+} uptake by boar spermatozoa. Biochem Biophys Res Commun 114:28–33

Phillips DM and Olson G (1975) Mammalian sperm motility-structure in relation to function. In: The Functional Anatomy of the Spermatozoon. Ed BA Afzelius, Pergamon Press, Oxford. pp 117–126

Pholpramool C and Chaturapanich G (1979) Effect of sodium and potassium concentration and pH on the maintenance of motility of rabbit and rat epididymal spermatozoa. J Reprod Fert 57:245–251

Pholpramool C, Lea OA, Burrow PV, Dott HM and Setchell BP (1983) The effects of acidic epididymal glycoprotein (AEG) and some other proteins on the motility of rat epididymal spermatozoa. Int J Androl 6:240–248

Purvis K, Cusan L, Attramadal H, Ege A and Hansson V (1982) Rat sperm enzymes during epididymal transit. J Reprod Fert 65:381–387

Raff EC and Blum JJ (1968) A possible role for adenylate kinase in cilia: concentration profiles in a geometrically constrained dual enzyme system. J Theoret Biol 18:53–71

Rama Sastry BV and Sadavongvivad C (1978) Cholinergic systems in non-nervous tissues. Pharmacol Rev 30:65–132

Rama Sastry BV, Jensen BV and Chaturvedi AK (1981) Inhibition of human sperm motility by inhibitors of choline acetyltransferase. J Pharm exp Therap 216:378–384

Reed PW and Lardy HA (1972) Antibiotic A23187 as a probe for the study of calcium and magnesium function in biological systems. In: The Role of Membranes in Metabolic Regulation. Eds MA Mehlmnan and RW Hanson, Acad Press, London. pp 111–131

Reyes A and Chavarria M (1981) Interference with epididymal physiology as possible site of male contraception. Archs Androl 7:159–168

Reyes A, Mercado E, Goicoechea B and Rosado A (1976) Participation of membrane sulfydryls groups in the epididymal maturation of human and rabbit spermatozoa. Fert Steril 27:1452–1458

Reyes-Fuentes A, and Chavarria-Olarte ME (1980) Interferencia con la maduracion epididimaria del espermatozoide. Gaceta Medicale de Mexico 116:312–318

Rikmenspoel R (1965a) The inhibition by amytal of respiration and motility of bull spermatozoa. Exp Cell Res 37:312–326

Rikmenspoel R (1984) Movements and active moments of bull sperm flagella as a function of temperature and viscosity. J exp Biol 108:205–230

Rikmenspoel R, Sinton S and Janick JJ (1969) Energy conservation in bull sperm flagella. J Gen Physiol 54:782–805

Rikmenspoel R, Orris SE and O'Day PM (1978) Ionic-requirements of impaled bull spermatozoa driven by external ADP and ATP. Exp Cell Res 111:253–259

Rikmenspoel R, Orris SE and Isles CA (1981) Effects of vanadate, Mg^{2+} and electric current injection on the stiffness of impaled bull spermatozoa. J Cell Sci 51:53–61

Rossman I (1937) Uterine contractions and the transport of sperm in the rat. Anat Rec 69:133–149

Rufo GA, Schoff PK and Lardy HA (1984) Regulation of calcium content in bovine spermatozoa. J Biol Chem 259:2547–2552

Saling PM (1982) Development of the ability to bind to zonae pellucidae during epididymal maturation: reversible immobilisation of mouse spermatozoa by lanthanum. Biol Reprod 26:429–436

Salisbury GW (1956) The function of the epididymis of the bull.I. A theory for the activation of spermatozoan motility. Tijdschr. Diergeneesk 87:616–623

Sanyal RK and Khanna SK (1971) Action of cholinergic drugs on motility of spermatozoa. Pharmacol Rev 30:65–132

Satir P (1979) Basis of flagellar motility in spermatozoa: current status. In: The Spermatozoon. Eds DW Fawcett and JM Bedford, Urban and Schwarzenberg, Baltimore. pp 81–90

Sattayasai N and Panyim S (1982) Nature of the proteins which form disulfide bonds during the maturation of rat spermatozoa. Int J Androl 5:337–344

Sekine T (1951) Choline esterase in pig spermatozoa. J Biochem (Tokyo) 38:171–179

Serres C and Kann ML (1984) Motility induction in hamster spermatozoa from caput epididymidis: effects of forward motility protein (FMP) and calmodulin inhibitor. Reprod Nutr Develop 24:81–94

Setchell BP and Hinton BT (1981) The effects on the spermatozoa of changes in the composition of luminal fluid as it passes along the epididymis. In: Epididymis and Fertility: Biology and Pathology. Eds C Bollack and A Clavert, Karger, Munich. pp 58–66

Setchell BP, Scott TW, Voglmayr JK and Waites GMH (1969) Characteristics of testicular spermatozoa and the fluids which transport them into the epididymis. Biol Reprod Suppl 1:40–66

Sexton TJ, Amann RP and Flipse RJ (1971) Free amino acids and protein in rete testis fluid, vas deferens plasma, and accessory sex gland fluid and seminal plasma of the conscious bull. J Dairy Sci 54:412–416

Shilon M, Paz(Frenkel) G, Homonnai ZT and Schoenbaum M (1978) The effect of caffeine on guinea pig epididymal spermatozoa: motility and fertilizing capacity. Int J Androl 1:416–423

Singh JP, Babcock DF and Lardy HA (1983) Motility activation, respiratory stimulation and alteration of Ca^{2+} transport in bovine sperm treated with amine local anaesthetics and calcium transport antagonists. Archs Biochem Biophys 221:291–303

Smith MB, Babcock DF and Lardy HA (1985) A ^{31}P NMR study of the epididymis and epididymal sperm of the bull and hamster. Biol Reprod 33:1029–1040

Smithwick EB and Young LG (1977) Isolation of plasmalemmae-free, mitochondria-free bull sperm flagella. Biol Reprod 17:443–452

Smithwick EB and Young LG (1978) Ultrastructural evaluation of the isolation of mitochondria-free bull sperm flagella. Biol Reprod 19:280–290

Stengel D and Hanoune J (1981) The catalytic unit of ram sperm adenylate cyclase can be activated through the guanine nucleotide regulatory component and prostaglandin receptors of human erythrocyte. J Biol Chem 256:5394–5398

Stengel D, Guenet L, Desmier M, Insel P and Hanoune J (1982) Forskolin requires more than the catalytic unit to activate adenylate cyclase. Mol Cell Endocr 28:681–690

Stephens DT, Wang JL and Hoskins DD (1979) The cyclic AMP phosphodiesterase of bovine spermatozoa: multiple forms, kinetic properties and changes during development. Biol Reprod 20:483–491

Stephens DT, Acott TS and Hoskins DD (1981) A cautionary note on the determination of forward motility protein activity with bovine epididymal spermatozoa. Biol Reprod 25:945–949

Stewart TA and Forrester IT (1976) Choline acetyltransferase activity in ram spermatozoa. Proc U Otago Med Sch 54:53–54

Stewart TA and Forrester IT (1978a) Acetylcholinesterase and choline acetyltransferase in ram spermatozoa. Biol Reprod 19:271–279

Stewart TA and Forrester IT (1978b) Identification of a cholinergic receptor in ram spermatozoa. Biol Reprod 19:965–970

Stewart TA and Forrester IT (1978c) Acetylcholine-mediated control of spermatozoan motility. Proc U Otago Med Sch 56:27

Stewart TA and Forrester IT (1979) Acetylcholine-induced calcium movements in hypotonically washed ram permatozoa. Biol Reprod 21:109–115

Storey BT and Kayne FJ (1975) Energy metabolism of spermatozoa.V. The Embden-Meyerhof pathway of glycolysis: activities of pathway enzymes in hypotonically treated rabbit epididymal spermatozoa. Fert Steril 26:1257–1265

Storey BT and Kayne FJ (1980) Properties of pyruvate kinase and flagellar ATPase in rabbit spermatozoa: relation to metabolic strategy of the sperm cell. J exp Zool 211:361–367

Summers K (1974) ATP-induced sliding of microtubules in bull sperm flagella. J Cell Biol 609:321–324

Tabor CW and Rosenthal SM (1956) Pharmacology of spermine and spermidine. Some effects on animals and bacteria. J Pharm exp Therap 116:139–155

Tamblyn TM (1980) Identification of actin in boar epididymal spermatozoa. Biol Reprod 22:727–734

Tamblyn TM (1981) Evidence for nonmuscle myosin in bovine ejaculated spermatozoa. Gamete Res 4:499–506

Tang FY and Hoskins DD (1975) Phosphoprotein phosphatase of bovine epididymal spermatozoa. Biochem Biophys Res Commun 62:328–385

Tash JS (1976) Investigations on adenosine 3′,5′-monophosphate phosphodiesterase in ram semen and initial characterization of a sperm-specific isoenzyme. J Reprod Fert 47:63–72

Tash JS and Means AR (1982) Regulation of protein phosphorylation and motility of sperm by cyclic adenosine monophosphate and calcium. Biol Reprod 26:745–763

Tash JS and Means AR (1983) Cyclic adenosine 3′,5′monophosphate, calcium and protein phosphorylation in flagellar motility. Biol Reprod 28:75–104

Tash JS, Kakar SS and Means AR (1984) Flagellar motility requires the cAMP-dependent phosphorylation of a heat-stable NP-40-soluble 56kd protein, axokinin. Cell 38:551–559

Terner C, MacLaughlin J and Smith BR (1975) Changes in lipase and phosphatase activities of rat spermatozoa in transit from the caput to the cauda epididymidis. J Reprod Fert 45:1–8

Terner C, Holtz A, Brennan RG and Battista D (1980) Androgen control of phosphodiesterase modulator in epididymis and prostate of rat. Fed Proc 39:Abst 3765

Tongkao D and Chulavatnatol M (1979) Phosphorylation of microtubules of rat spermatozoa during epididymal maturation. In: The Spermatozoon. Eds DW Fawcett and JM Bedford, Urban and Schwarzenberg, Baltimore. pp 129–134

Treetipsattit N and Chulavatnatol M (1982) Effects of ATP, cAMP and pH on the initiation of flagellar movement in demembranated models of rat epididymal spermatozoa. Exp Cell Res 142:495–499

Turner TT and Giles RD (1981) The effects of carnitine, glycerylphosphorylcholine, caffeine, and egg yolk on the motility of rat epididymal spermatozoa. Gamete Res 4:283–295

Turner TT and Giles RD (1982) The effects of cyclic adenine nucleotides, phosphodiesterase inhibitors, and cauda epididymal fluid on the motility of rat epididymal spermatozoa. J Androl 3:134–139

Vermouth NT, Carriazo CS, Ponce RH and Blanci A (1986) Lactate dehydrogenase-X, malate dehydrogenase and total protein in rat spermatozoa during epididymal transit. Comp Biochem Physiol B83:381–384

Vijayaraghavan S and Hoskins DD (1985) Forskolin stimulates bovine epididymal sperm motility and cyclic AMP levels. J Cyclic Nucleotide Res 10:499–510

Vijayaraghavan S, Critchlow LM and Hoskins DD (1985) Evidence for a role of cellular alkalinization in the cyclic adenosine 3′,5′-monophosphate-mediated initiation of motility in bovine caput spermatozoa. Biol Reprod 32:489–500

Virtanen I, Badley RA, Paasivuo R and Lehto VP (1984) Distinct cytoskeletal domains revealed in sperm cells. J Cell Biol 99:1083–1091

Voglmayr JK (1975) Metabolic changes in spermatozoa during epididymal transit. In: Handbook of Physiology, Sect 7 Endocrinology, Vol V Male Reproductive System. Eds DW Hamilton and RO Greep, Amer Physiol Soc Washington DC. pp 437–451

Voglmayr JK and Gandhi JS (1978) Survival of ram testicular spermatozoa in vitro: effects of glucose, glucose metabolites, rete testis fluid proteins, selected androgens and phospholipids. Theriogeneology 9:463–478

Voglmayr JK, Scott TW, Setchell BP and Waites GMH (1967) Metabolism of testicular spermatozoa and characteristics of testicular fluid collected from conscious rams. J Reprod Fert 14:87–99

Voglmayr JK, White IG and Quinn PJ (1969) A comparison of adenosine triphosphatase activity in testicular and ejaculated spermatozoa of the ram. Biol Reprod 1:121–129

Voglmayr JK, Larson LH and White IG (1970) Metabolism of spermatozoa and composition of fluid collected from the rete testis of living bulls. J Reprod Fert 21:449–460

Voglmayr JK, Musto NA, Saksena SK, Brown-Woodman PDC, Marley PB and White IG (1977) Characteristics of semen collected from the cauda epididymidis of conscious rams. J Reprod Fert 49:245–251

Voglmayr JK, White IG and Parks RP (1978) The fertilizing capacity of ram testicualr spermatozoa, freshly collected and after storage in cauda epididymal fluid. Theriogenology 10:313–321

Warren MR (1938) Observations of the uterine fluid of the rat. Am J Physiol 122:602–608

Wasco WM and Orr GA (1984) Function of calmodulin in mammalian sperm: presence of a calmodulin-dependent cyclic nucleotide phosphodiesterase associated with demembranted rat caudal epididymal sperm. Biochem Biophys Res Commun 118:636–642

White IG and Voglmayr JK (1986) ATP-induced reactivation of ram testicular, cauda epididymal, and ejaculated spermatozoa extracted with Triton X-100. Biol Reprod 34:183–193

Williams JA (1970) Origin of transmembrane potentials in non-excitable cells. J Theoret Biol 28:287–296

Wong PYD and Lee WM (1983) Potassium movement during sodium-induced motility initiation in the rat caudal epididymal spermatozoa. Biol Reprod 28:206–212

Wong PYD and Lee WM (1985) Ionic mechanisms of sperm motility initiation. In: Male Fertility and its Regulation. Eds TJ Lobl and ESE Hafez, MTP Press Lancaster. pp 411–416

Wong PYD, Tsang AYF, Lee WM and Li CM (1980) Secretion of the rat cauda epididymidis. Archs Androl 5:327–336

Wong PYD, Lee WM and Tsang AYF (1981a) The effects of extracellular sodium on acid release and motility initiation in rat caudal epididymal spermatozoa vitro. Exp Cell Res 131:97–104

Wong PYD, Lee WM and Tsang AYF (1981b) The effects of sodium and amiloride on the motility of the caudal epididymal spermatozoa of the rat. Experientia 37:69–71

Wong PYD, Tsang AYF and Lee WM (1982) Secretion of macromolecules by the rat epididymis. Int J Androl Suppl 5:34–47

Wright MI, Bradley MP and Forrester IT (1981) Acetylcholinesterase and the maturation of mammalian sperm. Proc U Otago Med Sch 59:28–29

Wyker R and Howards SS (1977) Micropuncture studies on the motility of rete testis and epididymal spermatozoa. Fert Steril 28:108–112

Yamanaka HS and Soderwall AL (1960) Transport of spermatozoa through the female genital tract of hamsters. Fert Steril 11:470–474

Yasazumi G, Yamaguchi S, Takahashi Y, Nishimura Y, Yamagishi N, Nakai Y, Naito N and Iwashita T (1975) The structural and cytochemical bases for vertebrate and invertebrate sperm motility. In: The Functional Anatomy of the Spermatozoon. Ed BA Afzelius, Pergamon Press, Oxford. pp 151–159

Yeung CH (1984) Effects of cyclic AMP on the motility of mature and immature hamster epididymal spermatozoa studied by reactivation of the demembranated cells. Gamete Res 9:99–114

Young LG and Nelson L (1968) Viscometric analysis of the contractile proteins of mammalian spermatozoa. Exp Cell Res 51:34–44

Young LG and Nelson L (1969) Divalent cation activation of flagellar ATP-phosphohydrolase from bull sperm. J Cell Physiol 74:315–322

Young LG and Smithwick EB (1975a) Characterization of the ATP-phosphohydrolase activity of bovine spermatozoa flagellar extracts. J Cell Physiol 85:143–150
Young LG and Smithwick EB (1975b) Extraction and ATP-phosphohydrolase activity from epididymal bull sperm and ultrastructure of flagellar degradation. Exp Cell Res 90:223–236
Young LG and Smithwick EB (1976) Effects of mechanical and chemical disruption on the ATP-phosphohydrolase activity and ultrastructure of sperm flagella. Exp Cell Res 102: 179–190
Young LG and Smithwick EB (1983) Studies on the flagellar ATPase of bull spermatozoa: extraction and characterization. J exp Zool 226:459–466
Young WC (1931) A study of the function of the epididymis. III. Functional changes undergone by spermatozoa during their passage through the epididymis and vas deferens in the guinea pig. J exp Biol 8:151–162
Zamboni L (1972) Fertilization in the mouse. In: Biology of Mammalian Fertilization and Implantation. Eds KS Moghissi and ESE Hafez, Charles C Tomas, Springfield Illinois. pp 213–262
Zimmerman KJ, Crabo BG, Moore R, Weisberg S, Diebel FC and Graham EF (1979) Movements of sodium and potassium into epididymal boar spermatozoa. Biol Reprod 21:173–179

B. Capacitation

1. Detection of Capacitation

Capacitation is an obligatory pre-fertilisation event, before which even mature spermatozoa are incapable of fertilising eggs, *i.e.* are incapable of binding to or penetrating the zona pellucida or vitellus (see Sections II.C, D, E). Capacitation was originally detected as the period of residence needed by sperm in the female tract before they could fertilise freshly ovulated eggs (see Chang, 1984). Its reversibility was shown by the ability of seminal plasma to "decapacitate", *i.e.* contribute a "decapacitation factor" (DF) to, sperm previously capacitated in the female tract. Much work on DF from seminal plasma has shown it to have characteristics of peptides, glycoproteins, lipids and membranous vesicles (see Robertson et al., 1971; Eng and Oliphant, 1978; Davis and Davis, 1983; Oliphant et al., 1985).

That mature epididymal spermatozoa must also reside for some time in the female tract before they can fertilise eggs (Austin, 1951; Chang 1951) implies that they also need to be capacitated, even without contact with seminal DF, and this has been confirmed in numerous studies in-vitro. A proteinaceous decapacitation activity is present in epididymal extracts from the hamster (Morton and Chang, 1973), mouse (Iwamatsu and Chang, 1971; Oliphant and Brackett, 1973b; Fraser and Drury, 1976) and DF activity is demonstrable in epididymal fluid from rabbits (Davis, 1973; Weinman and Williams, 1974), boars (Hunter et al., 1978) and guinea pigs (Aonuma et al., 1973, 1976; Hyne and Garbers 1982). Whereas a high molecular weight carbohydrate was suggested to be the active principal in mouse epididymal fluid (Shur and Hall, 1982), a low molecular weight factor was considered to be active in the guinea pig (Hyne and Garbers 1982). In the rabbit DF is a well characterised glycoprotein (Oliphant et al., 1985), but it is uncertain if the membrane vesicles which can be observed in centrifuged epididymal fluid from the rat (Davis, 1973), and which can decapacitate sperm reversibly (Davis, 1974), constitute an epididymal DF. Their apposition to sperm lacking membranes in the centrifuged samples suggests a sperm-derived, rather than epithelial cell origin.

2. Changes in the Sperm Surface During Capacitation

Many changes occur to sperm plasma membranes during capacitation (see Johnson, 1975; Austin, 1975). These include a reduction in surface charge, monitored by a decreased mobility in electric fields (Vaidya et al., 1971) and more subtle changes in the surface reflected in altered egg-binding properties (Hartmann and Gwatkin, 1971) and recognition by leucocytes (Bedford, 1965). They probably represent changes in exposed carbohydrate groupings demonstrated by parallel changes in lectin-binding to sperm heads and tails with capacitation in many species (see Koehler, 1976; Lambert and van Le, 1984). The susceptibility towards lectin-induced tail-tail agglutination (Talbot and Franklin, 1978a,b; Talbot and Chacon, 1981) and the extent of binding to lectin-coated beads also changes on capacitation (Ahuja, 1984).

Alterations to membrane proteins during capacitation are indicated by changes in distribution of antibodies raised to sperm antigens (Johnson and Hunter, 1972; Koehler 1976), rearrangements of intramembranous particles (Friend and Rudolf, 1974; Friend et al., 1977; Kinsey and Koehler, 1978; Friend, 1980, 1982; Yanagimachi and Suzuki, 1985) and losses of high molecular weight proteins from porcine sperm capacitated in-utero (Esbenshade and Clegg, 1980) and low molecular weight proteins from rat sperm under capacitating conditions in-vitro (Davis and Gergely, 1979).

While these effects may be related to capacitation, few experiments have proved that the conditions used to obtain the membrane changes were capable of supporting capacitation as defined by egg penetration. Tetracycline (which binds to sperm coating proteins: Ericsson and Baker, 1967) is lost from sperm on capacitation (Ericsson et al., 1967a,b) but the time-course does not parallel the acquisition of the capacitated state in-vivo (Vaidya et al., 1969) or in-vitro (Byrd, 1981).

3. Mechanisms of Capacitation

At least three mechanisms of capacitation have been advanced, each being thought to lead to destabilisation of sperm membranes which is deemed essential for subsequent post-capacitation events (see Clegg, 1983; Sections II.C,D,E).

a. Loss of Protein Coats

In most studies, proteins demonstrably lost from the sperm surface under capacitating conditions (Kohane et al., 1980b; Dravland and Joshi, 1981; Bearer and

Friend, 1982) were not shown to have DF activity. The loss of surface proteins in parallel with an increase of fertilising capacity is more suggestive evidence of a functional role of the lost protein, and has been demonstrated for epididymal spermatozoa from the rabbit (Oliphant and Brackett, 1973a; Oliphant and Singhas, 1979) and mouse (Fraser, 1984b). High ionic strength treatment of sperm, which should elute surface (peripheral) membrane proteins held electrostatically to the surface (Singer, 1974; see Section III.G.10.a.i), also promotes the loss of surface proteins from spermatozoa and under these conditions there are increases in fertilisation in-vitro by epididymal sperm from rabbits (Brackett et al., 1978), mice (Oliphant and Brackett, 1973b) and bulls (Lorton and First, 1979). High ionic strength may not be a mechanism which operates in-vivo (Bedford, 1979; Akruk et al., 1979; Viviyapanich and Bedford, 1981) although fertilisation occurs earlier in-vivo after high ionic strength treatment in-vitro (Oliphant and Eng, 1981).

b. Lipid Exchanges

Most evidence for this comes from rodents (see Davis, 1981; Go and Wolf, 1985). In these species albumin promotes fertilisation in-vitro by removing lipids from the membrane (Davis, 1976a): albumin is known to associate with sperm from rabbits (Blanck et al., 1976) and rats (Davis and Gergely, 1979), and whereas in-vitro fertilisation in rats and mice is promoted by defatted albumin, it is inhibited by cholesterol-rich albumin. The transfer of albumin-bound phospholipid and cholesterol to sperm and sperm-bound cholesterol to albumin have been demonstrated under capacitating conditions in the rat (Davis et al., 1979, 1980) and mouse (Go and Wolf, 1985).

Fertilisation of rabbit sperm is also inhibited by cholesterol (Davis, 1976b, 1980) and cholesterol can be transferred from sterol-rich liposomes to rabbit sperm (Davis and Byrne, 1980). A general relationship between the time required for capacitation and the cholesterol and phospholipid content of spermatozoa from a variety of species has been claimed to support the view that cholesterol is lost from sperm during capacitation in the female tract (Davis, 1981). How this is achieved is not known but rabbit uterine fluid can remove cholesterol from rabbit sperm (Davis, 1982) and human oviductal lipids may play a role in capacitation (Menezo et al., 1984). Phospholipid transfer from sperm is less extensive than that of cholesterol (Davis and Byrne, 1980) but changes in the concentration of phospholipid do occur during incubation of porcine spermatozoa in-utero (Snider and Clegg, 1975; Clegg et al., 1975; Evans et al., 1980).

A water-soluble lipid present in the post-acrosomal regions of rabbit sperm (Teichmann et al., 1972) identified as choline plasmalogen (Teichmann et al., 1974a) is lost from sperm in-utero, but not in-vitro (Cummins and Teichmann, 1974), and it may be involved in capacitation (Soupart et al., 1979). Steroids that prevent capacitation of spermatozoa by cumulus cells in-vitro (Gwatkin and Williams, 1970) and in-vivo (Briggs, 1973) may stabilise sperm cell membranes and vitamin A, which is present in epididymal fluid of rabbits (Velazquez et al., 1975) and in epithelial cells of the vas deferens of mice (Hirosawa, 1977) may act similarly if retained by spermatozoa in the female tract.

c. Cleavage of Sterol Sulphates

The evidence for this comes mainly from the hamster (see Langlais et al., 1981; Go and Wolf, 1983). The major sterol in spermatozoa from this species is not cholesterol but desmosterol and its sulphate (Bleau and vandenHeuvel, 1974; Legault et al., 1978, 1979a). This is a membrane-active sterol and a potent inhibitor of capacitation by cumulus cells in-vitro (Bleau et al., 1975). The presence of a sterol sulphatase in the secretion of the female hamster oviduct, the activity of which varies with the female cycle (Legault et al., 1980) and which could thus remove this inhibitor, *i.e.* promote capacitation at the time of ovulation, makes this an attractive hypothesis. Desmosterol and its sulphate bind to human spermatozoa (Langlais et al., 1981) and there is also a steroid sulphatase in the human female tract (LaLumiere et al., 1976).

4. Consequences for Membrane Fluidity

The loss of stabilising proteins, cholesterol or sterol sulphates is anticipated to produce more fluid membranes in capacitated sperm cells. In support of this, capacitation can be promoted by a "membrane-mobility" agent in guinea pig sperm (Fleming et al., 1982) and the activity of PCM, which modulates membrane fluidity in other membranes (Gagnon and Heisler, 1979), increases in activity during capacitation of hamster sperm (Bardin and Gagnon, 1982; Castaneda et al., 1983). The decreases in membrane fluidity accompanying capacitation (O'Rand, 1977, 1979) do not necessarily argue aginst this view, for the clustering of membrane proteins may be a secondary consequence of segregation of lipid domains. An increase in the diffusibility of membrane lipids over the mouse sperm tail has been observed after capacitation (Wolf et al., 1986).

5. Action of Decapacitation Factors

Because guinea pig sperm can be "arrested" in a capacitated state in Ca^{2+}-free media and Ca^{2+} can subsequently initiate synchronous Ca^{2+}-dependent events (Yanagimachi and Usui, 1974), capacitation may be involved in increasing the permeability of the sperm cell to this cation. Certainly, Ca^{2+} uptake into the guinea pig sperm cell increases during capacitation (Singh et al., 1978), although this is unlikely to be reversible and may strictly be the beginning of post-capacitation events. An increase in the permeability to K^+ and protons occurs during capacitation of hamster sperm (Working and Meizel, 1981, 1983).

Eluting coating proteins from spermatozoa does promote uptake of calcium ions into bovine spermatozoa (Triana et al., 1980) and capacitation can be bypassed when intracellular calcium ion concentrations are raised by the use of ionophores (see Yanagimachi, 1981). Thus in the bull, epididymal DF may prevent the uptake of Ca^{2+} into spermatozoa in a similar way to a seminal calcium-binding protein which reduces the uptake of this ion into ejaculated sperm (Babcock et al., 1979); however, the ability of bovine seminal plasma to reduce the uptake of Ca^{2+} into bovine epididymal spermatozoa (Rufo et al., 1982, 1984) would suggest that calcium transport in bovine epididymal spermatozoa is not normally inhibited by a component of epididymal fluid. Likewise in the rabbit, the histochemical demonstration of Ca^{2+}-ATPase activity on epididymal, but not ejaculated, sperm and the ability of seminal plasma to inhibit the histochemically demonstrable enzyme activity of epididymal sperm (Gordon, 1973; Gordon and Dandekar, 1977) argue that epididymal fluid does not reduce calcium binding to or transport in spermatozoa from this species.

6. Maintenance of Motility

In-vitro fertilisation at the low sperm/egg ratios found in the oviduct at the time of fertilisation (see Section II.E.8.a) is difficult, because of the decline in motility of spermatozoa at high dilution. Such problems can be overcome by including agents which are capable of maintaining sperm motility under capacitating conditions (see Bavister, 1979). The effect of adrenal extracts on capacitation (Bavister et al., 1976) may be an indirect one on the maintenance of sperm motility (see Section II.A.4.g.i). Adrenalin can also reduce the potassium ion requirement for capacitation (Bize and Santander, 1985; see Section II.B.7).

Other "sperm survival factors" (SSF) that protect the sperm cell against membrane damage at high dilution (see Bavister, 1979) can be mimicked by macromolecules with phosphodiesterase inhibitors (Morton et al., 1979a; see Section III.G.12.a.v).

7. Requirements for Capacitation

Capacitation requires K^+ (see Fraser, 1984a) and the increased activity of Na^+/K^+ ATPase of hamster sperm that occurs during capacitation (Mrsny et al., 1984) may mediate the necessary ion fluxes. The role of cyclic nucleotides in capacitation (see Garbers and Kopf, 1982) are more clearly seen in post-capacitation events (see Sections II.D,E), but the activity of adenylate cyclase increases during capacitation in hamsters (Morton and Albagli, 1973), guinea pigs (Hyne and Garbers, 1979a) and mice (Stein and Fraser, 1984), whereas cyclic nucleotide phosphodiesterase activity decreases (Stein and Fraser, 1984). However, imidazole, which activates cyclic AMP PDE, promotes capacitation of porcine spermatozoa in-utero (Hanada and Nagase, 1981).

8. Development in the Epididymis of the Ability of Spermatozoa to be Capacitated

Immature spermatozoa would be unable to fertilise eggs if they could not undergo capacitation. Immature rabbit spermatozoa were initially thought to require longer capacitation times than mature sperm cells (Orgebin-Crist, 1967a, 1968; Overstreet, 1970), but increasing the time they spent in the female tract before ovulation did not promote fertilisation of caput cells (Overstreet. 1970; Cummins and Orgebin-Crist, 1971) and Overstreet and Bedford (1974) found no difference in capacitation times of distal corpus and ejaculated spermatozoa in mixed inseminations.

It would appear now, however, that shorter capacitation times are required for immature spermatozoa. Upper corpus sperm from the boar epididymis can fertilise eggs 4–4.5 h after insemination into the oviducts, whereas at this time mature caudal sperm from the vas deferens have not fertilised eggs (Hunter et al., 1976). That this was so because the immature sperm cells had less DF associated with them, or held it less tightly, and therefore required less time to lose it, was confirmed when the amount of DF associated with upper corpus sperm was increased by inseminating them in cauda epididymal plasma (CEP) and early fertilisation was prevented (Hunter et al., 1978). This not only demonstrated the presence of DF in CEP, but also the ability of immature sperm cells to respond to it.

Reducing the capacitation time in-vivo failed to improve fertilisation by more immature (testicular or caput) rabbit sperm (see Table 7). This is compatible with the view that these immature cells are incapable of being capacitated as they have no DF to lose; certainly high ionic strength treatment of testicular sperm does not enable them to fertilise eggs in-vitro (Brackett et al., 1978). Thus, mature cells may need to be stabilised by DF, whereas immature sperm cells may not (see Section II.B.9.d; Part IV).

9. Involvement of the Epididymis in Permitting Capacitation

Spermatozoa from the hamster caput or corpus epididymidis do not display capacitation-dependent events in-vivo, even after previous ageing in the ligated epididymis (Cummins, 1976), but it is not known if this reflects their failure to be capacitated or a failure to undergo capacitation-dependent events (see Section II.D,E).

a. Secretion of Sperm Coating Proteins

Where more surface proteins are eluted by high ionic strength media from the surface of mature than immature cells in rats (Kohane et al., 1980) and hamsters (Gonzalez-Echeverria et al., 1982) they were not shown to have DF activity. Conversely, where less DF activity was associated with immature spermatozoa in the guinea pig (Aonuma et al., 1973), only simple washing was required for the proteins to be lost. Androgens promote the incorporation of amino acids into specific epididymal proteins which are secreted into the lumen and subsequently associate with the surface of maturing sperm cells in hamsters (Cuasnica et al., 1984) and humans (Tezon et al., 1985), but the function of the proteins may be related more to sperm-egg binding that decapacitation. The binding of these proteins to spermatozoa may be promoted by the low ionic strength of epididymal fluid, resulting from other epididymal functions, namely, the absorption of Na^+ and secretion of inositol (see Sections III.C,F).

An acrosome-stabilising, sperm-coating protein has been localised in the epithelium of the rabbit epididymis (Thomas et al., 1984; see Oliphant et al., 1985) and α-lactalbumin, by binding to galactosyltransferase on the sperm surface (Schur and Hall, 1982b), could also exhibit decapacitation factor activity (Byers et al., 1984).

b. Removal of Lipids

If capacitation reflects the time needed for the cholesterol/phospholipid ratio to reach a level compatible with fertilisation (Davis, 1981), immature spermatozoa would require longer capacitation times as a consequence of the higher content of cholesterol found in these spermatozoa from rams (Scott et al., 1967; Quinn and White, 1967), bulls (Lavon et al., 1979), rats (Terner et al., 1975; Adams and Johnson, 1977) and hamsters (Legault et al., 1978). Total sperm phospholipids also decrease on sperm maturation in rams (Poulos et al., 1975; Evans and Setchell, 1979a,b; Dacheux, 1977), boars (Grogan et al., 1966), bulls (Poulos et al., 1973) and rats (Dacheux, 1977) and it has recently been shown that these are lost from the membranes in the rat (Cooper, 1985), mouse (Toshimori et al., 1985), ram (Parks and Hammerstedt, 1985) and boar (Nikolopoulou et al., 1985).

If the epididymis is involved in removing sperm lipids, in one sense it could be considered to initiate capacitation-like events which are completed in the female tract. How the epididymis achieves this is not known, but the high concentration of proteins in epididymal fluid (Table 18) may be involved in mediating lipid transfer from spermatozoa. The epithelial clear cells may be involved since in the corpus epididymidis they contain lipid inclusions and in the cauda they can accumulate what are considered to be products of sperm disintegration under experimental conditions (see Section III.A.9).

Choline plasmalogen in sperm heads increases upon maturation of rabbit spermatozoa (Cummins and Teichmann, 1974) and appears in immature sperm after ageing them in occluded epididymal ducts, and so is probably an age-dependent, rather than an epididymal-dependent product (Cummins and Teichmann, 1974). Choline can be incorporated into this sperm lipid in-vivo (Teichmann et al., 1974b) but choline plasmalogen can accumulate in human sperm merely upon incubation (Soupart et al., 1979).

c. Secretion of Sterol Sulphates

The epididymis can synthesise sterols (Hamilton, 1971; see Section III.B.2.c.i) and other prenol polymers (Wenstrom and Hamilton, 1980) and could be the source of the vitamin A in epididymal fluid in the caput and cauda (Velasquez et al., 1975). Steroid sulphotransferase is present in the epididymis of hamsters (Bouthillier et al., 1981, 1984, 1985a,b), rats (Djoseland, 1976) and other species (Hamilton, 1971) and desmosterol sulphate is secreted distally into the epididymal lumen (Legault et al., 1979a). The greater uptake of both free and sulphated sterols into mature spermatozoa compared to immature cells demonstrated in-vitro (Legault et al., 1979b) may reflect the lower total sterol content of the mature cell membrane. Loss of cholesterol accompanies the rise in desmosterol sulphate in maturing hamster spermatozoa (Legault et al., 1978, 1979a) and it is lost from epididymal fluid in this species and the ram (Quinn and White, 1967). It is not known whether these sterols are transported out of the duct or incorporated into epithelial lipids.

d. Possible Sequence of Events

Epididymal maturation may occur in two sequential phases, necessitating one, or all, of the capacitation-associated events cited above. Firstly, a loss of sterols would tend to destabilise the sperm cell's membranes and this could lead to a precocious expression of fertility, possibly resulting in damage to the epididymal epithelium by the now potentially-fertilising spermatozoon, if it were not prevented by the addition of either proteinaceous decapacitation factors or sterol sulphates, acting to stabilise the membranes. It is not clear if some apparently conflicting evidence, the fluidity of immature sperm cell membranes, demonstrated by lectin-induced clustering of surface components of rabbit caput sperm (Nicolson and Yanagimachi, 1976) and the lack of observable change in membrane fluidity on maturation when NMR probes (Hammerstedt et al., 1979) or laser-induced dye quenching (Wolf and Voglmayr, 1984) are used, reflects the nature of capacitated or uncapacitated cells.

10. Summary

Capacitation reflects a phase of epididymal action involved in preventing untimely expression of the abilities that are acquired by spermatozoa as they mature. The stabilisation of the sperm cell membranes by adsorption of proteins or sterols secreted by the epididymis is reversed in the female tract so that postcapacitation events occur around the freshly ovulated egg.

11. References

Adams CS and Johnson AD (1977) The lipid content of epididymal spermatozoa of Rattus Norvegicus. Comp Biochem Physiol 58B:409–411

Ahuja KK (1984) Lectin-coated agarose beads in the investigation of sperm capacitation in the hamster. Develop Biol 104:131–142

Akruk SR, Humphreys WJ and Williams WL (1979) In vitro capacitation of ejaculated rabbit spermatozoa. Differentiation 13:125–131

Aonuma S, Mayumi T, Suzuki K, Noguchi T, Iwai I and Okabe M (1973) Studies on sperm capacitation. I. The relationship between a guinea-pig sperm-coating antigen and a sperm capacitation phenomenon. J Reprod Fert 35:425–432

Aonuma S, Okabe M, Suzuki K, Matsumoto K and Kawai Y (1976) Studies on sperm capacitation. V. Characterization of decapacitation from guinea-pig spermatozoa. Chem Pharm Bull 24:907–911

Austin CR (1951) Observations on the penetration of the sperm into the mammalian egg. Aust J Sci Res 34:581–596

Austin CR (1975) Membrane fusion events in fertilization. J Reprod Fert 44:155–166

Babcock DF, Singh JP and Lardy HA (1979) Alteration of membrane permeability to calcium ions during maturation of bovine spermatozoa. Develop Biol 69:85–93

Bardin CW and Gagnon C (1982) The possible role of protein-carboxyl methylation in sperm motility. In: Physiotherapy of Hypophysial Disturbance and Diseases of Reproduction. Eds A DeNicola, J Blaquier and RJ Soto, Alan Liss Inc, New York. pp 217–234

Bavister BD (1979) Elimination of the adverse effect of dilution on hamster sperm motility in vitro. In: The Spermatozoon. Eds DW Fawcett and JM Bedford, Urban and Schwarzenberg, Baltimore. pp 169–172

Bavister BD, Yanagimachi R and Teichman RJ (1976) Capacitation of hamster spermatozoa with adrenal gland extracts. Biol Reprod 14:219–221

Bearer EL and Friend DS (1982) Modifications of anionic- lipid domains preceding membrane fusion in guinea pig sperm. J Cell Biol 92:604–615

Bedford JM (1965) Effect of environment on phagocytosis of rabbit spermatozoa. J Reprod Fert 9:249–256

Bedford JM (1979) Some caveats of mammalian gamete research. In: Animal Models for Research on Contraception and Fertility. Ed NJ Alexander, Harper and Row, Hagerstown. pp 269–283

Bize I and Santander G (1985) Epinephrine decreases the potassium requirements of hamster sperm capacitation: furosemide blocks the effects of epinephrine. J exp Zool 235:261–267

Blank M, Soo L and Britten JS (1976) Adsorption of albumin on rabbit sperm membranes. J Memb Biol 29:401–409

Bleau G and vandenHeuvel WJA (1974) Desmosterol sulphate and desmosterol in hamster epididymal spermatozoa. Steroids 24:549–556

Bleau G, vandenHeuvel WJA, Anderson CF and Gwatkin RBL (1975) Desmosterol sulphate of hamster spermatozoa, a potent inhibitor of capacitation in vitro. J Reprod Fert 43:175–178

Bouthillier M, Chapdelaine A, Bleau G and Roberts KD (1981) Steroid sulfotransferase in hamster epididymis. Steroids 38:523–535

Bouthillier M, Bleau G, Chapdelaine A and Roberts KD (1984) Distribution of steroid sulfotransferase in the male hamster reproductive tract. Biol Reprod 31:936–941

Bouthillier M, Bleau G, Chapdelaine A and Roberts KD (1985) The assay and partial characterization of 3β-hydroxysteroid sulfotransferase of tha hamster epididymis. Can J Biochem Cell Biol 63:71–76

Brackett BG, Hall JL and Oh YK (1978) In-vitro fertilizing ability of testicular, epididymal, and ejaculated rabbit spermatozoa. Fert Steril 29:571–582

Briggs MH (1973) Steroid hormones and the fertilizing capacity of spermatozoa. Steroids 22:547–551

Byers SW, Qasba PK and Paulson HL (1984) Immunocytochemical localization of α-lactalbumin in the male reproductive tract. Biol Reprod 30:171–177

Byrd W (1981) In vitro capacitation and the chemically induced acrosome reaction in bovine spermatozoa. J exp Zool 215:35–46

Castaneda E, Bouchard P, Saling P, Phillips D, Gagnon C and Bardin CW (1983) Endogenous protein carboxyl methylation in hamster spermatozoa: changes associated with capacitation in vitro. Int J Androl 6:482–496

Chang MC (1951) Fertilizing capacity of spermatozoa deposited into the Fallopian tubes. Nature (Lond) 168:697–698

Chang MC (1984) The meaning of sperm capacitation: a historical perspective. J Androl 5:45–50

Clegg ED (1983) Mechanism of mammalian sperm capacitation. In: Mechanism and Control of Animal Fertilization. ED JF Hartmann, Acad Press New York. pp 177–212

Clegg ED, Morre DJ and Lunstra DD (1975) Porcine sperm membranes: in vivo phospholipid changes, isolation and electron microscopy. In: The Biology of the Male Gamete. Eds JG Duckett and PA Racey, Biol J Linnaen Soc, Suppl 1, Vol 7 Acad Press. pp 321–335

Cooper TG (1985) Prevention of hypo-osmotic swelling by detergents provides clues to the membrane structure of rat sperm. Int J Androl 8:159–167

Cuasnicú PS, González Echeverría F, Piazza A and Blaquier JA (1984) Addition of androgens to cultured hamster epididymis increases zona recognition by immature spermatozoa. J Reprod Fert 70:541–547

Cummins JM (1976) Effects of epididymal occlusion on sperm maturation in the hamster. J exp Zool 197:187–190

Cummins JM and Orgebin-Crist MC (1971) Investigations into the fertility of epididymal spermatozoa. Biol Reprod 5:13–19

Cummins JM and Teichmann RJ (1974) The accumulation of malachite green stainable phospholipid in rabbit spermatozoa during maturation in the epididymis, and its possible role in capacitation. Biol Reprod 10:555–564

Dacheux JL (1977) Reinvestigation of the variation in total phospholipid content of the spermatozoa of the rat and the ram during epididymal transit. IRCS Med Sci 5:18

Davis BK (1973) Occurrence of vesicles in rabbit seminal plasma. Experientia 29:1484–1487

Davis BK (1974) Decapacitation and recapacitation of rabbit spermatozoa treated with membrane vesicles from seminal plasma. J Reprod Fert 41:241–244

Davis BK (1976a) Influence of serum albumin on the fertilizing ability in vitro of rat spermatozoa. Proc Soc exp Biol Med 151:240–243

Davis BK (1976b) Inhibitory effect of synthetic phospholipid vesicles containing cholesterol on the fertilizing ability of rabbit spermatozoa. Proc Soc exp Biol Med 152:257–261

Davis BK (1980) Interaction of lipids with the plasma membrane of sperm cells. I. The antifertilization action of cholesterol. Archs Androl 5:249–254

Davis BK (1981) Timing of fertilization in mammals: sperm cholesterol/phospholipid ratio as a determinant of the capacitation interval. Proc Nat Acad Sci 78:7560–7564

Davis BK (1982) Uterine fluid proteins bind sperm cholesterol during capacitation in the rabbit. Experientia 38:1063–1064

Davis BK and Byrne R (1980) Interaction of lipids with the plasma membrane of sperm cells. II. Evidence of a membrane-thermotropic transition. Archs Androl 5:255–261

Davis BK and Davis NV (1983) Binding by glycoproteins of seminal plasma membrane vesicles accelerates decapacitation in rabbit spermatozoa. Biochem Biophys Acta 727:70–76

Davis BK and Gergely AF (1979) Studies on the mechanism of capacitation: changes in plasma membrane proteins of rat spermatozoa during incubation. Biochem Biophys Res Commun 88:613–618

Davis BK, Byrne R and Hungund B (1979) Studies on the mechanism of capacitation. II. Evidence for lipid transfer between plasma membrane of rat sperm and serum albumin during caacitation in vitro. Biochem Biophys Acta 558:257–266

Davis BK, Byrne R and Bedigian K (1980) Studies on the mechanism of capacitation. Albumin-mediated changes in plasma membrane lipids during in vitro incubation of rat sperm cells. Proc Nat Acad Sci 77:1546–1550

Djøseland O (1976) Androgen metabolism by rat epididymis. 4. The formation of conjugates. Steroids 27:617–636

Dravland E and Joshi MS (1981) Sperm-coating antigens secreted by the epididymis and seminal vesicle of the rat. Biol Reprod 25:649–658

Eng LA and Oliphant G (1978) Rabbit sperm reversible decapacitation by a membrane stabilization with a highly purified glycoprotein from seminal plasma. Biol Reprod 19:1083–1094

Ericsson RJ (1967a) A fluorimetric method for measurement of sperm capacitation. Proc Soc exp Biol Med 125:1115–1118

Ericsson RJ (1967b) Technology, physiology and morphology of spermatozoa capacitation. J Reprod Fert Suppl 2:65–74

Ericsson RJ and Baker VF (1967) Binding of tetracycline to mammalian spermatozoa. Nature (Lond) 214:403–404

Esbenshade KL and Clegg ED (1980) Surface proteins of ejaculated porcine sperm and sperm incubated in the uterus. Biol Reprod 23:530–537

Evans RW and Setchell BP (1979a) Lipid changes in boar spermatozoa during epididymal maturation with some observations on the flow and composition of boar rete testis fluid. J Reprod Fert 57:189–196

Evans RW and Setchell BO (1979b) Lipid changes during epididymal maturation in ram spermatozoa collected at different times of the year. J Reprod Fert 57:197–203

Evans RW, Weaver DE and Clegg ED (1980) Diacyl, alkenyl, and alkyl ether phospholipids in ejaculated, in utero-, and in vitro-incubated porcine spermatozoa. J Lipid Res 21:223–228

Fleming AD and Armstrong DT (1985) Effects of polyamines upon capacitation and fertilization in the guinea pig. J exp Zool 233:93–100

Fleming AD, Kosoner NS and Yanagimachi R (1982) Promotion of capacitation of guinea pig spermatozoa by the membrane mobility agent, A2C, and inhibition by the disulphide reducing agent, DTT. Gamete Res 5:19–33

Fraser LR (1984a) Mechanisms controlling mammalian fertilization. In: Oxford Reviews of Reproductiove Biology, Vol. 6. Ed JR Clarke, Clarendon Press, Oxford. pp 174–225

Fraser LR (1984b) Mouse sperm capacitation in vitro involves loss of a surface-associated inhibitory component. J Reprod Fert 72:373–384

Fraser LR and Drury LM (1976) Effect of removal of epididymal secretion on fertilization in vitro of mouse eggs. J Reprod Fert 48:125–128

Friend DS (1980) Freeze-fracture alterations in guinea pig sperm membranes preceding gamete fusion. In: Membrane-Membrane Interactions. Ed NB Gilula, Raven Press, New York. pp 153–165

Friend DS (1982) Plasma-membrane diversity in a highly polarised cell. J Cell Biol 93:243–249

Friend DS and Rudolf I (1974) Acrosomal disruption in sperm. Freeze-fracture of altered membranes. J Cell Biol 63:466–479

Friend DS, Orci L, Perrelet A and Yanagimachi R (1977) Membrane particle changes attending the acrosome reaction in guinea pig spermatozoa. J Cell Biol 74:561–577

Gagnon C and Heisler S (1979) Minireview. Protein carboxy-methylation: role in exocytosis and chemotaxis. Life Sci 25:993–1000

Garbers DL and Kopf GS (1980) The regulation of spermatozoa by calcium and cyclic nucleotides. In: Advances in Cyclic Nucleotide Research Vol 13. Eds P Greengard and GA Robison, Raven Press, New York. pp 251–306

Go KJ and Wolf DP (1983) The role of sterols in sperm capacitation. Adv Lipid Res 20:317–330

Go KJ and Wolf DP (1985) Albumin-mediated changes in sperm sterol content during capacitation. Biol Reprod 32:145–153

González Echeverría FM, Cuasnicú PS and Blaquier JA (1982) Identification of androgen-dependent glycoproteins in the hamster epididymis and their association with spermatozoa. J Reprod Fert 64:1–7

Gordon M (1973) Localization of phosphatase activity on the membranes of the mammalian sperm head. J exp Zool 185:111–120

Gordon M and Dandekar PV (1977) Fine-structural localization of phosphatase activity on the plasma membrane of the rabbit sperm head. J Reprod Fert 49:155–156

Grogan DE, Mayer DT and Sikes JD (1966) Quantitative differences in phospholipids of ejaculated spermatozoa. J Reprod Fert 12:431–436

Gwatkin RBL and Williams DT (1970) Inhibition of sperm capacitation in vitro by contraceptive steroids. Nature (Lond) 227:182–183

Hamilton DW (1971) Steroid function in the mammalian epididymis. J Reprod Fert Suppl 13:89–97

Hammerstedt RH, Keith AD, Hay S, DeLuca N and Amann RP (1979) Changes in ram sperm membranes during epididymal transit. Archs Biochem Biophys 196:7–12

Hanada A and Nagase H (1981) Effects of sperm preincubation in rabbit uterus and of imidazole on the penetration of zona-free hamster eggs by bull and boar spermatozoa. Jap J Anim Reprod 27:113–118

Hartmann JF and Gwatkin RBL (1971) Alteration of sites on the mammalian sperm surface following capacitation. Nature (Lond) 234:479–481

Hunter RHF, Holtz W and Henfrey PJ (1976) Epididymal function in the boar in relation to the fertilizing ability of spermatozoa. J Reprod Fert 46:463–466

Hunter RHF, Holtz W and Hermann H (1978) Stabilizing role of epididymal plasma in relation to the capacitation tiome of boar spermatozoa. Anim Reprod Sci 1:161–166

Hyne RV and Garbers DL (1979) Calcium-dependent increase in adenosine 3′,5′-monophosphate and induction of the acrosome reaction in guinea pig spermatozoa. Proc Natn Acad Sci 76:5699–5703

Hyne RV and Garbers DL (1982) Inhibition of the guinea-pig sperm acrosome reaction by a low molecular weight factor(s) in epididymal fluid and serum. J Reprod Fert 64:151–157

Iwamatsu T and Chang MC (1971) Factors involved in the fertilization of mouse eggs in-vitro. J Reprod Fert 26:197–208

Johnson MH (1975) The macromolecular organization of membranes and its bearing on events leading up to fertilization. J Reprod Fert 44:167–184

Johnson WL and Hunter AG (1972) Seminal antigens: their alteration in the genital tract of female rabbits and during partial in vitro capacitation with beta-amylase and beta-glucuronidase. Biol Reprod 7:332–340

Kinsey WH and Koehler JK (1978) Cell surface changes associated with in vitro capacitation of hamster sperm. J Ultrastruct Res 64:1–13

Koehler JK (1976) Changes in antigenic site distribution on rabbit spermatozoa after incubation in "capacitating" media. Biol Reprod 15:444–456

Kohane AC, González Echeverría FMC, Pineiro L and Blaquier JA (1980) Interaction of proteins of epididymal origin with spermatozoa. Biol Reprod 23:737–742

LaLumiere G, Bleau G, Chapdelaine A and Roberts KD (1976) Cholesterol sulfate and sterol sulfatase in the human reproductive tract. Steroids 27:247–260

Lambert H and vanLe A (1984) Possible involvement of a sialylated component of the sperm plasma membrane in the sperm-zona interaction in the mouse. Gamete Res 10:153–163

Langlais J, Zollinger M, Plante L, Chapdelaine A, Bleau G and Roberts KD (1981) Localization of cholesteryl sulfate in human spermatozoa in support of a hypothesis for the mechanism of capacitation. Proc Natn Acad Sci US 78:7266–7270

Lavon U, Volcani R and Danon D (1970) The lipid content of bovine spermatozoa during maturation and ageing. J Reprod Fert 23:215–222

Legault Y, vandenHeuvel WJA, Arison BH, Bleau G, Chapdelaine A and Roberts KD (1978) 5α-cholesta-7,24-dien-3β-ol as a major sterol of the male hamster reproductive tract. Steroids 32:649–658

Legault Y, Bouthillier M, Bleau G, Chapdelaine A and Roberts KD (1979a) The sterol and sterol sulfate content of the male hamster reproductive tract. Biol Reprod 20:1213–1219

Legault Y, Bleau G, Chapdelaine A and Roberts KD (1979b) The binding of sterol sulfates to hamster spermatozoa. Steroids 34:89–99

Legault Y, Bleau G, Chapdelaine A and Roberts KD (1980) Steroid sulfatase activity of the hamster reproductive tract during the estrous cycle. Biol Reprod 23:720–725

Lorton SP and First NL (1979) Hyaluronidase does not disperse the cumulus oophorus surrounding bovine ova. Biol Reprod 21:301–308

Menezo Y, Testart J, Khatchadourian C and Frydman R (1984) Human preovulatory follicular fluid: the lipids. Are they the trigger for capacitation? Int J Fert 29:61–64

Morton B and Albagli L (1973) Modification of hamster sperm adenyl cyclase by capacitation in vitro. Biochem Biophys Res Commun 50:697–703

Morton B and Chang TSK (1973) The effect of fluid from the cauda epididymidis, serum components and caffeine upon the survival of diluted epididymal hamster spermatozoa. J Reprod Fert 35:255–263

Morton BE, Fraser CF and Albagli L (1979) Studies on factors in hamster caudal epididmal plasma and other sources which inhibit sperm dilution damage. Fert Steril 32:99–106

Mrsny RJ, Siiteri JE and Meizel S (1984) Hamster sperm Na^+, K^+-adenosine triphosphatease: increased activity during capacitation in vitro and its relationship to cyclic nucleotides. Biol Reprod 30:573–584

Nicolson GL and Yanagimachi R (1976) Mobility and the restriction of mobility of plasma membrane lectin-binding components. Science 184:1294–1296

Nikolopoulou M, Souck DA and Vary JC (1985) Changes in the lipid content of boar sperm plasma membranes during epididymal maturation. Biochem Biophys Acta 815:486 498

Oliphant G and Brackett BG (1973a) Immunological assessment of surface changes of rabbit sperm undergoing capacitation. Biol Reprod 9:404–414

Oliphant G and Brackett BG (1973b) Capacitation of mouse spermatozoa in media with elevated ionic strength and reversible decapacitation with epididymal extracts. Fert Steril 24:948–955

Oliphant G and Eng LA (1981) Collection of gametes in laboratory animals and preparation of sperm for in vitro fertilization. In: Fertilization and Embryonic Development In Vitro. Eds L Mastroianni and JD Biggers, Plenum Press, London. pp 11–26

Oliphant G and Singhas CA (1979) Iodination of rabbit sperm plasma membrane: relationship of specific surface proteins to epididymal function and sperm capacitation. Biol Reprod 21:937–944

Oliphant G, Reynolds AB and Thomas, TS (1985) Sperm surface components involved in the control of the acrosome reaction. Am J Anat 174:269–283

Orgebin-Crist MC (1967) Maturation of spermatozoa in the rabbit epididymis: fertilizing ability and embryonic mortality in does inseminated with epididymal spermatozoa. Ann biol Anim Biochem Biophys 7:373–389

Orgebin-Crist MC (1968) Maturation of spermatozoa in the rabbit epididymis: Delayed fertilization in does inseminated with epididymal spermatozoa. J Reprod Fert 16:29–33

Overstreet JW (1970) Fertilizing capacity of epididymal spermatozoa. J Reprod Fert 21:423–429

Overstreet JW and Bedford JM (1974) Transport, capacitation and fertilizing ability of epididymal spermatozoa. J exp Zool 189:203–214

O'Rand MG (1977) Restriction of a sperm surface antigen's mobility during capacitation. Develop Biol 55:260–270

O'Rand MG (1979) Changes in sperm surface properties correlated with capacitation. In: The Spermatozoon. Eds DW Fawcett and JM Bedford, Urban and Schwarzenberg, Baltimore. pp 185–204

Parks JE and Hammerstedt RH (1985) Developmental changes occurring in the lipids of ram epididymal spermatozoa plasma membrane. Biol Reprod 32:653–668

Poulos A, Voglmayr JK and White IG 1973) Phospholipid changes in spermatozoa during passage through the genital tract of the bull. Biochem Biophys Acta 306:194–202

Poulos A, Brown-Woodman PDC, White IG and Cox RI (1975) Changes in phospholipids of ram spermatozoa during migration through the epididymis and possible origin of prostaglandin $F_{2\alpha}$ in testicular and epididymal fluid. Biochem Biophys Acta 388:12–18

Quinn PJ and White IG (1967) Phospholipid and cholesterol content of epididymal and ejaculated spermatozoa and seminal plasma in relation to cold shock. Aust J Biol Sci 20:1205–1215

Robertson RT, Bhalla VK and Williams WL (1971) Purification and the peptide nature of decapacitation factor. Biochem Biophys Res Commun 45:1331–1336

Rufo GA, Singh JP, Babcock DF and Lardy HA (1982) Purification and characterization of a calcium transport inhibitor protein from bovine seminal plasma. J Biol Chem 257:4627–4632

Rufo GA, Schoff PK and Lardy HA (1984) Regulation of calcium content in bovine spermatozoa. J Biol Chem 259:2547–2552

Scott TW, Voglmayr JK and Setchell BP (1967) Lipid composition and metabolism in testicular and ejaculated ram spermatozoa. Biochem J 102:456–461

Shur BD and Hall NG (1982) A role for mouse sperm surface galactosyltransferase in sperm binding to the egg zona pellucida. J Cell Biol 95:574–579

Singer SJ (1974) The molecular organisation of membranes. Ann Rev Biochem 43:805–833

Singh JP, Babcock DF and Lardy HA (1978) Increased calcium-ion influx is a component is capacitation of spermatozoa. Biochem J 172:549–556

Snider DR and Clegg ED (1975) Alteration of phospholipids in porcine spermatozoa during in vivo uterus and oviduct incubation. J Anim Sci 40:269–274

Soupart P, Anderson ML, Albert DH, Coniglio JG and Repp JE (1979) Accumulation, nature, and possible function of the malachite green affinity material in ejaculated human spermatozoa. Fert Steril 32:450–454

Stein DM and Fraser LR (1984) Cyclic nucleotide metabolism in mouse epididymal spermatozoas during capacitation in vitro. Gamete Res 10:283–299

Talbot P and Franklin LE (1978a) Surface modification of guinea pig sperm during in vitro capacitation: an assessment using lectin-induced agglutination of living sperm. J exp Zool 203:1–14

Talbot P and Franklin LE (1978b) Trypsinization increases lectin-induced agglutinability of uncapacitated guinea pig sperm. J exp Zool 204:291–297

Talbot P and Chacon R (1981) Detection of modificiations in the tail of capacitated guinea pig sperm using lectins. J exp Zool 216:435–444

Teichmann RJ, Fujimoto M and Yanagimachi R (1972) A previously unrecognized material in mammalian spermatozoa as revealed by malachite green and pyronine. Biol Reprod 7:73–81

Teichmann RJ, Cummins JM and Takei GH (1974a) The characterization of a malachite green stainable, glutaraldehyde extractable phospholipid in rabbit spermatozoa. Biol Reprod 10:565–577

Teichmann RJ, Cummins JM and Takei GH (1974b) The in vivo incorporation of (^{14}C) and (^{3}H) choline into phospholipids of maturing rabbit spermatozoa and reproductive tract tissues. Biol Reprod 11:644–653

Terner C, MacLaughlin J and Smith BR (1975) Changes in lipase and phosphatase activities of rat spermatozoa in transit from the caput to the cauda epididymidis. J Reprod Fert 45:1–8

Tezon JG, Ramella E, Cameo MS, Vazquez MH and Blaquier, JG (1985) Immunocytochemical localization of secretion antigens in the human epididymis and their association with spermatozoa. Biol Reprod 32:591–597

Thomas TS, Reynolds AL and Oliphant G (1984) Evaluation of the site of synthesis of rabbit sperm acrosome stabilizing factor using immunocytochemical and metabolic labeling techniques. Biol Reprod 30:693–705

Toshimori K, Higashi R and Oura C (1985) Distribution of intramembranous particles and filipin-sterol complexes in mouse sperm membranes: polyene antibiotic filipin treatment. Am J Anat 174:455–470

Triana LR, Babcock DF, Lorton SP, First NL and Lardy HA (1980) Release of acrosomal hyaluronidase follows increased membrane permeability to calcium in the presumptive capacitation sequence for spermatozoa from the bovine and other mammalian species. Biol Reprod 23:47–59

Vaidya RA, Bedford JM, Glass RH and Morris JM (1969) Evaluation of the removal of tetracycline fluorescence from spermatozoa as a test for capacitation in the rabbit. J Reprod Fert 19:483–489

Vaidya RA, Glass RH, Dandekar P and Johnson K (1971) Decrease in the electrophoretic mobility of rabbit spermatozoa following intra-uterine incubation. J Reprod Fert 24:299–301

Velazquez A, Rosado A and Rosado A (1975) Participation of vitamin A in the maturation of rabbit spermatozoa. Int J Fert 20:151–155

Viriyapanich P and Bedford JM (1981) The fertilization performance in vivo of rabbit spermatozoa capacitated in vitro. J exp Zool 216:169–174

Weinman DE and Williams WL (1964) Mechanism of capacitation of rabbit spermatozoa. Nature (Lond) 203:423–424

Wenstrom JC and Hamilton DW (1980) Dolichol concentration and biosynthesis in rat testis and epididymis. Biol Reprod 23:1054–1059

Wolf DE and Voglmayr JK (1984) Diffusion and regionalization in membranes of maturing ram spermatozoa. J Cell Biol 98:1678–1684

Wolf DE, Hagopian SS and Ishijima S (1986) Changes in sperm plasma membrane lipid diffusibility after hyperactivation during in vitro capacitation in the mouse. J Cell Biol 102:1372–1377

Working PK and Meizel S (1981) Evidence that an ATPase functions in the maintenance of the acidic pH of the hamster sperm acrosome. J Biol Chem 256:4708–4711

Working PK and Meizel S (1983) Correlation of increased intraacrosomal pH with the hamster sperm acrosome reaction. J exp Zool 227:97–108

Yanagimachi R (1981) Mechanisms of Fertilization in Mammals. In: Fertilization and Embryonic Development In Vitro. Eds L Mastroianni and JD Biggers, Plenum Press, London. pp 81–182

Yanagimachi R and Usui N (1974) Calcium dependence of the acrosome reaction and activation of guinea pig spermatozoa. Exp Cell Res 89:161–174

Yanagimachi R and Suzuki F (1985) A further study of lysolecithin-mediated acrosome reaction of guinea pig spermatozoa. Gamete Res 11:29–40

C. Consequences of Capacitation. I. Sperm-Egg Binding

1. Sperm-Egg Binding

Greater numbers of capacitated sperm bind to fertilised eggs than uncapacitated sperm bind to unfertilised eggs (Bedford, 1967; Overstreet and Bedford, 1974) and during this association there is a change from "loose attachment" to a more "firm binding" to the egg surface (Hartmann and Hutchison, 1974a; see Hartmann et al., 1983). It is possible to separate these two events by observing the zona: only bound spermatozoa display lytic activity on the surface and penetrate the zona pellucida (Jedlicki and Barros, 1985), an event intimately linked to the expression of the acrosome reaction (see Section II.D).

Uncapacitated porcine spermatozoa can bind to eggs Peterson et al., 1970, 1981, 1982, 1984, 1985) but generally, association of uncapacitated sperm to eggs is not species-specific (Bedford, 1977), in contrast, capacitated sperm bind only to homologous eggs (see Yanagimachi, 1977, 1978; 1980; Gulyas and Schmell, 1981; Swenson and Dunbar, 1982). Solubilised zonae from the hamster and bovine eggs only display receptor activity towards capacitated sperm cells (Gwatkin and Williams, 1978). A complicated sequence of prepenetration events in the hamster (see Hartmann and Hutchison, 1974) has not been confirmed in the mouse (Saling et al., 1978).

2. Relationship to Capacitation

A role for polycarbohyrate substrates of galactosyl-transferase in epididymal fluid as decapacitation factors has been proposed, their loss from the sperm during capacitation exposing to the zona pellucida the sperm surface enzyme which is considered to be the sperm zona receptor (Shur, 1981; Shur and Hall, 1982a).

3. Relationship to the Acrosome Reaction

The state of the acrosome at the time of binding to the zona is an area currently under active study. In-vitro, only acrosome-reacted guinea pig spermatozoa bind to the zona (Huang et al., 1981). The fact that spermatozoa recovered from the perivitelline space may fertilise eggs in the rabbit (Kuzan et al., 1984) suggests that acrosome-reacted sperm cells are also able to bind to and penetrate the zona in this species. In the hamster both acrosome-reacted and non acrosome-reacted spermatozoa bind to the zona pellucida, but only the former bind at the inner acrosomal membrane; the latter pivoting at the equatorial region and generating no thrust for penetration of this structure (Suarez et al., 1984). The acrosomal "ghost" may be important as an "anchor" to bind sperm temporarily to the egg surface (Talbot, 1985) and this implies that the acrosome reaction must occur close to the zona.

Determining the state of the acrosome in spermatozoa undergoing fertilisation is difficult. Cells observed after in-vivo insemination show that the acrosome reaction of ovine sperm occurs on the surface of the zona (Crozet and Dumont, 1984) and studies in which fertilisation is effected at low hamster sperm/egg ratios in-vitro (Corselli and Talbot, 1986) have supported this view.

Many studies of the binding of mouse sperm to cumulus-free eggs show that contact is made with the plasma membrane of intact (*i.e.* not acrosome reacted) spermatozoa (Saling and Storey, 1979; Saling et al., 1979; Florman and Storey, 1982; Bleil and Wassarman, 1983). This also implies that the acrosome reaction of relevance to fertilisation occurs on the surface of the zona pellucida (but see Section II.D.8) and the visualisation of the acrosome with chlortetracycline has confirmed that the majority of mouse sperm bound to the zona pellucida of even cumulus-invested eggs have intact acrosomes (Storey et al., 1984) that preclude the entry of ions (Lee and Storey, 1985). The zona intself may promote the zona (Florman and Storey, 1982; Florman et al., 1983; Bleil and Wasserman, 1983). "Zona Receptors" are also present on the plasma membrane of uncapacitated porcine spermatozoa and antibodies to them prevent sperm-egg binding (Peterson et al., 1979, 1981, 1982, 1985).

With the exception of the guinea-pig, then, current thinking contradicts the earlier views that the acrosome reaction and subsequent release of enzymes are needed for penetration of the normal egg investments (see Section II.D.9).

4. Mechanisms of Binding

Treatment of acrosome-reacted spermatozoa with trypsin (Huang and Yanagimachi, 1982) or DTT (Yanagimachi et al., 1983) prevents binding of sperm to eggs and implicates surface proteins and sulphydryl groups in the binding process. Binding of non-acrosome reacted mouse sperm is sensitive to trypsin inhibitors (Saling, 1981) and cold (Heffner and Storey, 1982) and Ca^{2+} is necessary for binding in mice (Saling et al., 1978; Heffner et al., 1980) but not hamsters (Yanagimachi, 1982) or guinea pigs (Huang and Yanagimachi, 1982). The ability of multi-charged spermine^{4+} or lanthanum^{3+} to promote sperm-binding in the mouse at lower concentrations than divalent calcium implicates electrostatic forces in binding (Saling et al., 1978).

a. Surface Carbohydrates

Evidence that surface carbohydrates are important in gamete recognition comes from hapten inhibition tests where incubation with low molecular weight aminosugars prevents both binding and in-vitro fertilisation in hamsters (Hartmann and Hutchison, 1977b; Ahuja, 1982, 1985), guinea pigs (Huang et al., 1982; Huang and Yanagimachi, 1984), mice (Lambert, 1984; Lambert and vanLe, 1984) and rats (Matityaha et al., 1984). In the guinea pig these receptors have been found on the inner and equatorial membranes of the acrosome (Huang and Yanagimachi, 1984), *i.e.* the carbohydrates do not become surface moieties until the acrosome reaction has occurred. As these carbohydrates are not species-specific in their effects and have no effect on sperm motility, it is suggested that they compete with the egg for the surface zona binding sites on the sperm.

b. Surface Enzymes

Sialyl- and galactosyl-transferases are involved in cell-cell contact and recognition in many cell types (see Shur and Roth, 1973; Pierce et al., 1980) and sialyltransferases have been implicated in sperm-egg contact by an association of the enzyme activity related to ova binding (Durr et al., 1977) and the different activities of the enzymes in sperm from strains of mice differing in fertility (Shur and Bennett, 1979; Shur, 1981). Purified galactosyltransferase competitively inhibits binding of sperm to eggs (Lopez et al., 1985) stressing the importance of substrates (N-acetylglucosamine) on the zona surface. The importance of sperm surface galactosyl transferase in sperm binding is also supported by studies with α-lactalbumin. This protein alters the specificity of galactosyltransferase away from N-acetylglucosamine towards glucose as galactose acceptor, so the decline in binding of sperm to eggs induced by α-lactalbumin in-vitro (Schur and Hall, 1982b) suggests that binding involves recognition N-acetylglucosamine on the zona pellucida by an enzyme in the sperm surface. That galactosylation of the zona by addition of UDPgalactose to the enzme depresses sperm-egg binding suggests that the sperm-zona recognition process does not involve a catalytic event; rather the enzyme is being used as a specific receptor (Lopez et al., 1985).

5. Development in the Epididymis of the Ability of Sperm to Bind to Eggs

Rabbit testicular spermatozoa are able to adhere to eggs after in-vivo insemination (Lambiase and Amann, 1973; Cooper and Orgebin-Crist, 1975; Table 9), but in-vitro testicular spermatozoa from the ram (Fournier-Delpech et al., 1982, 1983a,b) and rat (Orgebin-Crist and Fournier-Delpech, 1982) are unable to bind to rat eggs. With zona-intact, but cumulus-free eggs, sperm binding capacity appears in the corpus of boars, rams, hamsters and rats but in the cauda of mice (Table 8).

Increased attachment of gametes does not reflect the increased motility of, and the anticipated greater rate of collision with eggs by, more distal spermatozoa, because mature sperm rendered immobile by low temperature (Orgebin-Crist and Fournier-Delpech, 1982; Fournier-Delpech et al., 1984a) or lanthanum (Saling, 1982) still bind to eggs. Furthermore, agitating the medium containing mouse caput sperm (Hoppe, 1975) or initiating forward motility in porcine testicular spermatozoa (Dacheux and Paquignon, 1980b, 1983a) or stimulating caput sperm with caffeine (Peterson et al., 1984) does not improve the binding of the immature spermatozoa.

A variety of other changes occurring to the sperm surface during their stay in the epididymis, including charge, agglutination and antigenicity (see Table 19) may be involved in sperm-egg recognition. These changes are known to reflect alteration in membrane constituent glycoproteins (see Brooks and Tiver, 1984), proteins (see Vierula and Rajaniemi, 1982), lipid (see Hammerstedt et al., 1982) and carbohydrate (see Nicolson and Yanagimachi, 1979), but only in a few cases are these changes coincident with sperm-egg binding capacity. Development of agglutinating ability occurs more proximally in the epididymis than the ability of sperm to bind to zona-free eggs (Dacheux et al., 1983), but the more relevant interaction with zona-intact eggs was not studied. In the hamster aquisition of increasing amounts of a specific epididymal protein (Gonzáles Echeverría et al., 1972) parallels that of sperm-zona recognition (Cuasnicú et al., 1984a). Some men who ejaculate spermatozoa with abnormal surface charge are infertile (Moore, 1979), and this may reflect poor contact of the sperm with the egg.

6. Involvement of the Epididymis in the Development of the Sperm Surface

Allowing caput or corpus sperm to age in the hamster epididymis does not improve their ability to bind to eggs in-vivo (Cummins, 1976). The ability of mature sperm to bind to eggs is androgen-dependent in rams (Fournier-Delpech et al., 1983a) and rats (Fournier-Delpech et al., 1984) and in-vitro androgens can promote sperm-egg binding by immature sperm from the hamster upper corpus epididymidis (Cuasnicú et al., 1984a), an action prevented by inhibition of protein synthesis (Cuasnicú et al., 1984b). The epididymis may be involved by adding proteins to the sperm surface or by modifying constituent proteins of the membrane.

It is easier to envisage how the epididymis influences the external surface of the sperm (the plasma membrane) than, for example, the inner acrosomal membranes, yet the latter have been implicated in binding to the zona pellucida (see Sections II.C.4.a and II.D.8). Transmembrane effects of surface proteins (see Ji and Nicolson, 1974) or modification of sub-surface membranes by integral enzymes in the plasma membrane may be involved.

a. Secretion of Sperm-Coating Glycoproteins

That the epididymis is involved in secretion of sperm coating proteins is demonstrated by the failure of sperm aged in the efferent ducts to aquire the protein coats (Vernon et al., 1982) or membrane charge (Bedford et al., 1972) they usually gain on maturation. The secretion of specific epididymal proteins continues after inflow of testicular fluid into the epididymis is interrupted by ligation of testicular efferent ducts (Kohane et al., 1980a; Barker and Amann, 1970, 1971; see Section III.G.2.c) and androgens maintain the secretion of specific epididymal coating proteins in-vivo (Moore, 1981a) and in-vitro (Cuasnicú et al., 1984a). Androgen-dependent components of cauda epididymal fluid compete for lectin-binding sites on the sperm surface (Fournier-Delpech et al., 1983b).

i. The Role of Coating Proteins in Sperm-Egg Binding

Testicular sperm will bind to eggs in-vitro if the sperm are previously incubated with specific lectins (rams: Fournier-Delpech et al., 1983b) or specific epididymal sialoproteins (rats: Orgebin-Crist and Fournier-Delpech, 1982), although in the latter study other proteins were not tested as non-specific controls. Hamster sperm within the upper corpus epididymidis maintained in culture with androgens develop the ability to bind to eggs in-vivo and they also become coated with specific androgen-dependent epididymal glycoproteins that can be eluted from them with high ionic strength (Cuasnicú et al., 1984a). The addition of these extracted proteins to cultures of proximal corpus tubules increases the – egg binding capacity of the contained spermatozoa (Cuasnicú et al., 1984b).

ii. Relationship to Capacitation

It is important to establish if the immature spermatozoa that are induced to bind to eggs by epididymal secretions in-vitro are capacitated, if they are to provide information about normal epididymal processes. Normally only capacitated mature sperm cells can achieve fertilisation and the binding of uncapacitated sperm to eggs may differ from that of capacitated sperm because of the presence of epididymal fluid normally diluted under capacitation conditions. For example, Hartmann and Hutchison (1977b) found that an oligosaccharide in epididymal fluid mediated the binding of uncapacitated sperm to eggs through association with galactose residues on the sperm surface.

Since decapacitation factors should be lost during capacitation and those mediating sperm-egg binding must remain to be exposed to the zona, decapacitation factors may be associated less firmly with the membrane (peripheral proteins?) than proteins associated with sperm-egg binding, which may be inserted more firmly into the membrane (see Section III.G.10.b), but the exact relationship between the two remains unexplored. It is interesting that in the hamster, where sperm-egg binding is associated with a coating proteins (see Section II.C.6.a.i), capacitation is thought to involve modification of membrane sterols (see Section II.B.3.c).

iii. Relationship to Fertilisation

Active immunisation against a specific prealbumin involved in sperm-egg binding renders male rats infertile (Fournier-Delpech et al., 1985). Adding specific glycoproteins to immature hamster spermatozoa promotes their fertilising ability in-vitro and in-vivo (Gonzáles Echeverría et al., 1984) and culture of rat epididymal tubules in media containing an antibody raised to specific rat epididymal proteins reduces the fertilising ability of mature spermatozoa contained within the tubule (Cuasnicú et al., 1984c).

b. Changes in Integral Membrane Proteins of Sperm on Maturation

Several workers have found one major sialoglycoprotein on the plasma membrane of cauda sperm (Brooks and Tiver, 1984) which is not present on caput spermatozoa from the rat (Olson and Hamilton, 1978; Zeheb and Orr, 1984) and which first appears in the proximal cauda (Olson and Orgebin-Crist, 1982). Others have found more than one protein (Jones et al., 1981; Brown et al., 1983; Toowicharanont and Chulavatnatol, 1983c). The major protein appears to be an integral membrane protein and is only lost on disruption of the membrane (Olson and Orgebin-Crist, 1982; Brown et al., 1983; Zeheb and Orr, 1984).

i. Action and Control of Sperm Surface Glycosylation

Glycosylation of endogenous proteins in the immature sperm cell membrane within the epididymal lumen could be responsible for the appearance of galactose residues on mature spermatozoa. This could be mediated by the galactosyltransferases found in testicular and epididymal fluid from rats (Hamilton, 1980, 1981a) and epididymal fluid from mice (Letts et al., 1974; Shur and Bennett, 1979). Rat caput sperm, which lack galactose residues over their flagellum as judged from fluorescent lectin probes, develop them on incubation with UDP-galactose, especially in the presence of an additional source of galactosyltransferase (rete testis fluid; Hamilton and Gould, 1982). A fucosidase in the mouse epididymis can fucosylate mouse sperm in-vitro (Cossu and Boitani, 1984).

That galactose does not normally appear on the surface of the sperm until they reach the cauda suggests that mere contact of immature sperm with enzyme in-vivo is not sufficient for glycosylation to take place, and some form of control is implied. This could be a role for α-lactalbumin, which is present in testicular and epididymal fluids (Hamilton, 1981b; Byers et al., 1984) and binds to spermatozoa (Jones and Brown, 1982; Byers et al., 1984; Klinefelter and Hamilton, 1984). In the presence of this protein galactosyl transferase in RTF transfers galactose residues from UDP-galactose to glucose or inositol in-vitro (Hamilton, 1981b; Jones and Brown, 1982). As inositol is present along the length of the epididymis (see Section III.F) it could help to modulate the extent of sperm membrane glycosylation, possibly by competition, although the disaccharide product galactinol has not yet been detected in epididymal fluid.

c. Development in the Epididymis of the Ability of Sperm to be Glycosylated

Galactosyltransferase is associated with immature spermatozoa (Hamilton, 1981a) but it is easily washed off and is not associated with washed mature sperm in rats (Tadolini et al., 1977; Hamilton, 1981a) or mice (Letts et al., 1974). By contrast, high enzyme activity of testicular and caput sperm in the mouse (Shur and Bennett, 1979) may be reflect an integral membrane protein (Shur and Hall, 1982a,b) like the sialytransferase on mature sperm in the mouse (Durr et al., 1977) and rat (Bernals et al., 1980).

Rat caput sperm clearly have acceptors for galactose (Hamilton and Gould, 1982) and they have more acceptors for sialic acid than mature sperm (Bernals et al., 1980), so the extent of glycosylation on maturation would appear to depend on the supply of cosubstrates or activators of the enyzme.

i. Role of Glycosylation in Sperm-Egg Binding

A role for galactosyltransferase in sperm-egg binding has been postulated in the mouse (Shur and Hall, 1982a; see Section II.C.4.b), but the lack of egg binding by immature sperm is inconsistent with the higher activity of the membrane enzyme in these cells (Shur and Bennett, 1979). In rats few caput sperm bear galactose residues over the surface of the head (Hamilton and Gould, 1982) and they do not bind to eggs (Table 8), but whether this sugar is involved in sperm-egg binding in this species remains to be determined. Additionally, the accumulation of galactose or the masking of N-acetylglucosamine residues on the flagellum may be associated with changes in motility (see Section II.A.3.b).

7. Summary

The ability of sperm to adhere to eggs is a prerequisite of penetration and in some species reflects the uptake onto the sperm surface of specific epididymal proteins that themselves recognise proteins in the zona. Because these proteins must remain on the sperm surface after capacitation, proteinaceous decapacitation factors in these species must be held to the sperm surface by a different mechanism from the egg-binding proteins; alternatively, decapacitation may involve a change in the lipid component of the membrane. Where sperm-egg binding occurs on membranes normally hidden beneath the plasma membrane until exposure after the acrosome reaction it is less clear how the epididymis influences the binding ability of the maturing sperm.

8. References

Ahuja KK (1982) Fertilization studies in the hamster. The role of cell-surface carbohydrate. Exp Cell Res 140:353–362

Ahuja KK (1985) Carbohydrate determinants involved in mammalian fertilization. Am J Anat 174:207–223

Barker LDS and Amann RP (1970) Epididymal physiology. I. Specificity of antisera against bull spermatozoa and reproductive fluids. J Reprod Fert 22:441–452

Barker LDS and Amann RP (1971) Epididymal physiology.II. Immunofluorescent analyses of epithelial secretion and absorption and of bovine sperm maturation. J Reprod Fert 26:319–332

Bedford JM (1967) The importance of capacitation for establishing contact between eggs and sperm in the rabbit. J Reprod Fert 13:365–367

Bedford JM (1977) Sperm/Egg interaction: the specificity of human spermatozoa. Anat Rec 188:477–488

Bedford JM, Cooper GW and Calvin HI (1972) Post-meiotic changes in the nucleus and membranes of mammalian spermatozoa. In: The Genetics of the Spermatozoon. Eds RA Beatty and S Glucksohn-Waelsh, Bogtrykkeriet Forum, Copenhagen. pp 69–89

Bernals A, Torres J, Reyes A and Rosado A (1980) Presence and regional distribution of sialyltransferase in the epididymis of the rat. Biol Reprod 23:290–293

Bleil JD and Wassarman PM (1983) Sperm-egg interaction in the mouse: sequence of events and induction of the acrosome reaction by a zona pellucida glycoprotein. Develop Biol 95:317–324

Brooks DE and Tiver K (1984) Analysis of surface proteins of rat spermatozoa during epididymal transit and identification of antigens common to spermatozoa, rete testis fluid and cauda epididymal plasma. J Reprod Fert 71:249–257

Brown CR, von Glos KI and Jones R (1983) Changes in plasma membrane glycoproteins of rat spermatozoa during maturation in the epididymis. J Cell Biol 96:256–264

Byers SW, Qasba PK and Paulson HL (1984) Immunocytochemical localization of α-lactalbumin in the male reproductive tract. Biol Reprod 30:171–177

Cooper TG and Orgebin-Crist MC (1975) The effect of epididymal and testicular fluids on the fertilising capacity of testicular and epididymal spermatozoa. Andrologia 7:85–93

Corselli J and Talbot P (1986) An in vitro technique to study penetration of hamster oocyte-cumulus complexes by using physiological numbers of sperm. Gamete Res 13:293–308

Cossu G and Boitani C (1984) Lactosaminoglycans synthesised by mouse male germ cells are fucosylated by an epididymal fucosyltransferase. Develop Biol 102:402–408

Crozet N and Dumot M (1984) The site of the acrosome reaction during in vivo penetration of the sheep oocyte. Gamete Res 10:97–105

Cuasnicú PS, González Echeverría F, Piazza A and Blaquier JA (1984a) Addition of androgens to cultured hamster epididymis increases zona recognition by immature spermatozoa. J Reprod Fert 70:541–547

Cuasnicú PS, González Echeverría F, Piazza, Piñeiro L and Blaquier JA (1984b) Epididymal proteins that mimic the androgenic effect on zona pellucida recognition by immature hamster spermatozoa. J Reprod Fert 71:427–431

Cuasnicú PS, Gonzáles Echeverría F, Piazza AD, Cameo MS and Blaquier JA (1984c) Antibodies against epididymal glycoproteins block fertilizing ability in rat. J Reprod Fert 72:467–471

Cummins JM (1976) Effects of epididymal occlusion on sperm maturation in the hamster. J exp Zool 197:187–190

Dacheux JL and Paquignon M (1980) Relations between the fertilizing ability, motility and metabolism of epididymal spermatozoa. Reprod Nutr Develop 20:1085–1099

Dacheux JL and Paquignon M (1983) Influence of initiation of forward motility on the fertilizing ability of immature boar spermatozoa in in vivo homologous and in vitro heterologous systems of insemination. In: The Sperm Cell Ed J Andre, Martinus-Nijhoff, The Hague. pp 99–102

Dacheux JL, Paquignon M and Combarnous Y (1983) Head-to-head agglutination of ram and boar epididymal spermatozoa and evidence for an epididymal antagglutinin. J Reprod Fert 67:181–189

Durr R, Shur B and Roth S (1977) Sperm-associated sialyltransferase activity. Nature (Lond) 265:547–548

Florman HM and Storey BT (1982) Mouse gamete interactions: the zona pellucida is the site of the acrosome reaction leading to fertilization in vitro. Develop Biol 91:121–130

Florman HM, Saling PM and Storey BT (1983) Sperm/zona pellucida reactions leading to fertilization of mouse eggs in vitro. In: The Sperm Cell. Ed J Andre, Martinus-Nihjoff, The Hague. pp 111–114

Fournier-Delpech S, Courtens JL, Pisselet CL, DeLaleu B and Courot M (1982) Acquisition of zona binding by ram spermatozoa during epididymal passage, as revealed by interaction with rat oocytes. Gamete Res 5:403–408

Fournier-Delpech S, Hamamah S, Colas G and Courot M (1983) Aquisition of zona binding structures by ram spermatozoa during epididymal passage. In: The Sperm Cell. Ed J Andre, Martinus-Nijhoff, The Hague. pp 103–106.

Fournier-Delpech S, Hamamah S, Tananis-Anthony C, Courot M and Orgebin-Crist MC (1984a) Hormonal regulation of zona-binding ability and fertilizing ability of rat epididymal spermatozoa. Gamete Res 9:21–30

Fournier-Delpech S, Hamah S, Courot M and Kuntz C (1984b) Androgenic control of zona binding capacity of ram spermatozoa. In: The Male in Farm Animal Reproduction. Ed M Courot, Martinus-Nijhoff, Boston. pp 103–107

Fournier-Delpech S, Courot M and Dubois MP (1985) Decreased fertility and motility of spermatozoa from rats immunized with a prealbumin epididymal specific glycoprotein. J Androl 6:246–250

González Echeverría FM, Cuasnicú PS and Blaquier JA (1982) Identification of androgen-dependent glycoproteins in the hamster epididymis and their association with spermatozoa. J Reprod Fert 64:1–7

González Echeverría F, Cuasnicú PS, Piazza A, Piñeiro L and Blaquier JA (1984) Addition of an androgen-free epididymal protein extract increases the ability of immature hamster spermatozoa to fertilize in vivo and in vitro. J Reprod Fert 71:432–437

Gulyas BJ and Schmell ED (1981) Sperm-egg recognition and binding in mammals. In: Bioregulators of Reproduction. Eds G Jagiello and HJ Vogel, Acad Press, London. pp 499–519

Gwatkin RBL and Williams DT (1978) Bovine and hamster zona solutions exhibit receptor activity for capacitated but not for noncapacitated sperm. Gamete Res 1:259–263

Hamilton DW (1980) UDP-galactose: N-acetylglucosamine galactosyltransferase in fluids from rat rete testis and epididymis. Biol Reprod 23:377–385

Hamilton DW (1981a) Future trends in cell biological research on the male reproductive system. In: Goals in Male Reproductive Research. Eds S Boyarsky and K Polakoski, Pergamon Press, Oxford. pp 19–27

Hamilton DW (1981b) Evidence for α-lactalbumin-like activity in reproductive tract fluids of the male rat. Biol Reprod 25:385–392

Hamilton DW and Gould RP (1982) Preliminary observations on enzymatic galactosylation of glycoproteins on the surface of rat caput epididymal spermatozoa. Int J Androl Suppl 5:73–80

Hammerstedt RH, Hay SR and Amann RP (1982) Modification of ram sperm membranes during epididymal transit. Biol Reprod 27:745–754

Hartmann JF (1983) Mammalian fertilization: gamete surface interactions in vitro. In: Mechanism and Control of Animal Fertilization. Ed JF Hartmann, Acad Press, New York. pp 325–364

Hartmann JF and Hutchison CF (1974) Nature of the pre-penetration contact interactions between hamster gametes in vitro. J Reprod Fert 36:49–57

Hartmann JF and Hutchison CF (1977a) Involvement of two carbohydrate-containing components in the binding of uncapacitated spermatozoa to eggs of the golden hamster in vitro. J exp Zool 201:383–390

Heffner LJ and Storey BT (1982) Cold lability of mouse sperm binding to zona pellucida. J exp Zool 219:155–161

Heffner LJ, Saling PM and Storey BT (1980) Separation of calcium effects on motility and zona binding ability in mouse spermatozoa. J exp Zool 212:53–57

Hoppe PC (1975) Fertilizing ability of mouse sperm from different epididymal regions and after washing and centrifugation. J exp Zool 192:219–222

Huang TTF and Yanagimachi R (1982) Some factors controlling attachment of guinea pig spermatozoa to zona pellucida. Proc Japan Acad 58B:101–104

Huang TTF and Yanagimachi R (1984) Fucoidan inhibits attachment of guinea pig spermatozoa to the zona pellucida through binding to the inner acrosomal membrane and equatorial domains. Exp Cell Res 153:363–373

Huang TTF, Fleming AD and Yanagimachi R (1981) Only acrosome-reacted spermatozoa can bind to and penetrate the zona pellucida: a study using the guinea pig. J exp Zool 217:287–290

Huang TTF, Ohzu E and Yanagimachi R (1982) Evidence suggesting that L-fucose is part of a recognition signal for sperm-zona pellucida attachment in mammals. Gamete Res 5:355–361

Jedlicki A and Barros C (1985) Scanning electron microscope study of on vitro prepenetration gamete interactions. Gamete Res 11:121–131

Ji TH and Nicolson GL (1974) Lectin binding and perturbation of the outer surface of the cell membrane induces a transmembrane organizational alteration at the inner surface. Proc Natn Acad Sci US 71:2212–2216

Jones R and Brown CR (1982) Association of epididymal secretory proteins showing α-lactalbumin-like activity with the plasma membrane of rat spermatozoa. Biochem J 206:161–164

Jones R, Pholpramool C, Setchell BP and Brown CR (1981) Labelling of membrane glycoproteins in rat spermatozoa collected from different regions of the epididymis. Biochem J 200:457–460

Klinefelter GR and Hamilton DW (1984) Organ culture of rat caput epididymal tubules in a perifusion chamber. J Androl 5:243–258

Kohane AC, Cameo MS, Piñeiro L, Garberi JC and Blaquier JA (1980) Distribution and site of production of specific proteins in the rat epididymis. Biol Reprod 23:181–187

Kuzan F, Fleming A and Seidel G (1984) Successful fertilization in vitro of fresh oocytes by perivitelline (acrosome-reacted) spermatozoa of the rabbit. Fert Steril 41:766–770

Lambert H (1984) Role of sperm-surface glycoproteins in gamete recognition in two mouse species. J Reprod Fert 40:281–284

Lambert H and vanLe A (1984) Possible involvement of a sialylated component of the sperm plasma membrane in the sperm-zona interaction in the mouse. Gamete Res 10:153–163

Lambiase JT and Amann RP (1973) Infertility of rabbit testicular spermatozoa collected in their native environment. Fert Steril 24:65–67

Lee MA and Storey NT (1985) Evidence for plasma membrane impermeability to small ions in acrosome-intact mouse spermatozoa bound to mouse zonae pellucidae, using an amino-acridine fluorescent pH probe; time course of the zona-induced acrosome reaction monitored by both chlortetracycline and pH probe fluorescence. Biol Reprod 33:235–246

Letts PJ, Meistrich ML, Bruce WR and Schacter H (1974) Glycoprotein glycosyl-transferase levels during spermatogenesis in mice. Biochem Biophys Acta 343:192–207

Lopez LC, Bayna EM, Litoff D, Shaper NL, Shaper JH and Shur BD (1985) Receptor function of mouse sperm surface galactosyltransferase during fertilization. J Cell Biol 101:1501–1510

Matityahu A, Shalgi R and Nebel L (1984) The role of carbohydrates in sperm-egg attachment in rats. Am J Reprod Immunol 6:60 (abst).

Moore HDM (1979) The net surface charge of mammalian spermatozoa as determined by isoelectric focussing. Changes following sperm maturation, ejaculation, incubation in the female tract, and after enzyme treatment. Int J Androl 2:449–462

Moore HDM (1981) Effects of castration on specific glycoprotein secretions of the epididymis in the rabbit and hamster. J Reprod Fert 61:347–354

Nicolson GJ and Yanagimachi R (1979) Cell surface changes associated with the epididymal maturation of mammalian spermatozoa. In: The Spermatozoon. Maturation, Motility, Surface Properties and Comparative Aspects. Eds DW Fawcett and JM Bedford, Urban and Schwarzenberg, Baltimore-Munich. pp 187–194

Olson GE and Hamilton DW (1978) Characterization of the surface glycoproteins of rat spermatozoa. Biol Reprod 19:26–35

Olson GE and Orgebin-Crist MC (1982) Sperm surface changes during epididymal maturation. Ann NY Acad Sci 383:372–391

Orgebin-Crist MC and Fournier-Delpech S (1982) Sperm-egg interactions. Evidence for maturational changes during epididymal transit. J Androl 3:429–433

Overstreet JW and Bedford JM (1974) Importance of sperm capacitation for gamete contact in the rabbit. J Reprod Fert 39:393–398

Peterson RN, Russell L, Bundman D and Freund M (1979) Sperm-egg interaction: evidence for boar sperm plasma membrane receptors for porcine zona pellucida. Nature (Lond) 207:73–74

Peterson RN, Russell LD, Bundman D, Conway M and Freund M (1981) The interaction of living boar sperm and sperm plasma membrane vesicles with the porcine zona pellucida. Develop Biol 84:144–156

Peterson RN, Robl JM, Dziuk PJ and Russell LD (1982) The effects of antisperm plasma membrane antibodies on sperm-egg binding, penetration and fertilization in the pig. J exp Zool 223:79–81

Peterson RN, Russell LD, and Hunt WP (1984) Evidence for specific binding of uncapacitated boar spermatozoa to porcine zonae pellucidae in vitro. J exp Zool 231:137–147

Peterson RN, Henry L, Hunt W, Saxena N and Russell LD (1985) Further characterization of boar sperm plasma membrane proteins with affinity for the porcine zona pellucida. Gamete Res 12:91–100

Pierce M, Turley EA and Roth R (1980) Cell surface galactosyltransferase activities. Int Rev Cytol 65:1–47

Roth S (1973) A molecular model for cell interactions. Quart Rev Biol 48:541–563

Saling PM (1981) Involvement of trypsin-like activity in binding of mouse spermatozoa to zonae pellucidae. Proc Nat Acad Sci US 78:6231–6235

Saling PM (1982) Development of the ability to bind to zonae pellucidae during epididymal maturation: reversible immobilisation of mouse spermatozoa by lanthanum. Biol Reprod 26:429–436

Saling PM and Storey BT (1979) Mouse gamete interactions during fertilization in vitro: chlortetracycline as fluorescent probe for the sperm acrosome reaction. J Cell Biol 83:544–555

Saling PM, Storey BT and Wolf DP (1978) Calcium-dependent binding of mouse epididymal spermatozoa to the zona pellucida. Develop Biol 65:515–525

Saling PM, Sowinski J and Storey BT (1979) An ultrastructural study of epididymal mouse spermatozoa binding to zonae pellucidae in vitro: sequential relationship to the acrosome reaction. J exp Zool 209:229–238

Shur BD (1981) Galactosyltransferase activities on mouse sperm bearing multiple t lethal and t viable haplotypes of the T/t-complex. Genet Res Cambridge 38:225–236

Shur BD and Bennett D (1979) A specific defect in galactosyltransferase regulation on sperm bearing mutant alleles of the T/t locus. Develop Biol 71:243–259

Schur BD and Hall NG (1982a) Sperm surface galactosyltransferase activities during in vitro capacitation. J Cell Biol 95:567–573

Shur BD and Hall NG (1982b) A role for mouse sperm surface galactosyltransferase in sperm binding to the egg zona pellucida. J Cell Biol 95:574–579

Storey BT, Lee MA, Muller C, Ward CR and Wirtshafter DG (1984) Binding of mouse spermatozoa to the zona pellucida of mouse eggs in cumulus: evidence that the acrosome remains substantially intact. Biol Reprod 31:1119–1128

Suarez SS, Katz DF and Meizel S (1984) Changes in motility that accompany the acrosome reaction in hyperactivated hamster spermatozoa. Gamete Res 10:253–265

Swenson CE and Dunbar RS (1982) Specificity of sperm-zona interaction. J exp Zool 219:97–104

Tadolini B, Wilson J, Reddy PRK and Williams-Ashman HG (1977) Charactertistics and hormonal control of some glycoprotein glycosyltransferase reactions in male reproductive organs. Advances in Enzyme Regulation, Ed G Weber, Volume 15. pp 319–336

Talbot P (1985) Sperm penetration through oocyte investments in mammals. Am J Anat 174:331–346

Toowicharanont P and Chulavatnatol M (1983) Characterization of sialoglycoproteins of rat epididymal fluid and spermatozoa by periodate-tritiated borohydride. J Reprod Fert 67:113–141

Vernon RB, Mueller CH, Herr JC, Feuchter FA and Eddy EM (1982) Epididymal secretion of a mouse sperm surface component recognized by a monoclonal antibody. Biol Reprod 26:523–535

Vierula M and Rajaniemi H (1982) Epididymal maturation of the surface protein structure of mammalian spermatozoa. Medical Biology 60:323–327

Yanagimachi R (1977) Specificity of sperm-egg interaction. In: Immunobiology of Gametes. Eds M Edinin and MH Johnson, CUP Cambridge. pp 255–289

Yanagimachi R (1978) Sperm-Egg Association in Mammals. In: Current Topics in Developmental Biology, Volume 12 Fertilization. Eds AA Moscona and A Monroy, Academic Press, New York. pp 83–105

Yanagimachi R (1982) Requirement of extracellular calcium ions for various stages of fertilization and fertilization-related phenomena in the hamster. Gamete Res 5:323–344

Yanagimachi R, Huang TTF, Fleming AD, Kosower NS and Nicolson GL (1983) Dithiothreitol, a disulphide-reducing agent, inhibits capacitation, acrosome reaction and interaction with eggs by guinea pig spermatozoa. Gamete Res 7:145–154

Zeheb R and Orr GA (1984) Characterization of a maturation-associated glycoprotein on the plasma membrane of rat caudal epididymal sperm. J Biol Chem 259:839–848

D. Consequences of Capacitation. II. The Acrosome Reaction

1. The Acrosome Reaction

The "Acrosome Reaction" refers to a series of changes in the acrosome leading to its eventual loss, with the concomitant release of its content of hydrolytic enzymes and the exposure of a new limiting membrane over the anterior portion of the sperm head, both of which are essential for fertilisation (see Austin and Bavister, 1975; Austin, 1977; Bedford and Cooper, 1978). It proceeds by fusion of the sperm plasma membrane with the outer acrosomal membrane directly beneath it, giving rise to membrane vesicles which have been clearly shown in the boar to arise from two membranes with different structures (Russell et al., 1979a). The pores so created allow the escape of soluble enzymes (see Section II.D.9).

Loss of the acrosomal cap normally follows dispersion of the matrix, but if matrix dispersal is inhibited the sperm cell may appear to have retained its acrosome in the light microscope, yet electron micrographs would reveal that vesiculation has occurred. The mechanisms underlying the fusion is unclear; while several workers found that inhibitors of trypsin could prevent acrosomal matrix dispersal (see Huang et al., 1985), others (see Dravland et al., 1984) considered that the inhibitor acted on the vesiculation process. Dispersion of the acrosomal matrix leads to loss of the membrane vesicles, leaving the inner acrosomal membrane exposed. This membrane differs from the plasma membrane in surface and internal structures (Yanagimachi and Noda, 1972; Kinsey and Koehler, 1976), carbohydrates (Schwarz and Koehler, 1979), antigenicity (Tung et al., 1982) and fluidity (Huang and Yanagimachi, 1985): characteristics that may be important in the recognition of the zona pellucida and vitellus (see Section II.C.3).

2. Mechanism of the Acrosome Reaction

The cause of the membrane changes during the acrosome reaction is not established and detailed theories (Meizel, 1978; Green, 1978d) have been reviewed (Yanagimachi, 1981; Meizel, 1984). Vesiculation of sperm head membranes occurs because the membranes involved are close enough for fusion (see Section II.D.3.a.i) but the extent to which microtubules are involved in this process is unclear (Stambaugh and Smith, 1978; Peterson et al., 1978; Russell et al., 1979a,b). There is evidence that changes in the composition of the membrane make them more susceptible to fusion. This would be better understood if more were known of the composition of the membranes participating in the fusion events, but it is hard to prepare uncontaminated preparations of particular sperm head membranes, although progress is being made in this direction (Parks and Hammerstedt, 1985; Nikolopolou et al., 1985). It is known from biochemical studies that the outer acrosomal membrane has more cholesterol (Elias et al., 1978; Vijayasarathy et al., 1982) and less Ca^{2+}-ATPase activity (Vijayasarqthy et al., 1980) than the plasma membrane and its fluidity is less (Vijayasarathy et al., 1982). "Freeze-fracture cytochemistry" has revealed differences in phospholipid and sterol domains in fusinogenic and non-fusinogenic areas of sperm from the guinea pig (Elias et al., 1978, 1979; Bearer and Friend, 1980, 1982; Friend and Bearer, 1981; Friend, 1982), bulls (Bradley et al., 1979) and rams (Bradley et al., 1980). In mice the sterol-filipin complexes were found only in the plasma membrane and in neither acrosomal membrane (Toshimori et al., 1985).

Changes to lipids in fusogenic areas that occur before the acrosome reaction include clearance of sterols (Friend and Bearer, 1981; Bearer and Friend, 1982), increased acrosomal binding of cationic polymyxin B sulphate and increased incorporation of phosphate into cardiolipin and phosphatidic acid (Bearer and Friend, 1982). Intramembranous particles are also cleared from the acrosomal membrane before fusion (Friend and Rudolf, 1974; Kinsey and Koehler, 1978; Friend et al., 1977, 1978; Flechon et al., 1986) and this occurs after the clearance of sterols (Friend 1980; Friend and Bearer, 1981). Segregation of membrane lipids during or preceding the acrosome reaction may explain the reduced mobility of antibodies to sperm plasma membrane glcoproteins observed after capacitation (see Section II.B.4). The importance of these events for the fusion process is highlighted by the fusion of liposomes composed of phosphatidic acid with only the protein-free areas of sperm membranes (Bearer and Friend, 1982).

3. Requirements for the Acrosome Reaction

In-vitro experiments have established the requirements for acrosome reactions to occur in many species (see Meizel, 1985), but whole cell analysis of metabolites or enzymes does not provide clues to the mechanism of action at the level of the membranes involved. Some parameters that alter during capacitation or during the acrosome reaction are cited below.

a. Calcium

The acrosome reaction requires calcium ions in many species and the entry of extracellular calcium into the cell is implied by the rapid and synchronous induction of acrosome reactions in spermatozoa provided with calcium and a divalent metal ion ionophore (see Yanagimachi, 1981). Ca^{2+}-activated ATPases can be localised histochemically on the plasma membrane and outer acrosomal membranes of rabbit and guinea pig spermatozoa (see Gordon et al., 1978; Santos-Sacchi and Gordon, 1982); initially assumed to be inwardly-directed pumps they are now considered to exclude calcium from the cell (Santos-Sacchi and Gordon, 1984), in line with the transport ATPase from porcine spermatozoa membranes (Ashraf et al., 1982). Thus, calcium entry into the cell is passive and increases in intracellular Ca^{2+} are mediated by effects on the membrane permeability or pumps. The entry of calcium into the sperm cell is not thought to be *via* the usual calcium channels (Roldban et al., 1986). Magnesium ions can competitively inhibit Ca^{+}-dependent acrosome reactions in-vitro (Rogers and Yanangimachi, 1976). As the intracellular location of the ion is not certain the roles of calcium are not yet elucidated, but calcium can bind to acrosomal membranes in the human (Roomans, 1965).

i. Possible Roles of Calcium

A direct effect of calcium in the periacrosomal space could be to bring about contact of the outer acrosomal and plasma membranes by neutralising negative surface charges on both membranes and forming a Ca^{2+} bridge between them. Support for this has come from the cytochemical localisation of calcium at discrete points in the acrosome (Friend, 1977), especially areas of contact between the plasma membrane and the outer acrosomal membrane during acrosome reactions induced by the ionophore A23187 (Watson and Plummer, 1986). The length of this bridge (50nm) is such that contact with membranes is only likely to be achieved when mutual repulsion between the surfaces is overcome. Such conditions can be provided by protruberances with radii of curvature of 0.1 μm (Bangham and Pethica, 1960; Pethica, 1961: Bangham, 1964) and intramembranous particles could provide these points (Gingell, 1976) as these are known to move during capacitation within the membranes of interest (see Section II.B.2).

A role for calmodulin, indicated by the prevention of acrosome reactions by inhibitors of calmodulin action (Lenz and Cormier, 1980,1982), could be *via* an action on microfilaments and microtubules, which are affected by calmodulin in other tissues (see Cheung, 1980; Means and Dedman, 1980) and which are observed around vesiculated acrosomes (see Section II.D.2).

Indirect actions of calcium could stem from the ability of this cation to activate or inhibit enzymes. A Mg^{2+}-ATPase found on the outer acrosomal and plasma membrane of guinea pig sperm (Yanagimachi and Usui, 1974) was considered to maintain the osmotic properties of the acrosome. Inhibition of this enzyme by calcium (Usui and Yanagimachi, 1986) would upset these forces and may precipitate the acrosome reaction. A Mg-ATPase with characteristics of a proton translocase is released from hamster sperm during the acrosome reaction (Working and Meizel, 1980, 1982). In this species such an enzyme could be involved in generating the low pH within the acrosome (see Section II.D.3.c) and inhibition would dissipate this gradient across the acrosome with consequences for the activity of acrosomal enzymes. This enzyme may not be universally involved in the acrosome reaction, since in the bull the enzyme is only found on the plasma membrane (Vijayasarathy et al., 1980) and porcine spermatozoa membrane Mg^{2+}-ATPase does not have proton transocase activity (Ashraf et al., 1982).

Phospholipase A is activated by calcium in the mouse and it is lost from mouse sperm at the time of the acrosome reaction (Thakkar et al., 1983). The activity of the enzyme increases during the acrosome reaction of hamster sperm (Llanos et al., 1982) and both products liberated by the enzymes, fatty acids (see Meizel and Turner, 1984; Fleming and Yanagimachi, 1984) and lysolipids (see Llanos and Meizel, 1983; Yanagimachi and Suzuki, 1985), can promote acrosome reactions in-vitro. The activity of an associated sperm enzyme, transmethylase, increases in activity during capacitation of hamster sperm (Llanos and Meizel, 1983) and one product of this enzyme, lecithin, is a substrate for phospholipase A. However, transmethylase acts independently of lysolipid formation, since inhibitors of the enzyme prevent the lyolecithin-induced stimulation of acrosome reactions (Meizel, 1981), possibly by modifying membrane rigidity. The very lytic nature of lysolipids has been considered to argue against their mediation of physiological events, and one enzyme providing an alternative pathway for phospholipid degradation (phospholipase C: Hostetler and Hall, 1980) has been found in rabbit sperm heads (Srivastava et al., 1982).

While most unsaturated fatty acids act directly to destabilise cell membranes and induce acrosome reactions, the action of arachidonic acid can be prevented by inhibitors of prostaglandin synthesis (Meizel and Turner, 1984). The prostaglandin synthetase (Johnson and Ellis, 1977) and cyclo-oxygenase activity of sperm may thus be of importance.

A range of hydrolytic enzymes in the acrosome (see Meizel, 1984) may be activated by calcium entering the acrosome and contribute to the induction of the acrosome reaction.

b. Cyclic Nucleotides

The requirement for cyclic nucleotides in acrosome reactions in-vitro varies between species. In hamsters (Morton and Albagli, 1973) and mice (Stein and Fraser, 1984) the activity of adenylate cyclase inceases at capacitation and this is responsible for increased intracellular cAMP (Rogers and Morton, 1973) and cGMP/cAMP ratio (Stein and Fraser, 1984) in capacitated sperm cells.

Cyclic AMP is not essential, however: acrosome reactions can be induced under circumstances in which cAMP is not raised (Singh et al., 1980) and in the guinea pig cAMP prevents acrosome reactions whereas activators of phosphodiesterase (imidazole) promotes acrosome reactions (Rogers and Garcia, 1979; Rogers, 1981). An inhibitory role of cAMP on the expression of the acrosome reaction receives support from time-course studies where acrosome reactions are seen to occur as cAMP levels in the cell fall (Hyne and Garbers, 1979a).

Artificially increasing intracellular cAMP by inhibitors of PDE or permeant cAMP analogues can promote acrosome reactions in mice (Fraser, 1981) and hamsters (see Mrsny and Meizel, 1980). Permeant cGMP analogues promote acrosome reactions in both hamster and guinea pig sperm (Mrsny et al., 1984; Santos-Sacchi and Gordon, 1980), so the relative amounts of each may control this phenomenon.

i. Possible Roles of Cyclic Nucleotides

It has been argued that the formation of cAMP could raise the concentration of free Ca^{2+} by depleting the cell of the chelator ATP (Gordon et al., 1978; Santos-Sacchi and Gordon, 1980). ATP is known to be lost from capacitating hamster sperm (Rogers and Morton, 1973), although this may be more related to motility than the acrosome reaction.

An Na^{+}, K^{+}-ATPase in hamster sperm heads, whose activity increases during capacitation is activated by cGMP and little effected or inhibited by cAMP (Mrsny and Meizel, 1983; Mrsny et al., 1984), so that intracellular ions may be responsible for the effects of cyclic nucleotides (see Section II.D.3.c).

Despite reports of cAMP binding to human sperm (Rosado et al., 1975) and protein kinases on the surfaces of rat spermatozoa (Majumder, 1978; Horowitz et al., 1984), the role of ecto-enzymes in sperm function is not clear, although transmembrane effects of extracellular modifiers are possible (Atherton et al., 1985). There exist in spermatozoa protein kinases dependent upon cAMP (see Section II.A.4.c.i) and cGMP (Spruill et al., 1981) for which the substrates are as yet unknown, although human sperm membranes can be phosphorylated in the presence of PDE inhibitors (Huacuja et al., 1977). cAMP- and cGMP-dependent protein kinase from hamster sperm act synergistically with the cyclic nucleotides to reduce, or increase, repectively, the activity of hamster sperm head Na^+, K^+-ATPase (Mrsny and Meizel, 1983; Mrsny et al., 1984) with consequences for ion balance (see Section II.D.3.c).

c. Ions

There is increasing evidence for the involvement of monovalent cations in the acrosome reaction (see Hyne, 1984; Hyne et al., 1984). Potassium ions promote the acrosome reaction in hamsters (Mrsny and Meizel, 1981) and mice (Fraser, 1983b) and monovalent ionophores promote the acrosome reaction in spermatozoa from guinea pigs (Hyne et al., 1984) and hamsters (Mrsny and Meizel, 1981). However, high potassium concentrations in-vitro can retard or prevent acrosome reactions in the guinea pig (Rogers et al., 1971) and acrosome reactions do not occur in guinea pig sperm held at acid pH (Murphy and Yanagimachi, 1984). In the hamster acrosomal pH is kept acidic (Meizel and Deamer, 1978) and dissipation of the gradient precedes the acrosome reaction (Working and Meizel, 1983). The pH gradient can be dissipated by protonophores if potassium is supplied as counter-ion (Working and Meizel, 1981, 1983). The rise in acrosomal pH may be important in activation of various acrosomal enzymes (see Meizel, 1984) and any agents influencing the ion balance within the sperm head may thus affect the acrosome reaction.

Because they could find no Na^+, K^+-ATPase in membranes from the porcine sperm head, Ashraf et al. (1982) argued that ion concentrations in the sperm head of this species were regulated by the ATPases of the flagellum.

Polyamines can promote acrosome reactions (Fleming and Armstrong, 1985) but are thought to do so by promoting capacitation.

d. Catecholamines

Adrenal extracts that promote acrosome reactions in-vitro contain both catecholamines and taurine (see Meizel et al., 1980), but to what extent this reflects increased survival, capacitation (see Section II.B.6) or improved motility (see Section II.A.4.g.i) is not yet established.

e. Possible Sequence of Events

Despite these advances made on acrosome reactions in-vitro, an overall view of the process is not yet possible for even one species. Conditions within the female tract vary and different species may have employed different approaches to the same problem. Binding proteins and sodium may regulate entry into the cell of calcium ions; calcium may act directly to bring together the relevant membranes and indirectly *via* synthetic or degradatative enzymes that may control the relative content of cyclic nucleotides; cyclic nucleotides and their proteins kinases may modulate concentrations of Na^+ and K^+; the former may promote further entry of calcium and the latter activate the movement of protons from the acrosome; the acrosomal pH attendant upon these variations may activate enzymes within the acrosome and liberate membrane-active agents that finally destabilise the membrane and permit fusion with the plasma membrane.

4. Stimulus to the Acrosome Reaction

Capacitation *per se* is not stimulus enough for the acrosome reaction because sperm can be "arrested" in a capacitated state, when they do not undergo the acrosome reaction in-vitro until suitable challenge, *e.g.* addition of Ca^{2+} to sperm capacitated in Ca^{2+}-free media (Yanagimachi and Usui, 1974). Synchronous acrosome reactions follow the addition of follicular fluid to sperm capacitated in high ionic strength medium (Oliphant, 1976; Singhas and Oliphant, 1978). Thus, capacitation "prepares" sperm cells to respond to conditions capable of triggering the acrosome reaction. The trigger may be calcium ions, the uptake of which precedes or occurs in parallel with the acrosome reaction (Singh et al., 1978, 1980) and is promoted by follicular fluid after high ionic strength treatment (Triana et al., 1980).

The agent in follicular fluid that induces vesiculation and an acrosome reaction indistinguishable from that found in-vivo (Singhas and Oliphant, 1978) may be a proteoglycan (see Handrow et al., 1982; Lenz et al., 1982, 1983a,b) or one of several high molecular weight serum proteins that can enter oviductal fluid from blood (Oliphant et al., 1977), including complement (Cabot and Oliphant, 1978; Suarez and Oliphant, 1982) and albumin. The action of albumin in promoting acrosome reactions (Lui et al., 1977; Lui and Meizel, 1977; Fraser, 1985) may be in disturbing membrane lipids by removing fatty acids or promoting lysolipid formation (Singleton and Killian, 1983).

The ability of membrane vesicles from rabbit seminal plasma and cholesterol to prevent the acrosome reaction of rat and guinea pig sperm, respectively (Davis and Niwa, 1974; Fleming and Yanagimachi, 1981), suggests that these changes in the membrane may be a continuation of those membrane events occurring during capacitation (see Section II.B.3.b).

5. Relationship to Capacitation

The acrosome reaction is now thought to be a separate event following capacitation (Bedford, 1970; Yanagimachi and Usui, 1974; Saling et al., 1978) rather than being an integral part (Austin and Bishop, 1958; Hartree, 1977) or the final stage of it (Yanagimachi, 1969a). This is because, unlike capacitation, the acrosome reaction is irreversible and capacitated sperm do not necessarily undergo the acrosome reaction, *e.g.* if they are inhibited by Ca^{2+}-free media (see Section II.D.4) or by DTT (Yanagimachi et al., 1983). Although proteins are lost from the sperm surface during capacitation (see Section II.D.3.a) only in a few cases has loss of proteins been clearly associated with an increase in acrosome reactions (Oliphant, 1976; Eng and Oliphant, 1978). The promotion of acrosome reactions by proteases (Shinohara et al., 1985) may relate to the digestion of coating proteins. Decapacitation factors that prevent acrosome reactions in-vitro are present in epididymal fluid from hamsters (Bavister et al., 1978), rabbits (Thomas et al., 1984; Reynolds and Oliphant, 1984), guinea pigs and pigs (Hyne and Garbers, 1982).

6. Relationship to Hyperactivated Motility

Hyperactivation precedes the acrosome reaction (Fraser, 1982), but acrosome-reacted hyperactivated spermatozoa display lower beat frequencies and greater flagellar bends than non-acrosome-reacted cells (Suarez et al., 1984; see Section II.E.6).

7. Relationship to Fertilisation

The inhibition of the acrosome reaction in parallel with inhibition of fertilisation by seminal plasma glycoproteins (rabbit: Eng and Oliphant, 1978), lipid vesicles (rats: Davis and Niwa, 1974), the absence of Ca^{2+} (Fraser, 1983a), glucose (Fraser and Quinn, 1981) or the presence of muscarinic agonists (Florman and Storey, 1982a) in mice demonstrates an association between these two events. Acrosome-reacted sperm will not fertilise eggs, however, unless they are also hyperactivated (see Section II.E.6).

8. Site of the Acrosome Reaction

Although acrosome reacted spermatozoa may be recovered from the uterus of mated animals, these are probably "false" acrosome reactions associated with degenerating sperm cells (Bedford, 1970; Talbot and Franklin, 1976). Where the site of the acrosome reaction in-vivo has been studied carefully, in pigs (Szollosi and Hunter, 1978), rabbits (Overstreet and Cooper, 1979b) and sheep (Crozet and Dumont, 1984), it occurs in the oviducts, a little before or during interaction with the eggs. Although follicular fluid can induce the acrosome reaction in-vitro (see Yanagimachi, 1981; Section II.D.4) it is not thought that the release of this fluid at ovulation triggers the acrosome reaction in rabbits (Overstreet and Cooper, 1979b). Evidence that the stimulus to the acrosome reaction resides in the cumulus (Wolf et al., 1977; see Bavister, 1980), the zona pellucida (Bleil and Wassarman, 1983; Ward and Storey, 1984) or vitellus (Nicosia et al., 1977) has been challenged (Bedford, 1983).

One prevalent view is an old one (Austin and Bishop, 1958) that the acrosome reaction of relevance to fertilisation occurs within the cumulus mass (Bedford, 1968, 1972). Acrosome-reacted spermatozoa are observed within the cumulus mass (Yanagimachi and Phillips, 1984) and if sperm are acrosome reacted outside the cumulus mass they fail to penetrate into it, and merely adhere to the outside. In contrast, non-acrosome-reacted spermatozoa penetrate into the cell layers (Suarez et al., 1984), but whether the acrosome reaction occurs on the surface of the zona pellucida within the cumulus mass has been debated (see Section II.C.3).

9. Consequences of the Acrosome Reaction

a. Exposure or Release of Hyaluronidase

Hyaluronidase disperses the cumulus cell layer from eggs in-vitro and this has been used as a bioassay for the enzyme (Talbot and Franklin, 1974a,b). Other acrosomal enzymes (arylsulphatase, hexosaminidase) can also disperse the cumulus (Farooqi and Srivastava, 1979a) and may be important where hyaluronidase cannot achieve cumulus cell dispersal (Lorton and First, 1979). However, in species where cumulus cells are present on fertilised eggs only the passage between the follicular cells is facilitated by hyaluronidase released from the fertilising spermatozoon.

Hyaluronidase is released from sperm cells upon induction of the acrosome reaction with Ca^{2+} (Talbot and Franklin, 1974a; Rogers and Yanagimachi, 1975), calcium transport inhibitors (Singh et al., 1980), Ca^{2+}-ionophore (Primakoff et al., 1980; Shams-Borhan and Harrison, 1981) and high salt treatment with follicular fluid (Triana et al., 1980). Hyaluronidase binds with variable affinity to sperm from different species (Triana et al., 1980) and in the rat the bound form may be active (Lewin et al., 1982).

The importance of the enzyme in fertilisation is shown by the prevention of fertilisation in-vitro by capacitated rabbit sperm treated with an antibody to the enzyme (Dunbar et al., 1976) and use of an enzyme inhibitor (mycrisin) that prevents fertilisation of cumulus-intact eggs by mouse and hamster sperm (Reddy et al., 1980; Perreault et al., 1980); however, sperm from invertebrate species which lack hyaluronidase can also reach the zona through cumulus layers (Talbot, 1985) and this observation supports the alternative view that the acrosome reaction does not occur until the zona is reached (see Section II.C.3), in which case hyaluronidase would not appear to be important for the penetration of the egg investments.

b. Exposure or Release of Corona Penetrating Enzyme (CPE)

The follicular cells remaining adherent to the ovum after treatment with hyaluronidase comprise the "Corona Radiata", but they are not present in all species (Bradford et al., 1976a). Although antibodies to spermatozoa which prevent dispersion of the corona (anti-CPE?) also prevent fertilisation (Russo and Metz, 1974), other sites of inhibition by other antibodies are not excluded. Corona Penetrating Enzyme is an esterase (Bradford et al., 1976a,b) and the hyaluronidase inhibitor mycrisin, also an esterase inhibitor (Waibel et al., 1984), prevents fertilisation by cumulus-free eggs in-vitro (Perreault et al., 1980). Whether this enzyme has any relevance to fertilisation has also to be questioned if the acrosome reaction does not occur until the zona is reached.

c. Exposure or Release of Acrosin

Acrosin is a trypsin-like protease that can digest the zona pellucida in-vitro (see Zaneveld et al., 1975) but in-vivo only "penetration slits", equal in number to the number of penetrating spermatozoa, are produced in this structure (see Dziuk and Dickmann, 1965). The fact that sperm can penetrate lectin-treated, trypsin- and acrosin-resistant zona pellucidae (Bedford and Cross, 1978) can be explained by an action of other acrosomal enzymes such as arylsulphatase, that may be involved in penetration of the zona (Farooqi and Srivastava, 1980; Dudkiewicz, 1984).

Acrosin is present in mature epididymal spermatozoa as a precursor zymogen, proacrosin (Meizel and Mukerji, 1975, 1976) which may be located on the inner acrosomal membrane (Bradford et al., 1981), in the acrosomal matrix, or both (Johnson et al., 1983). It is liberated by autoactivation of proacrosin upon acrosomal disruption (Brown and Harrison, 1978; Harrison and Brown, 1979) or by an enzyme, acrolysin (McRorie et al., 1976; Bradford et al., 1981). Premature release of the enzyme is prevented by the low pH of acrosomal contents (*e.g.* hamster: Meizel and Deamer, 1978) or, where no low pH compartments exist in the sperm cell (bull: Babcock, 1983) by endogenous trypsin inhibitors (see Fink et al., 1971, 1973; Shams-Borhan and Harrison, 1981). Although trypsin inhibitors have been argued (Zaneveld et al., 1969) and demonstrated (Zimmerman and Burck, 1978) to be lost at capacitation, no proacrosin is converted during capacitation (Goodpasture et al., 1981).

Gentle removal of the acrosome from the sperm leaves much acrosin within the sperm (Brown and Hartree, 1974), but it is released to the medium during a physiological acrosome reaction (Goodpasture et al., 1981) or one induced by ionophores (Primakoff et al., 1980). Support for the view that the enzyme which is bound on the freshly exposed inner acrosomal membrane is responsible for the penetration slits in the zona pellucida comes from the silver proteinate staining method for protease activity which reveals an enzyme network localised there (Faltas et al., 1975; Stambaugh et al., 1975) and in the penetration slits of fertilised eggs (Stambaugh and Smith, 1976). Labelled exogenous inhibitors of acrosin are also localised on the inner acrosomal membrane (see Green and Hockaday, 1978). Endogenous trypsin inhibitors, on the other hand, act only on the soluble enzyme (Brown and Hartree, 1976), and this emphasises the importance of the non-inhibited, bound, form of the enzyme in fertilisation.

Trypsin inhibitors prevent fertilisation both in-vivo and in-vitro in mice (Suominen et al., 1973; Fraser, 1982b; Beyler and Zaneveld, 1982) and rabbits (Zaneveld et al., 1971). The cleavage of desmosterol sulphate at capacitation (see Section II.B.3.c) would remove a potent inhibitor of acrosin (Burck and Zimmerman, 1980) and there is also a loss of trypsin inhibitors from sperm at capacitation (Schill et al., 1971).

10. Development in the Epididymis of the Ability of Spermatozoa to Undergo the Acrosome Reaction

In the rabbit few testicular spermatozoa display an acrosome reation in-vivo, and most of these are "false" ones (Bedford, 1970), but some caput sperm can undergo acrosome reactions in-vitro (Brackett et al., 1978). Neither caput nor corpus sperm from the hamster epididymis are acrosome reacted in the female tract (Cummins, 1976).

The failure of immature spermatozoa to display acrosome reactions is more than their failure to undergo capacitation, because if proteins stabilising the acrosome were the only factor involved, then immature sperm with less DF (see Section II.B.8) would be anticipated to undergo the acrosome reaction more readily that mature cells. If the low Ca^{2+}-ATPase activity of immature sperm membranes (Bradley and Forrester, 1985) resides in the sperm head, then greater entry of calcium into the cell than occurs in mature cells would be anticipated and, again, more ready acrosome reactions would be anticipted.

Freeze-fracture cytochemmistry in the mouse shows that there are fewer sterol-filipin complexes in the plasma membrane overlying the acrosome of mature spermatozoa compared to immature cells (Toshimori et al., 1985). Such a decrease in membrane sterol, in the face of low sterol in the outer acrosomal membrane, could make the acrosome reaction more likely to occur by reducing the difference in sterol content between the two membranes that partake of fusion.

Two enzymes implicated in the acrosome reaction, phospholipase and lysophospholipase, have greater activity in intact sperm from the caput than the cauda when assayed with exogenous substrates (Atreja and Anand, 1985), although the activity towards endogenous substrates, possibly more relevant to acrosome reactions, was not examined.

Perhaps the "triggers" to the acrosome reaction or the ability to respond to them are limiting in immature spermatozoa. Histochemistry fails to demonstrate prostaglandin synthetase in caput sperm in the rat, but once released from the male tract they are as active as mature cells (Johnson and Ellis, 1977). This argues that an inhibitor to prostaglandin synthetase is present in the proximal epididymis which could reduce the effectiveness of endogenous arachidonate as a stimulus to the acrosome reaction within the epididymis.

a. Acrosomal Structure

There are few changes in the structure of the acrosome of maturing spermatozoa in most species (Nicander and Bane, 1962; Bedford, 1963, 1965; Bedford and Nicander, 1971; Jones, 1971) and where the plasma membranes are closer to the acrosome in immature spermatozoa (Bedford, 1965; Fawcett and Phillips, 1969), fixation artefacts have not been ruled out (Zamboni and Stefanini, 1968). Spermatozoa from the initial segment of the rat epididymis contain a proteinaceous material between the outer acrosomal membrane and the plasma membrane which disappears as the sperm mature (Suzuki and Nagano, 1980) and this may act physically to prevent the two membranes contacting each other.

There is a dramatic change in acrosomal morphology in maturing sperm from the guinea pig (Fawcett and Hollenberg, 1963; Blaquier et al., 1972) and chinchilla (Fawcett and Phillips, 1969). The acrosomal contents of the guinea pig may be at the limit of their compressibility owing to the greater osmotic pressure of the periacrosomal cytoplasm (Green, 1978c, 1978d), so changes in the osmotic pressure of different cellular compartments during maturation may be responsible for the changes in cell shape. It is known that the specific gravity of spermatozoa increases on maturation in bulls (Lindahl and Kihlstrom, 1952; Lavon et al., 1966, 1968) and rabbits (Benedict et al., 1967) and that immature sperm differ from mature sperm in their response to osmotic conditions (see Section II.A.4.d). Basic proteins, which bind to phosphotungstic acid (Scott, 1965) are lost from the periacrosomal regions on maturation in several species including primates (Holt, 1979; Moore et al., 1984), and it has been suggested that these are involved in acrosomal remodelling.

b. Acrosomal Enzymes

Immature spermatozoa from the rat are as able as mature cells to disperse cumulus cells from eggs, but do it rather more slowly (Fournier-Delpech et al., 1984), suggesting that there is a slower release of hyaluronidase from immature sperm. Rabbit testicular spermatozoa can activate eggs in-vitro, albeit abnormally, despite their absence from the perivitelline space (Brackett et al., 1978); this could be explained by the release of hyaluronidase, which can activate eggs parthogenetically (Graham, 1970; Graham and Deussen, 1974). Hyaluronidase activity from cauda spermatozoa is greater than that present in caput and testicular spermatozoa in the goat (Bhattacharya and Bhattacharya, 1985).

Hamster testicular spermatozoa contain as much acrosin as mature cells judging from silver proteinate staining (Yanagimachi and Teichmann, 1972). The creation of "haloes" around the head of air-dried sperm smeared on gelatin plates is a simple test of proteolytic activity (Gaddum and Blandau, 1970, 1971; Gaddum-Rosse and Blandau, 1971); only the soluble enzyme is detected because no haloes are formed around acrosome-reacted sperm (Gaddum-Rosse and Blandau, 1977). This technique reveals smaller haloes surrounding heads of immature spermatozoa from rats and rabbits (Gaddum and Blandau, 1971; Penn et al., 1972; Allen et al., 1974). If this technique reveals only post-mortem release of enzyme (Allen et al., 1974) then the results may reflect merely the greater stability of the membranes of immature spematozoa to the stresses of air-drying, rather than a lower total enzyme activity or a greater amount of acrosin inhibitor.

There is evidence for more stable membranes in immature cells (see Section II.C.9.b), but there is also both greater enzyme activity in mature (caudal) goat spermatozoa (Bhattacharrya and Bhattacharrya, 1985) and greater amounts of acrosin inhibitor in immature cells from the ram (Harrison and Brown, 1979). Human testicular sperm also contain acrosin, although the amounts relative to mature cells was not studied (Schill et al., 1981). Testicular spermatozoa contain slightly less proacrosin than mature cells, although a firmer binding of the enzyme to immature sperm cells cannot be ruled out (Harrison and Brown, 1979).

11. Involvement of the Epididymis in Permitting the Acrosome Reaction

Ageing spermatozoa in the epididymis does not make them capable of displaying the acrosome reaction in-vivo (Cummins, 1976) and changes in the shape of the acrosome that occur on maturation are androgen-dependent (Blaquier et al., 1972). The material separating the outer acrosomal and plasma membranes of immature rat spermatozoa disappears at the time a variable glycocalyx material (VGM) is deposited on the surface of the cell (Suzuki and Nagano, 1980), but it is not known if the events are causally related, or if VGM is a secretory product of the epididymis.

The high osmolarity of epididymal fluid (Table 5) may help to dehydrate the sperm with consequences for acrosomal shape (see Section II.D.10.a) and loss of cholesterol from the sperm membrane (see Section II.B.9.b) would also increase the potential for membrane fusion.

Mature cell membranes are fully capable of vesiculating and a small percentage of them do so immediately upon release in-vitro (Yanagimachi, 1975; Florman and Storey, 1982a). That they normally do not display acrosomes reactions in the epididymis (Jones, 1973) implies that they are prevented from doing so. Sperm in the epididymis are not capacitated, and decapacitation factors (see Section III.G.12.b.iii) may be the proteins in epididymal fluid that inhibit acrosome reactions in-vitro (see Section II.D.5) Other normal constituents of epididymal fluid that can prevent the acrosome reaction in-vitro are K^+ (Rogers et al., 1981) and Mg^{2+} (Rogers and Yanagimachi, 1976). Vitamin A in epididymal fluid (Velzquez et al., 1975) may influence the vitamin content of the acrosome (Gambhir and Abluwalia, 1975) and stabilise the acrosomal membrane.

Physical factors may also play a part in preventing inappropriate acrosome reactions. In the guinea pig mature spermatozoa stack in rouleaux (Fawcett and Hollenberg, 1963; Fawcett and Phillips, 1969; Hoffer and Greenberg, 1978; Williamson et al., 1980). The decline in rouleaux formation of newly maturing sperm soon after castration (Shepherd and Martan, 1976, 1979) or anti-androgen administration (McGlinn et al., 1976, 1979) implicates the epididymis in the adhesion. The ability of trypsin (Shepherd et al., 1974), antibodies to epididymal spermatozoa (Tung et al., 1980) and sulphydryl reducing agents (Yanagimachi et al., 1983) to dissociate the cells suggests that epididymal sulphydryl-containing proteins are involved in the stacking, and these have been localised at the site of sperm-sperm attachment (Tung et al., 1982). Sperm rouleaux are present in ejaculates (Martan, 1966) and the lower female tract (Martan and Shepherd, 1973; McGlinn et al., 1979) but there is some restriction to rouleaux at the uterotubal junction (Martan and Shepherd, 1973).

Since sperm cells in rouleaux display greater fertilisation than single cells when inseminated into the uterus (Williamson et al., 1980) it has been implied that sperm in rouleaux are the more mature form. However, single cells from androgen-deprived epididymides may fertilise eggs (McGlinn et al., 1979; Shepherd and Martan, 1979) and single cells predominate in the oviduct at fertilisation (Martan and Shepherd, 1973), so the phenomenon of stacking in rouleaux may be less to do with the state of maturity of the sperm cell than a mechanism to protect the acrosome in female tract secretions. For example, leucocytes which enter the uterus and scavenge isolated spermatozoa do not enter the oviduct (Martan and Shepherd, 1973). Rouleaux may also prevent premature acrosin loss and maintain cell viability (Tung et al., 1980). There is evidence for similar, but less extensive, association of spermatozoa within the epididymal canal in other species (Martan, 1970a,b; Martan and Hruban, 1970; Martan et al., 1970, 1971), and where no visible changes in the acrosome are noticeable in the epididymis, other mechanisms, *e.g.* complement activation, may be involved in protection of the acrosome (Witkin et al., 1983).

Desmosterol sulphate in epididymal fluid (see Section III.B) is a potent inhibitor of acrosin in pigs (Burck and Zimmerman, 1980), rabbits (Burck et al., 1982) and hamsters (Bouthillier et al., 1984). Other trypsin inhibitors are in epididymal fluid (Suominen and Setchell, 1972; Zaneveld et al., 1975) and in the mouse these are androgen-dependent (Poirier and Jackson, 1981; Poirier and Nicolson, 1984). Their presence may be related to the prevention of damage to the epididymal epithelium caused by acrosin leaking from damaged spermatozoa.

12. Summary

The increasing ability of maturing sperm to perform the acrosome reaction is due to changes in the relevant membranes of the spermatozoa. These probably reflect removal of elements that render the membrane rigid or insertion of labilising agents by the epididymis but the exact mechanism is unclear. It is made difficult to interpret by the simultaneous addition of decapacitation factors designed to prevent the expression of the acrosome reaction in the epididymis. The acrosomal enzymes released or exposed during the acrosome reaction are present in immature spermatozoa but some may be inhibited by other epididymal secretion products, subsequently lost during capacitation.

13. References

Allen GJ, Bishop MWH and Thompson TE (1974) Lysis of photographic emulsions by mammalian and chicken spermatozoa. J Reprod Fert 36:249–252

Ashraf M, Peterson RN and Russell LD (1982) Activity and localization of cation-dependent ATPase on the plasma membranes of boar spematozoa. Biochem Biophys Res Commun 107:1273–1278

Atherton RW, Khatoon S, Schoff PK and Hadley BE (1985) A study of rat epididymal sperm adenosine 3″,5″-monophosphate-dependent protein kinase: maturation differences and cellular location. Biol Reprod 32:155–171

Atreja SK and Anand SR (1985) Phospholipase and lysophospholipase activities of goat spermatozoa in transit from the caput to the cauda eididymidis. J Reprod Fert 74:687–691

Austin CR (1977) Spermatozoa and ova: the role of membranes in the fertilization process. In: Mammalian Cell Membranes, Vol. 3 Surface Membranes of Specific Cell types. Eds GA Jamieson and DM Robinson, Butterworths, London. pp 206–230

Austin CR and Bavister BD (1975) Preliminaries to the acrosome reaction in mammalian spermatozoa. In: Functional Anatomy of the Spermatozoon. Ed BA Afzelius, Pergamon Press, Oxford. pp 83–87

Austin CR and Bishop MWH (1958) Role of the rodent acrosome and perfatorium in fertilization. Proc Roy Soc (Lond) 149B:241–248

Babcock DF (1983) Examination of the intracellular ionic environment and of ionophore action by nullpoint measurements employing the fluorescent chromophore. J Biol Chem 258:6380–6389

Bangham AD (1964) The adhesiveness of leukocytes with special reference to zeta potential. Ann New York Acad Sci 116:945–949

Bangham AD and Pethica BA (1960) The adhesiveness of cells and the nature of the chemical groups at their surfaces. Proc Roy Phys Soc 28:43–52

Bavister BD (1980) Recent progress in the study of early events in mammalian fertilization. Develop Growth Diff 22:385–402

Bavister BD, Rogers BJ and Yanagimachi R (1978) The effects of cauda epididymal plasma on the motility and acrosome reaction of hamster and guinea pig spermatozoa in vitro. Biol Reprod 19:358–363

Bearer EL and Friend DS (1980) Anionic lipid domains: correlation with functional topography in a mammalian cell membrane. Proc Nat Acad Sci US 77:6601–6605

Bearer EL and Friend DS (1982) Modifications of anionic- lipid domains preceding membrane fusion in guinea pig sperm. J Cell Biol 92:604–615

Bedford JM (1963) Morphological changes in rabbit spermatozoa during passage through the epididymis. J Reprod Fert 5:169–177

Bedford JM (1965) Changes in fine structure of the rabbit sperm head during passage through the epididymis. J Anat 99:891–906

Bedford JM (1968) Ultrastructural changes in the sperm head during fertilization in the rabbit. Am J Anat 123:329–358

Bedford JM (1970) Sperm capacitation and fertilization in mammals. Biol Reprod Suppl 2:128–158

Bedford JM (1983) Significance of the need for capacitation before fertilization in Eutherian mammals. Biol Reprod 28:108–120

Bedford JM and Cooper GW (1978) Membrane fusion events in the fertilization of vertebrate eggs. In: Membrane Fusion. Eds G Poste and GL Nicolson. Volume 5 Cell Surface Reviews, Elsevier/North Holland, Amsterdam. pp 65–125

Bedford JM and Cross NL (1978) Normal penetration of rabbit spermatozoa through a trypsin- and acrosin-resistant zona pellucida. J Reprod Fert 54:385–392

Bedford JM and Nicander L (1971) Ultrastructural changes in the acrosome and sperm membranes during maturation of spermatozoa in the testis and epididymis of the rabbit and monkey. J Anat 108:527–543

Benedict RC, Schumaker VN and Davis RE (1967) The buoyant density of bovine and rabbit spermatozoa. J Reprod Fert 13:237–249

Beyler SA and Zaneveld LJD (1982) Inhibition of in-vitro fertilization of mouse gametes by proteinase inhibitors. J Reprod Fert 66:425–431

Bhattacharya A and Bhattacharya AK (1985) Biochemical changes in some acrosomal enzymes of spermatozoa during maturation. Experientia 41:62–63

Blaquier JA, Cameo MS and Burgos MH (1972) The role of androgens in the maturation of epididymal spermatozoa in the guinea pig. Endocrinology 90:839–842

Bleil JD and Wassarman PM (1983) Sperm-egg interaction in the mouse: sequence of events and induction of the acrosome reaction by a zona pellucida glycoprotein. Develop Biol 95:317–324

Bouthillier M, Bleau G, Chapdelaine A and Roberts KD (1984) Distribution of steroid sulfotransferase in the male hamster reproductive tract. Biol Reprod 31:936–941

Brackett BG, Hall JL and Oh YK (1978) In-vitro fertilizing ability of testicular, epididymal, and ejaculated rabbit spermatozoa. Fert Steril 29:571–582

Bradford MM, McRorie RA and Williams WL (1976a) Involvement of esterases in sperm penetration of the corona radiata of the ovum. Biol Reprod 15:102–106

Bradford MM, McRorie RA and Williams WL (1976b) A role for esterases in the fertilization process. J exp Zool 197:297–301

Bradford MM, Dudkiewicz AB, Penny GS, Dyckes DF, Burleigh BD, Wooley RE and McRorie RA (1981) Localization of proacrosin on the inner acrosomal membrane of spermatozoa in rabbits and hamsters. Am J Vet Res 42:1082–1083

Bradley and MP and Forrester IT (1985) Sperm calcium homeostasis during maturation. In: Male Fertility and its Regulation. Eds TJ Lobl and ESE Hafez, MTP Press, Lancaster. pp 437–449

Bradley MP, Forrester IT and Rayns DG (1979) Cholesterol distribution in bull sperm plasma membrane: effects of filipin and digitonin. Proc U Otago Med Sch 57:3–5

Bradley MP, Rayns DG and Forrester IT (1980) Effects of filipin, digitonin and polymyxin B on plasma membrane of ram spermatozoa – an EM study. Archs Androl 4:195–204

Brown CR and Harrison RAP (1978) The activation of proacrosin in spermatozoa from ram, bull and boar. Biochem Biophys Acta 526:202–217

Brown CR and Hartree EF (1976) Effects of acrosin inhibitors in the soluble and membrane-bound forms of ram acrosin and a reappraisal of the role of the of the enzyme in fertilization. Hoppe Selye's Z Physiol Chem 357:57–65

Burck PJ and Zimmerman RE (1980) The inhibition of acrosin by sterol sulphates. J Reprod Fert 58:121–125

Cabot CL and Oliphant G (1978) The possible role of immunological complement in induction of rabbit sperm acrosome reaction. Biol Reprod 19:666–672

Cheung WY (1980) Calmodulin plays a pivotal role in cellular regulation. Science 207:19–27

Crozet N and Dumont M (1984) The site of the acrosome reaction during in vivo penetration of the sheep oocyte. Gamete Res 10:97–105

Cummins JM (1976) Effects of epididymal occlusion on sperm maturation in the hamster. J exp Zool 197:187–190

Davis BK and Niwa K (1974) Inhibition of mammalian fertilization in vitro by membrane vesicles from seminal plasma. Proc Soc exp Biol Med 146:11–16

Dravland JE, Llanos MN, Munn RJ and Meizel S (1984) Evidence for the involvement of a sperm trypsinlike enzyme in the membrane events of the hamster sperm acrosome reaction. J exp Zool 232:117–128

Dudkiewicz AB (1984) Purification of boar acrosomal arylsulfatase A and possible role in the penetration of cumulus cells. Biol Reprod 30:1005–1014

Dunbar BS, Munoz MG, Cordle CT and Metz CB (1976) Inhibition of fertilization in vitro by treatment of rabbit spermatozoa with univalent isoantibodies to rabbit sperm hyaluronidase. J Reprod Fert 47:381–384

Dziuk PJ and Dickmann Z (1965) Sperm penetration through the zona pellucida of the sheep egg. J exp Zool 158:237–240

Elias PM, Goerke J, Friend DS and Brown BE (1978) Freeze-fracture identification of sterol-digitonin complexes in cell and liposome membranes. J Cell Biol 78:577–596

Elias PM, Friend DS and Goerke J (1979) Membrane sterol heterogeneity. Freeze-fracture detection with saponins and filipin. J Histo Cyto 27:1247–1260

Eng LA and Oliphant G (1978) Rabbit sperm reversible decapacitation by a membrane stabilization with a highly purified glycoprotein from seminal plasma. Biol Reprod 19:1083–1094

Faltas S, Smith M and Stambaugh R (1975) The organised distribution of acrosomal proteinase within the acrosome. Fert Steril 26:1070–1074

Farooqui AA and Srivastava PN (1979) Isolation, characterization and the role of rabbit testicular arylsulphatase A in fertilization. Biochem J 181:331–337

Farooqui AA and Srivastava PN (1980) Isolation of β-N-acetylhexosaminidase from rabbit semen and its role in fertilization. Biochem J 191:827–834

Fawcett DW and Hollenberg RD (1963) Changes in the acrosome of guinea-pig spermatozoa during passage through the epididymis. Z Zellforsh mikr Anat 60:276–292

Fawcett DW and Phillips DM (1969) Observations on the release of spermatozoa and on changes in the head during passage through the epididymis. J Reprod Fert Suppl 6:405–418

Fink E, Klein G, Hammer F, Muller-Bardorff G and Fritz H (1971) Protein proteinase inhibitors in male sex glands. Proc Int Res Conf Proteinase Inhibitors, Munich 1971. Walter de Gruyter, New York. pp 225–235

Fink E, Fritz H, Jaumann E, Schissler H, Forg.Brey B and Werle E (1973) Protein proteinase inhibitors in male sex glands and their secretions. In: Protides of the Biological Fluids. Ed H Peters, Pergamon Press, Oxford. pp 425–431

Flechon JE, Harrison RAP, Flechon B and Escaig J (1986) Membrane fusion events in the Ca^{2+}/ionophore-induced acrosome reaction of ram spermatozoa. J Cell Sci 81:43–63

Fleming AD and Armstrong DT (1985) Effects of polyamines upon capacitation and fertilization in the guinea pig. J exp Zool 233:93–100
Fleming AD and Yanagimachi R (1981) Effects of various lipids on the acrosome reaction and fertilizing capacity of guinea pig spermatozoa with special reference to the possible involvement of lysophospholipids in the acrosome reaction. Gamete Res 4:253–273
Fleming AD and Yanagimachi R (1984) Evidence suggesting the importance of fatty acid moieties of sperm membrane phospholipids in the acrosome reaction of guinea pig spermatozoa. J exp Zool 229:485–489
Florman HM and Storey BT (1981) Inhibition of in vitro fertilization of mouse eggs: 3-quinuclidinyl benzilate specifically blocks penetration of zonae pellucidae by mouse spermatozoa. J exp Zool 216:159–167
Florman HM and Storey BT (1982a) Mouse gamete interactions: the zona pellucida is the site of the acrosome reaction leading to fertilization in vitro. Develop Biol 91:121–130
Florman HM and Storey BT (1982b) Characterization of cholinomimetic agents that inhibit in vitro fertilization in the mouse. Evidence for a sperm-specific binding site. J Androl 3:157–164
Fournier-Delpech S, Hamamah S, Tananis-Anthony C, Courot M and Orgebin-Crist MC (1984) Hormonal regulation of zona-binding ability and fertilizing ability of rat epididymal spermatozoa. Gamete Res 9:21–30
Fraser LR (1981) Dibutyryl cyclic AMP decreases capacitation time in vitro in mouse spermatozoa. J Reprod Fert 62:63–72
Fraser LR (1982a) Potential contraception by interference with capacitation. In: Progress towards a male contraceptive. Eds SL Jeffcoate and M Sandler, John Wiley and Sons, Chichester. pp 185–208
Fraser LR (1982b) p-Aminobenzamidine, an acrosin inhibitor, inhibits mouse sperm penetration of the zona pellucida but not the acrosome reaction. J Reprod Fert 65:185–194
Fraser LR (1983a) Ca^{2+} is required for mouse sperm capacitation and fertilization in vitro. J Androl 3:412–419
Fraser LR (1983b) Potassium ions modulate expression of mouse sperm fertilizing ability, acrosome reaction and hyperactivated motility in vitro. J Reprod Fert 69:539–553
Fraser LR (1985) Albumin is required to support the acrosome reaction but not capacitation in mouse spermatozoa in vitro. J Reprod Fert 74:185–196
Fraser LR and Quinn PJ (1981) A glycolytic product is obligatory for initiation of the sperm acrosome reaction and whiplash motility required for fertilization in the mouse. J Reprod Fert 61:25–35
Friend DS (1977) The organization of the spermatozoal membrane. In: The Immunobiology of Gametes. Eds M Edinin and MH Johnson, Cambridge University Press. pp 5–30
Friend DS (1980) Freeze-fracture alterations in guinea pig sperm membranes preceding gamete fusion. In: Membrane-Membrane Interactions. Ed NB Gilula, Raven Press, New York. pp 153–168
Friend DS (1982) Plasma-membrane diversity in a highly polarised cell. J Cell Biol 93:243–249
Friend DS and Bearer EL (1981) β-hydroxysterol distribution as determined by freeze fracture cytochemistry. Histochem J 13:535–546
Friend DS and Rudolf I (1974) Acrosomal disruption in sperm. Freeze-fracture of altered membranes. J Cell Biol 63:466–479
Friend DS, Orci L, Perrelet A and Yanagimachi R (1977) Membrane particle changes attending the acrosome reaction in guinea pig spermatozoa. J Cell Biol 74:561–577
Gaddum P and Blandau RJ (1970) Proteolytic reaction of mammalian spermatozoa on gelatin membranes. Science 170:749–751
Gaddum P and Blandau RJ (1971) Comparative studies on the proteolytic activity of sperm acrosomes during development, maturation and capacitation. Anat Rec 169:321a
Gaddum-Rosse P and Blandau RJ (1972) Comparative studies on the proteolysis of fixed gelatin membranes by mammalian sperm acrosomes. Am J Anat 134:133–144
Gaddum-Rosse P and Blandau RJ (1977) Proteolytic activity of guinea pig spermatozoa after induction of the acrosomal reaction in vitro. Am J Anat 149:423–430
Gambhir KK and Ahluwalia BS (1975) Vitamin A in bovine sperm acrosomes. J Reprod Fert 43:129–132

Gingell D (1976) Electrostatic control of membrane permeability via intramembranous particle aggregation. In: Mammalian Cell Membranes Vol 1 General Concepts. Eds GA Jamieson and DM Robinson, Butterworths, London, pp 198–223

Gordon M, Dandekar PV and Eager PR (1978) The identification of phosphatase on the membranes of guinea pig sperm. Anat Rec 191:123–134

Goodpasture JC, Reddy JM and Zaneveld LJD (1981) Acrosin, proacrosin, and acrosome inhibitor of guinea pig spermatozoa capacitated and acrosome reacted. Biol Reprod 25:44–45

Graham CF (1970) Parthogenetic mouse blastocysts. Nature (Lond) 226:165–167

Graham CF and Deussen ZA (1974) In vitro activation of mouse eggs. J Embryol exp Morph 31:497–512

Green DPL (1978a) The osmotic properties of the acrosome of guinea pig sperm. J Cell Sci 32:165–176

Green DPL (1978b) The mechanism of the acrosome reaction. In: Development of Mammals. Ed MH Johnson, North Holland Publishing Co, Amsterdam. pp 65–81

Green DPL and Hockaday AR (1978) The histochemical localization of acrosin in guinea pig sperm after the acrosome reaction. J Cell Sci 32:177–184

Handrow RR, Lenz RW and Ax RL (1982) Structural comparisons among glycosaminoglycans to promote an acrosome reaction in bovine spermatozoa. Biochem Biophys Res Commun 107:1326–1332

Harrison RAP and Brown CR (1979) The zymogen form of acrosin in testicular, epididymal and ejaculated spermatozoa from the ram. Gamete Res 2:75–87

Hartree EF (1977) Spermatozoa, eggs and proteinases. Biochem Soc Transactions 5:375–394

Hoffer AP and Greenberg JA (1978) The structure of the epididymis, efferent ductules and ductus deferens of the guinea pig: a light microscopical study. Anat Rec 190:659–678

Holt WV (1979) Development and maturation of the mammalian acrosome. J Ultrastruct Res 68:58–71

Horowitz, Toeg H and Orr GA (1984) Characterization and localization of cAMP-dependent protein kinases in rat caudal epididymal sperm. J Biol Chem 259:832–838

Hostetler KY and Hall LB (1980) Phospholipase C activity of rat tissues. Biochem Biophys Res Commun 96:388–393

Huacuja L, Delgado NM, Merchant H, Pancardo RM and Rosado A (1977) Cyclic AMP induced incorporation of ^{33}Pi into human spermatozoa membrane components. Biol Reprod 17:89–96

Huang TTF and Yanagimachi R (1985) Inner acrosomal membrane of mammalian spermatozoa: its properties and possible functions in fertilization. Am J Anat 174:249–268

Huang TTF, Hardy D, Yanagimachi H, Teuscher C, Tung K, Wild G and Yanagimachi R (1985) pH and protease control of acrosomal content status and release during the guinea pig sperm acrosome reaction. Biol Reprod 32:451–462

Hyne RV (1984) Bicarbonate- and Ca-dependent induction of rapid guinea pig sperm acrosome reactions by monovalent ionophores. Biol Reprod 31:312–323

Hyne RV and Garbers DL (1979) Calcium-dependent increase in adenosine 3′,5′-monophosphate and induction of the acrosome reaction in guinea pig spermatozoa. Proc Natn Acad Sci US 76:5699–5703

Hyne RV and Garbers DL (1982) Inhibition of the guinea-pig sperm acrosome reaction by a low molecular weight factor(s) in epididymal fluid and serum. J Reprod Fert 64:151–157

Hyne RN, Higginson RE, Hohlman D and Lopata A (1984) Sodium requirement for capacitation and membrane fusion during the guinea pig acrosome reaction. J Reprod Fert 70:83–94

Johnson JM and Ellis LC (1977) The histochemical localization of prostaglandin synthetase activity in reproductive tract of the male rat. J Reprod Fert 51:17–22

Johnson LA, Garner DL, Truitt-Gilbert AJ and Lessley BA (1983) Immunocytochemical localization of acrosin on both acrosomal membranes and in the acrosomal matrix of porcine spermatozoa. J Androl 4:222–229

Jones RC (1971) Studies of the structure of the head of boar spermatozoa from the epididymis. J Reprod Fert Suppl 13:51–64

Jones RC (1973) Changes opccurring in the head of boar spermatozoa: vesiculation or vacuolation of the acrosome? J Reprod Fert 33:113–118
Kinsey WH and Koehler JK (1976) Fine structural localization of concanavalin A binding sites on hamster spermatozoa. J Supramolecular Struct 5:185–198
Kinsey WH and Koehler JK (1978) Cell surface changes associated with in vitro capacitation of hamster sperm. J Ultrastruct Res 64:1–13
Koehler JK (1976) Changes in antigenic site distribution on rabbit spermatozoa after incubation in "capacitating" media. Biol Reprod 15:444–456
Lavon U, Volcani R, Amir D and Danon D (1966) The specific gravity of bull spermatozoa from different parts of the reproductive tract. J Reprod Fert 12:597–599
Lavon U, Volcani R and Danon D (1968) Determination of dry matter percentage of ejaculated and epididymal bull spermatozoa after differential flotation. J Reprod Fert 17:151–168
Lenz RW and Cormier MJ (1980) Effects of phenathiozine drugs on the acrosome reaction in mammalian spermatozoa. In: Calcium-Binding Proteins: Structure and Function. Eds FL Siegel, E Carafoli, RH Kretsinger, DH MacLennon and RH Wasserman, Elsevier North Holland Inc. pp 227–228
Lenz RW and Cormier MJ (1982) Effects of calmodulin-binding drugs on the guinea pig spermatozoon acrosome reaction and the use of these drugs as vaginal contraceptive agents in rabbits. Ann New York Acad Sci 383:85–97
Lenz RW, Ax RL, Grimek HJ and First NL (1982) Proteoglycan from bovine follicular fluid enhances an acrosome reaction in bovine spermatozoa. Biochem Biophys Res Commun 106:1092–1098
Lenz RW, Bellin ME and Ax RL (1983a) Rabbit spermatozoa undergo an acrosome reaction in the presence of glycosaminoglycans. Gamete Res 8:11–19
Lenz RW, Ball GD, Lohse JK, First NL and Ax RL (1983b) Chondroitin sulfate facilitates an acrosome reaction in bovine spermatozoa as evidenced by light microscopy, electron microscopy and in vitro fertilization. Biol Reprod 28:683–690
Lewin LM, Nevo Z, Gabsu A and Weissenberg R (1982) The role of sperm-bound hyaluronidase in the dispersal of the cumulus oophorus surrounding rat ova. Int J Androl 5:37–44
Lindahl PE and Kihlstrom JE (1952) Alterations in specific gravity during the ripening of bull spermatozoa. J Dairy Sci 35:393–402
Llanos MN and Meizel S (1983) Phospholipid methylation increases during capacitation of golden hamster sperm in vitro. Biol Reprod 28:1043–1051
Llanos MN, Lui CW and Meizel S (1982) Studies on phospholipase A2 related to the hamster sperm acrosome reaction. J exp Zool 221:107–117
Lorton SP and First NL (1979) Hyaluronidase does not disperse the cumulus oophorus surrounding bovine ova. Biol Reprod 21:301–308
Lui CW and Meizel S (1977) Biochemical studies of the in vitro acrosome reaction inducing activity of bovine serum albumin. Differentiation 9:59–66
Lui CW and Meizel S (1979) Further evidence in support of a role for hamster sperm hydrolytic enzymes in the acrosome reaction. J exp Zool 207:173–186
Lui CW, Cornett LE and Meizel S (1977) Identification of the bovine follicular fluid protein involved in the in vitro induction of the hamster sperm acrosome reaction. Biol Reprod 17:34–41
Majumder GC (1978) Occurence of a cyclic AMP-dependent protein kinase on the outer surface of rat epididymal spermatozoa. Biochem Biophys Res Commun 83:829–836
Martan J (1966) Occurrence of the intact spermatozoa in spontaneous ejaculations of isolated male guinea pigs. Transactions of the Illinois State Academy of Science 59:78–80
Martan J (1970a) Cohesive properties of mammalian epididymal spermatozoa. J Morphol 132:389–396
Martan J (1970b) Epididymal spermatozoa in chinchilla. Anat Rec 166:354 abstr
Martan J and Hruban Z (1970) Unusual spermatozoan formation in the epididymis of the flying squirrel (Glaucomys volans). J Reprod Fert 21:1678–170
Martan J and Shepherd BA (1973) Spermatozoa in rouleaux in the female genital tract. Anat Rec 175:625–630
Martan J, Adams CS and Perkins BL (1970) Epididymal spermatozoa of two species of squirrel. J Mammal 51:376–378

Martan J, Hruban Z and Brown P (1971) The formation and development of the cylindrical bodies in the mammalian epididymis. J Reprod Fert 27:115–117
McGlinn SM, Shepherd BA and Martan J (1976) Effects of cyproterone acetate on associations of spermatozoa and seminal vesicles in the guinea pig. Contraception 14:183–191
McGlinn SM, Shepherd BA and Martan J (1979) Rouleau formation and fertility of spermatozoa in guinea pigs treated with cyproterone. Archs Androl 2:197–202
McRorie RA, Turner RB, Bradford MM and Williams WL (1976) Acrolysin, the aminoproteinase catalysing the initial conversion of proacrosin to acrosin in mammalian fertilization. Biochem Biophys res Commun 71:492–498
Means AR and Dedman JR (1980) Calmodulin – an intracellular calcium receptor. Nature (Lond) 285:73–77
Meizel S (1978) The mammalian sperm acrosome reaction. A biochemical Approach. In: Development in Mammals. Vol 3 Ed MH Johnson, North Holland Pub Co, Amsterdam. pp 1–64
Meizel S (1981) Inhibition of the hamster sperm acrosome reaction by transmethylation inhibitors. J exp Zool 217:443–446
Meizel S (1984) The importance of hydrolytic enzymes to an exocytotic event, the mammalian sperm acrosome reaction. Biol Rev 59:125–157
Meizel S (1985) Molecules that initiate or help stimulate the acrosome reaction by their interaction with the mammalian sperm surface. Am J Anat 174:285–302
Meizel S and Deamer DW (1978) The pH of the hamster sperm acrosome. J Histo Cyto 26:98–105
Meizel S and Mukerji SK (1975) Proacrosin from rabbit epididymal spermatozoa partial purification and initial biochemical characterization. Biol Reprod 13:83–93
Meizel S and Mukerji SK (1976) Biochemical studies of proacrosin and acrosin from hamster cauda epididymal spermatozoa. Biol Reprod 14:444–450
Meizel S and Turner KO (1984) The effects of products and inhibitors of arachidonic acid metabolism on the hamster sperm acrosome reaction. J exp Zool 231:238–288
Meizel S and Working PK (1980) Further evidence suggesting the hormonal stimulation of hamster sperm acrosome reactions by catecholamines in vitro. Biol Reprod 22:211–216
Meizel S, Lui CW, Working PK and MRsny RJ (1980) Taurine and hypotaurine: their effects on motility, capacitation and the acrosome reaction of hamster sperm in vitro and their presence in sperm and reproductive tract fluids of several mammals. Develop Growth Diff 22:483–494
Moore HDM, Hartmann TD and Holt WV (1984) The structure and epididymal maturation of the spermatozoa of the common marmoset (Callithrix jacchus). J Anat 138:227–235
Morton B and Albagli L (1973) Modification of hamster sperm adenyl cyclase by capacitation in vitro. Biochem Biophys Res Commun 50:697–703
Mrsny RJ and Meizel S (1980) Evidence suggesting a role for cyclic nucleotides in acrosome reactions of hamster sperm in vitro. J exp Zool 211:153–157
Mrsny RJ and Meizel S (1981) Potassium ion influx and Na^+, K^+-ATPase activity are required for the hamster soerm acrosome reaction. J Cell Biol 91:77–82
Mrsny RJ and Meizel S (1983) Initial evidence for the modification of hamster sperm Na^+, K^+-ATPase activity by cyclic nucleotide-mediated processes. Biochem Biophys Res Commun 112:132–138
Mrsny RJ, Siiteri JE and Meizel S (1984) Hamster sperm Na^+, K^+-ATPase increases in activity during capacitation in vitro and its relationship to cyclic nucleotides. Biol Reprod 30:573–584
Murphy SJ and Yanagimachi R (1984) The pH dependence of motility and the acrosome reaction of guinea pig spermatozoa. Gamete Res 10:1–8
Nicander L and Bane A (1962) Fine structure of boar spermatozoa. Z fur Zellforschung 57:390–405
Nikolopoulou M, Soucke DA and Vary JC (1985) Changes in the lipid content of boar sperm plasma membranes during epididymal maturation. Biochem Biophys Acta 815:486–498
Nicosia SV, Wolf DP and Inoue M (1977) Cortical gfranule distribution and cell surface characteristics in mouse eggs. Dev Biol 57:56–74V
Ohzu E and Yanagimachi R (1982) Acceleration of acrosome reaction in hamster spermatozoa by lysolecithin. J exp Zool 224:259–263

Ono K, Yanagimachi R and Huang TTF (1982) Phospholipase A of guinea pig spermatozoa: its preliminary characterization and possible involvement in the acrosome reaction. Develop Growth Differentiation 24:305–310
Oliphant G (1976) Removal of sperm-bound seminal plasma components as a prerequisite to induction of the rabbit acrosome reaction. Fert Steril 27:28–38
Oliphant G, Randall P and Cabot CL (1977) Immunological components of rabbit Fallopian tubule fluid. Biol Reprod 16:463–469
Overstreet JW and Cooper GW (1979) The time and location of the acrosome reaction during sperm transport in the female rabbit. J exp Zool 209:97–104
Parks JE and Hammerstedt RH (1985) Developmental changes occurring in the lipids of ram epididymal spermatozoa plasma membrane. Biol Reprod 32:653–668
Penn A, Gledhill BL and Darzynkiewicz Z (1972) Modification of the gelatin substrate procedure for demonstration of acrosomal proteolytic activity. J Histo Cyto 20:499–506
Perreault S, Zaneveld LJD and Rogers BJ (1980) Inhibition of fertilization in the hamster by sodium aurothiomalate, a hyaluronidase inhibitor. J Reprod Fert 60:461–467
Peterson R, Russell L, Bundman D and Freund M (1978) Presence of microfilaments and tubular srtuctures in boar spermatozoa after chemically inducing the acrosome reaction. Biol Reprod 19:459–466
Pethica BA (1961) The physical chemistry of cell adhesion. Exp Cell Res Suppl 8:123–140
Poirier GR and Jackson J (1981) Isolation and characterization of two proteinase inhibitors from the male reproductive tract of mice. Gamete Res 4:555–569
Poirier CR and Nicholson N (1984) Distribution of a proteinase inhibitor of epididymal origin in the tissues and secretions of the male reproductive tract of mice. J exp Zool 230:465–472
Primakoff P, Myles DG and Bellve AR (1980) Biochemical analysis of the released products of the mammalian acrosome reaction. Develop Biol 80:324–331
Reddy JM, Joyce C and Zaneveld LJD (1980) Role of hyaluronidase in fertilization: the antifertility activity of mycrisin, a non toxic hyaluronidase inhibitor. J Androl 1:28–32
Reynolds AB and Oliphant G (1984) Production and characterization of monoclonal antibodies to the sperm acrosome stabilizing factror (ASF): utlization for purification and molecular analysis of ASF. Biol Reprod 30:775–786
Rogers BJ (1981) Factors affecting mammalian In Vitro fertilization. In: Bioregulators of Reproduction. Eds G Jagielle and HJ Vogel, Acad Press, London. pp 459–486
Rogers BJ and Garcia L (1979) Effect of cAMP on acrosome reactions and fertilization. Biol Reprod 21:365–372
Rogers BJ and Morton B (1973) ATP levels in hamster spermatozoa during capacitation in vitro. Biol Reprod 9:361–369
Rogers BJ and Yanagimachi R (1975) Release of hyaluronidase from guinea-pig spermatozoa through an acrosome reaction initiated by calcium. J Reprod Fert 44:135–138
Rogers BJ and Yanagimachi R (1976) Competitive effect of magnesium on the calcium-dependent acrosome reaction in guinea pig sperm. Biol Reprod 15:614–619
Rogers BJ, Ueno M and Yanagimachi R (1981) Fertilization by guinea pig spermatozoa requires potassium ions. Biol Reprod 25:639–648
Roldban ERS, Shibata S and Yanagimachi R (1986) Effect of Ca^{2+} channel antagonists on the acrosome reaction of guijnea pig and golden hamster spermatozoa. Gamete Res 13:281–292
Roomans GM (1975) Calcium binding to the acrosomal membrane of human spermatozoa. Exp Cell Res 96:23–30
Rosado A, Huacuja L, Delgado NM, Hicks JJ and Pancardo RM (1975) Cyclic-AMP receptors in the human spermaozoa membrane. Life Sci 17:1707–1714
Russell L, Peterson R and Freund M (1979a) Direct evidence for formation of hybrid vesicles by fusion of plasma and outer acrosomal membranes during the acrosome reaction in boar spermatozoa. J exp Zool 208:41–56
Russell L, Peterson RN and Freund M (1979b) Morphological characteristics of the chemically induced acrosome reaction in human spermatozoa. Fert Steril 32:87–92
Russo I and Metz CB (1974) Inhibition of fertilization in vitro by treatment of rabbit spermatozoa with univalent isoantibody. J Reprod Fert 38:211–215
Saling PM, Storey BT and Wolf DP (1978) Calcium-dependent binding of mouse epididymal spermatozoa to the zona pellucida. Develop Biol 65:515–525

Santos-Sacchi J and Gordon M (1980) Induction of the acrosome reactio in guinea pig spermatozoa by cGMP analogues. J Cell Biol 85:798–803

Santos-Sacchi J and Gordon ML (1982) The effect of ATP depletion upon the acrosome reaction in guinea pig sperm. J Androl 3:108–112

Schwartz MA and Koehler JK (1979) Alterations in lectin binding to guinea pig spermatozoa asccompanying in utero capacitation and the acrosome reaction. Biol Reprod 21:1295–1307

Schill WB, Feifel M, Fritz H and Hammerstein J (1981) Inhibitors of acrosomal protease as antifertility agents – a problem of acrosomal membrane permeability. Int J Androl 4:25–38

Scott JE (1975) Interactions of phosphotungstate with biological polymers. Acta Histochem Suppl 15:213–219

Shams-Borhan G and Harrison RAP (1981a) Production, characterization, and use of ionophore-induced calcium-dependent acrosome reactions in ram spermatozoa. Gamete Res 4:407–432

Shepherd BA and Martan J (1976) Differences in longevity of guinea pig spermatozoa following castration and hemicastration. Anat Rec 184:527–528 (Abst)

Shepherd BA and Martan J (1979) Morphology and fertility of guinea pig spermatozoa aged in vivo. Archs Androl 2:53–58

Shepherd BA, Martan J and Murphy R (1974) In vitro studies of guinea pig spermatozoa in rouleaux. Biol Reprod 11:470–474

Shinohara H, Yanagimachi R and Srivastava PN (1985) Enhancement of the acrosome reaction of hamster spermatozoa by the proteolytic enzymes, kallikrein, trypsin, and chymotrypsin. Gamete Res 11:19–28

Singleton CL and Killian GJ (1983) A study of phospholipase in albumin and its role in inducing the acrosome reaction of guinea pig spermatozoa in vitro. J Androl 4:150–156

Singh JP, Babcock DF and Lardy HA (1978) Increased calcium-ion influx is a component is capacitation of spermatozoa. Biochem J 172:549–556

Singh JP, Babcock DF and Lardy HA (1980) Induction of accelerated acrosome reaction in guinea pig sperm. Biol Reprod 22:566–570

Singhas CA and Oliphant G (1978) Ultrastuctural observations of the time sequence of induction of acrosome membrane alterations by ovarian follicular fluid. Fert Steril 29:194–203

Spruill WA, Koide Y, Huang HL, Levine SN, Ong SH, Steiner AL and Beavo JA (1981) Immunocytochemical localization of cyclic guanosine monophosphate-dependent protein kinase in endocrine tissues. Endocrinology 109:2239–2248

Srivastava PN, Breuer JM and White RA (1082) Hydrolysis of p-nitrophenylphosphorylcholine by alkaline phosphatase and phospholipase C from rabbit sperm acrosomes. Biochem Biophys Res Commun 108:1120–1125

Stambaugh R and Smith M (1976) Sperm proteinase release during fertilization of rabbit ova. J exp Zool 197:121–125

Stambaugh R and Smith M (1978) Tubulin and microtubules-like structures iun mammalian acrosomes. J exp Zool 203:135–141

Stambaugh R, Smith M and Faltas S (1975) An organized distribution of acrosomal proteinase in rabbit sperm acrosomes. J exp Zool 193:119–123

Stein DM and Fraser LR (1984) Cyclic nucleotide metabolism in mouse epididymal spermatozoas during capacitation in vitro. Gamete Res 10:283–299

Suarez SS and Oliphant G (1982) Interaction of rabbit spermatozoa and serum complement components. Biol Reprod 27:473–483

Suarez SS, Katz DF and Meizel S (1984) Changes in motility that accompany the acrosome reaction in hyperactivated hamster spermatozoa. Gamete res 10:253–265

Suominen J, Kaufman MH and Setchell BP (1972) Prevention of fertilization in vitro by an acrosin inhibitor from rete testis fluid of the ram. J Reprod Fert 34:385–388

Suzuki F and Nagano T (1980) Epididymal maturation of rat spermatozoa studies by thin sectioning and freeze fracture. Biol Reprod 22:1219–1231

Szollosi D and Hunter RHF (1978) The nature and occurrence of the acrosome reaction in spermatozoa of the domestic pig, Sus scrofa. J Anat 127:33–41

Talbot P (1985) Sperm penetration through oocyte investments in mammals. Am J Anat 174:331–346

Talbot P and Franklin LE (1974a) Hamster sperm hyaluronidase. I. A bioassay procedure based on cumulus dispersion rate. J exp Zool 189:311–320

Talbot P and Franklin LE (1974b) Hamster sperm hyaluronidase. II. Its release from sperm in vitro in relation to the degenerative and normal acrosome reaction. J exp Zool 189:321–332

Talbot P and Franklin LE (1976) Morphology and kinetics of the hamster sperm acrosome reaction. J exp Zool 198:163–176

Thakkar JK, East J, Seyler D and Franson RC (1983) Surface-active phospholipase A2 in mouse spermatozoa. Biochem Biophys Acta 754:44–50

Thomas TS, Reynolds AL and Oliphant G (1984) Evaluation of the site of synthesis of rabbit sperm acrosome stabilizing factor using immunocytochemical and metabolic labeling techniques. Biol Reprod 30:693–705

Toshimori K, Higashi R and Oura C (1985) Distribution of intramembranous particles and filipin-sterol complexes in mouse sperm membranes: polyene antibiotic filipin treatment. Am J Anat 174:455–470

Triana LR, Babcock DF, Lorton SP, First NL and Lardy HA (1980) Release of acrosomal hyaluronidase follows increased membrane permeability to calcium in the presumptive capacitation sequence for spermatozoa from the bovine and other mammalaina species. Biol Reprod 23:47–59

Tung KSK, Okada A and Yanagimachi R (1980) Sperm autoantigens and fertilization. I. Effects of antisperm autoantibodies on rouleaux formation, viability and acrosome reaction of guinea pig spermatozoa. Biol Reprod 23:872–886

Tung KSK, Yanagimachi H and Yanagimachi R (1982) Sperm autoantigens and fertilization. III. Ultrastructural localization of guinea pig autoantigens. Anat Rec 202:241–253

Usui N and Yanagimachi R (1986) Cytochemical localization of membrane-bound Mg^{2+}-dependent ATPase activity in guinea pig sperm head before and during the acrosome reaction. Gamete Res 113:271–280

Velazquez A, Rosado A and Rosado A (1975) Participation of vitamin A in the maturation of rabbit spermatozoa. Int J Fert 20:151–155

Vijayasarathy S, Shivaji S and Balaram P (1980) Plasma membrane bound Ca^{2+}-ATPase activity in bull sperm. FEBS Letts 114:45–47

Vijayasarathy S, Chivaji S and Balaram P (1982) Bull sperm plasma and acrosomal membranes: fluorescence studies of lipid phase fluidity. Biochem Biophys Res Commmun 108:585–591

Waibel R, Ginsberg LC and Ficsor G (1984) Cytochemical detection of hyaluronidase activity in single human and mouse sperm by an improved substrate-film technique. J Histo Cyto 32:63–66

Ward CR and Storey BT (1984) Determination of the time course of capacitation in mouse spermatozoa using a chlortetracycline fluorescence assay. Develop Biol 104:287–296

Watson PF and Plummer JM (1986) Relationship between calcium binding sites and membrane fusion during the acrosome reaction induced by ionophore in ram spermatozoa. J exp Zool 238:113–118

Williamson BR, Shepherd BA and Martan J (1980) Fertility of spermatozoa form the excurrent ducts of the guinea pig. J Reprod Fert 59:515–517

Witkin SS, Richards JM and Bedford JM (1983) Influence of epididymal maturation on the capacity of hamster and rabbit spermatozoa for complement activation. J Reprod Fert 69:517–521

Working PK and Meizel S (1981) Evidence that an ATPase functions in the maintenance of the acidic pH of the hamster sperm acrosome. J Biol Chem 256:4708–4711

Working PK and Meizel S (1982) Preliminary charactrization of a Mg^{2+}-ATPase in hamster sperm head membranes. Biochem Biophys Res Commun 104:1060–1065

Working PK and Meizel S (1983) Correlation of increased intraacrosomal pH with the hamster sperm acrosome reaction. J exp Zool 227:97–108

Yanagimachi R (1969) In vitro capacitation of hamster spermatozoa by follicular fluid. J Reprod Fert 18:275–286

Yanagimachi R (1975) Acceleration of the acrosome reaction and activation of guinea pig spermatozoa. Biol Reprod 13:519–526

Yanagimachi R (1981) Mechanisms of Fertilization in Mammals. In: Fertilization and Embryonic Development In Vitro. Eds L Mastroianni and JD Biggers, Plenum Press, London. pp 81–182

Yanagimachi R and Noda YD (1972) Scanning electron microscopy of golden hamster spermatozoa before and during fertilization. Experientia 28:69–72

Yanagimachi R and Phillips DM (1984) The status of acrosomal caps of hamster spermatozoa immediately before fertilization in vivo. Gamete Res 9:1–19

Yanagimachi Rand Suzuki F (1985) A further study of lysolecithin-mediated acrosome reaction of guinea pig spermatozoa. Gamete Res 11:29–40

Yanagimachi R and Teichman RJ (1972) Cytochemical demonstration of acrosomal proteinase in mammalaian and avian spermatozoa by a silver proteinate method. Biol Reprod 6:87–97

Yanagimachi R and Usui N (1974) Calcium dependence of the acrosome reaction and activation of guinea pig spermatozoa. Exp Cell Res 89:161–174

Yanagimachi R, Huang TTF, Fleming AD, Kosower NS and Nicolson GL (1983) Dithiothreitol, a disulphide-reducing agent, inhibits capacitation, acrosome reaction and interaction with eggs by guinea pig spermatozoa. Gamete Res 7:145–154

Zaneveld LJD, Srivastava PN and Williams WL (1969) Relationship of a trypsin-like enzyme in rabbit spermatozoa to capacitation. J Reprod Fert 20:337–339

Zaneveld LJD, Robertson RT, Kessler M and Williams WL (1971) Inhibition of fertilization in vivo by pancreatic and seminal plasma trypsin inhibitors. J Reprod Fert 25:387–392

Zaneveld LJD, Polakoski KL and Schumacher GFB (1975) The proteolytic enzyme systems of mammalian genital tract secretions and spermatozoa. In: Proteases and Biological Control. Eds E Reich, DB Rifkin and E Shaw, Cold Spring Harbour Lab. pp 683–706

Zamboni L and Stefanini M (1968) On the configuration of the plasma membrane of the mature spermatozoon. Fert Steril 19:570–579

Zimmerman RE and Burck PJ (1978) The loss of low molecular weight acrosin inhibitor from acrosomes during capacitation. Proc Soc exp Biol Med 158:490–495

E. Consequences of Capacitation. III. Hyperactivation

1. Hyperactivation

Hyperactivation (see Yanagimachi, 1981 for terminology) refers to the changed nature of sperm motility that occurs around the time of fertilisation both in-vivo and in-vitro, variously described as "bobbing" (Gwatkin and Anderson, 1969), "whiplash" (Yanagimachi, 1970b; Fraser, 1977; Katz and Yanagimachi, 1980) or "serpentine" (Yanagimachi, 1972a) in nature. The descriptive terms employed emphasise that a clearly recognisable change in motility occurs that is associated with increased flexibility of the sperm tail. Measurements of beat frequency of hyperactivated spermatozoa (Katz et al., 1978; Katz and Yanagimachi, 1980, 1981; Mohri and Yanagimachi, 1980; Dresdner and Katz, 1981; Johnson et al., 1981; Ishijima and Mohri, 1985) have not consistently confirmed the increase in frequency originally proposed (Yanagimachi, 1969a) and neither was an anticipated greater power output (Yanagimachi, 1970) confirmed from measurements in the guinea pig and hamster; this only applied to non-space-gaining hyperactivated spermatozoa from the rabbit (Katz et al., 1978; Dresdner and Katz, 1981; Johnson et al., 1981).

Tracings of sequential flagellar positions of hyperactivated spermatozoa reveal large amplitude bends in the mid-piece and principal-piece of golden hamsters (Yanagimachi, 1970b; 1981), mice (Fraser, 1977; Aonuma et al., 1980; Cooper and Woolley, 1982), guinea pigs (Yanagiimachi and Usui, 1974), bats (Lambert, 1981) and rams (Cummins, 1982) and a doubling in amplitude was measured by Ishijima and Mohri (1985). Hyperactivation by Chinese hamster sperm, however, is atypical (Yanagimachi et al., 1983) and may reflect the relatively large outer dense fibres in this species (Phillips, 1972; Phillips and Olson, 1975). This pattern of motility alternates between non-progressive and progressive movement in rabbits (Cooper et al., 1979; Johnson et al., 1981) and hamsters (Katz and Yanagimachi, 1980), although this was not confirmed by Suarez et al., (1983) who considered all hyperactivated cells to move in circles. The pattern observed depends on the state of the acrosome (Cummins, 1982) and is affected by the restrictions imposed by the egg investments (Corselli and Talbot, 1986) and contact with the zona (Katz and Yanagimachi, 1981).

Hyperactivation has been observed in-vitro in many species (see Fraser, 1984) but it is now no longer thought that the observed motility changes represent an artefactual manifestation in-vitro of the altered metabolic states of capacitated sperm that reveal themselves differently in-vivo (Bavister, 1980). Similar sperm movements have been observed within the female tract (see Section II.E.8) especially the oviducts (Katz and Yanagimachi, 1980). There is no clear evidence for hyaperactivation by human spermatozoa, although sperm moving in a similar way with large amplitude bends have been observed in-vitro (Burkman, 1984; see Fig. 2). Demembranated human sperm cells reactivated with ATP can display movements similar to those of hyperactivated intact cells (Mohri and Yanagimachi, 1980).

2. Requirements for Hyperactivated Motility

a. Calcium

The dependence of hyperactivation on extracellular Ca^{2+} and the rapid promotion of hyperactivation by Ca^{2+}-ionophores in the same species (see Yanagimachi, 1981) suggest that intracellular calcium plays an important part in the regulation of this flagellar activity. Support for this view comes from a very similar flexible flagellum which is observed when a form of "cycling" motility is induced in mature, forward progressing bovine spermatozoa by agents that displace membrane-bound, and presumably raise intracellular, calcium (Singh et al., 1983). Once initiated in the presence of Ca^{2+}, hyperactivation continues when extracellular calcium is withdrawn (Cooper, 1984). However, large amplitude bends resembling those of hyperactivated spermatozoa can be created in demembranated models by ATP and cAMP in the absence of calcium (Mohri and Yanagimachi, 1980; Yeung, 1984) and in these models calcium is inhibitory to motion (see Section II.A.4.d.i) so the action of Ca^{2+} is likely to be an indirect one.

b. Cyclic Nucleotides

Agents that should raise intracellular cAMP promote hyperactivated motility in sperm from mice (Fraser, 1979, 1981) and hamsters (Mrsny and Meizel, 1980), and permanent cGMP analogues promote hyperactivation in guinea pig sperm (Santos-Sacchi and Gordon, 1980). However, Cooper (1984) considered the effect of phosphodiesterase inhibitors in mice to be indirect since objective measurements revealed the proportion of non-hyeractivated cells was also increased by these agents.

c. Catecholamines

Catecholamines stimulate hyperactivation of hamster sperm, but need synergistic factors, probably taurine and hypotaurine for full expression (Mrsny et al., 1979; Meizel et al., 1980). The action of these compounds is not known, but sperm extracts are also active (Bavister and Yanagimachi, 1977) and this suggests that these factors enter the cell and so could have a direct action on the axoneme (see Section II.B.6).

d. Lipids

The onset of hyperactivation is promoted by lysolipids (Fleming and Yanagimachi, 1981) and prevented by inhibitors of phospholipase A (Llanos and Meizel, 1983). These agents may alter the nature of the membrane in some way: the fluidity of lipid domains is altered coincident with changes in motility (Wolf et al., 1986; see Section II.E.3).

3. Mechanisms of Hyperactivated Motility

Changes in the exposed carbohydrates of the membrane over the flagellum that occur during capacitation have been demonstrated by binding of lectins (Kinsey and Koehler, 1978; Talbot and Chacon, 1981; Ahuja, 1984, 1985) and increased lectin-mediated tail-tail agglutination. That these effects can be mimicked by trypsin (Talbot and Franklin, 1978a,b) suggests that previously masked sites are exposed. Changes also occur in the distribution of intramembranous particles of the plasma membrane over the mid-piece and tail during capacitation (Friend and Rudolf, 1974; Koehler and Gaddum-Rosse, 1975). These particle movements may relate to changes in motility brought about by ion fluxes, because a specialised region of intramembranous particles (the "zipper") is thought to be a site of ion exchange (Friend and Heuser, 1981) and cyanide inhibits both sperm motility and dispersion of the particles (Friend, 1977).

Hyperactivation is characterised by large bend angles which are created by greater sliding of axonemal doublets, but little is known at present about the control of this process in vertebrate sperm (Satir, 1979). If the normal extent of sliding during bend formation is limited by the curvature of the bend so created (Yeung and Woolley, 1983a,b) then there must be a change in the "set-point" of the switch controlling the activation of alternate half-axonemes (see Section II.A.4.a).

4. Stimulus to Hyperactivated Motility

Freshly ovulated cumulus cells can promote an "erratic" motion of mouse spermatozoa progressing towards them in capillary tubes (Bronson and Hamada, 1972) and forward progression in bovine epididymal spermatozoa in-vitro (Bradley and Garbers, 1983), but in neither case was hyperactivated motility specifically mentioned. In the rabbit hyperactivation may be triggered by changes in the composition of oviductal fluid occurring at ovulation (Overstreet et al., 1978, 1980; see Section II.E.8).

Hunter et al. (1983) considered that ovarian steroids and prostaglandins could act locally, directly or indirectly, to affect the movement of spermatozoa through the oviduct.

5. Relationship to Capacitation

High ionic strength media promote hyperactivated motility (Brackett and Oliphant, 1975; Fraser, 1984b; Cooper, 1984) so removal of surface proteins (see Section II.B.3.a) may precede the changes in motility. Capacitation itself, however, is insufficient to trigger hyperactivation in the absence of calcium ions (Yanagimachi and Usui, 1974; Cooper, 1984). The inability of agents that block capacitation to prevent hyperactivation once it has been initiated in their absence (Aonuma et al., 1980; Cooper, 1984) demonstrates not only the sequence of events but the irreversibility of the change in motility.

6. Relationship to the Acrosome Reaction

That some hyperactivated hamster spermatozoa do not bind to zona-free eggs is more likely to reflect the non-synchronous appearance of hyperactivation and the acrosome reaction following capacitation, rather than question the basic tenet that hyperactivated sperm are capacitated (Barros and Berrios, 1977). Hyperactivation precedes (and may help trigger, see Section II.E.9.d) the acrosome reaction (Mahi and Yanagimachi, 1973; Fraser, 1977, 1982) but it is possible to induce hyeractivation without the acrosome reaction and *vice-versa* by the use of certain drugs (Yanagimachi, 1975), different temperatures (Mahi and Yanagimachi, 1973) and different energy substrates (Fraser, 1981; Fraser and Quinn, 1981). The beat frequency of acrosome-reacted hyperactivated spermatozoa is lower than that of non-acrosome reacted hyperactivated spermatozoa (Suarez et al., 1984).

7. Relationship to Fertilisation

A role for hyperactivation in fertilisation is demonstrated in in-vitro studies where the acrosome reaction and hyperactivation can be separated. No fertilisation of eggs occurs if acrosome-reacted mouse spermatozoa do not display hyperactivated motility, and fertilisation follows the induction of hyperactivation (Fraser, 1981). In the guinea pig the decline in the fertilising capacity of sperm parallels the decline in the population of hyperactivated spermatozoa (Fleming and Yanagimachi, 1982).

Sperm motility within the confines of the periviteline space may be sluggish (mouse: Fraser, 1977) or more active (rabbit: Gaddum et al., 1968; Brackett, 1970; Katz and Overstreet, 1980) and it persists until engulfment of the sperm by the egg (Yanagimachi, 1978; Katz and Overstreet, 1980). Its role in fusion with the vitellus varies with species: zona-free eggs are fertilised by populations of non-hyperactivated sperm in the guinea pig (Fleming and Yanagimachi, 1982), whereas in the mouse fertilisation in-vitro is improved if the sperm cells display the exaggerated motility (Fraser, 1981). There is an absolute requirement for hyperactivated motility if bat sperm are to fuse with the zona-free hamster ovum (Lambert, 1981), but this may reflect the unphysiological nature of this particular gamete union. However, human sperm are able to fuse with the zona-free hamster egg even if they are immotile owing to lack of dynein arms (Aitken et al., 1984; Williamson et al., 1984), but this infertile disorder may also have expression in the fusing ability of their membranes.

8. Site of Hyperactivated Motility

It is not clear if the "undulating" motion of spermatozoa recovered from the mouse uterus after mating (Phillips, 1972) is that of a hyperactivated spermatozoon, but it is an inappropriate site for this to occur. Hyperactivated spermatozoa have been oberved in the oviducts of every species examined (see Fraser, 1977; Cooper et al., 1979; Yanagimachi et al., 1983), but it is difficult to tell if the movement of sperm in the perivitelline space is hyperactivated (Gaddum, 1968; Brackett, 1970).

Where the time-course of movement and the type of motility displayed has been studied (in the rabbit) no sperm were found in the tubal ampulla by 4 h *p.c.* and sperm cells in the isthmus at this time were poorly progressive. By 6 h *p.c.*, however, the motility displayed by oviductal spermatozoa was hyperactivated (Overstreet and Cooper, 1978a, 1979; Cooper et al., 1979).

a. Sperm Numbers at Fertilisation

Sperm motility may dictate the number of potentially fertilising spermatozoa arriving in the tubal ampulla at the time of ovulation. The artefacts asociated with manipulation of the tract in the determination of sperm numbers inside it have been discussed (Thibault, 1973) and fixation of the tract in-situ or immediately after removal at slaughter has been considered one way to reduce such errors.

When sections of such tissues are examined, a concentration gradient of sperm cells along the length of the oviduct is noted, with the majority accumulating just distal to the utero-tubal junction in cows (Thibault, 1973) and mice (Zamboni, 1972), confirming careful observations with more routine methods in hamsters (Yanagimachi and Chang, 1963) and rabbits (Overstreet et al., 1978). Where information on the number and fertilisation state of eggs was also recorded, a very low ratio of sperm to eggs (about 1:1) was found at the time of fertilisation, with greater numbers of spermatozoa being observed free in the lumen or associated with the eggs only when the majority of eggs had been fertilised. As a consequence, the original suggestion of Austin (1951), that the first sperm to arrive at the site of fertilisation around ovulation is the fertilising spermatozoon, has received recent support from work in both laboratory animals (see Shalgi and Kraicer, 1978; Cummins and Yanagimachi, 1982) and farm animals. In domestic species the spermatozoa progressing beyond the isthmus in the periovulatory period are known to be fertilising in sheep (Hunter et al., 1980, 1982; Hunter and Nichol, 1983), pigs (Hunter, 1981, 1984) and cows (Hunter and Wilmut, 1983, 1984; Wilmut and Hunter, 1983).

9. Consequences of Hyperactivated Motility

The change in swimming behaviour of spermatozoa have been linked to several aspects of fertilisation.

a. Ascent to Eggs

It has been argued that hyperactivated cells are responsible for the migration of sperm away from the isthmus to the site of fertilisation in rabbits and hamsters (Cooper and Katz, 1978; Katz and Yanagimachi, 1980). Reports in the rat (Blandau, 1978) and hamster (Battalia and Yanagimachi, 1979), however, reveal a concerted activity of the uterus and oviduct around the time of ovulation that permits even inert particles to move rapidly to the tubal ampulla at this time. The mass transport of spermatozoa to the egg, however, would not account for the 1-to-1 ratio of sperm to eggs observed (see Section II.E.8).

b. Sustained Release of Capacitated Spermatozoa

A display of circular movement (Suarez et al., 1983, 1984) would help retain sperm within the confines of the oviduct, preventing their premature loss and thus ensuring a population of capacitated cells for the freshly ovulated eggs.

c. Penetration of Egg Investments

The greater amplitude of beat of hyperactivted cells should generate more hydrodynamic thrust than non-hyperactivated cells (Katz et al., 1978; Katz and Yanagimachi, 1980) and have consequences for penetration of the egg investments (Gwatkin and Anderson, 1969; Yanagimachi, 1970b). Hyperactivated spermatozoa are found within the cumulus mass (Katz and Overstreet, 1980; Cummins and Yanagimachi, 1982) and on the zona pellucida (Katz and Yanagimachi, 1981) and direct observations of penetrating spermatozoa reveal that increased thrust is required for zona penetration (Yanagimachi, 1966). The existence of resistive forces in the visco-elastic zona pellucida to overcome on penetration is inferred from a deformed sperm head within this structure (Bedford et al., 1972; Calvin and Bedford, 1974), whereas lower species whose eggs have thinner investments may not require their spermatozoa to perform hyperactivated motility (Bedford, 1983).

The force generated by even non-hyperactivated spermatozoa is calculated to be greater than that required to break covalent bonds that presumably hold together the acellular zona (Green and Purves, 1984). If these calculations are valid and hyperactivated sperm generate even greater thrust, the role of acrosomal enzymes in zona penetration would need to be reassessed.

d. Diffusion of Ions

Greater flagellation of hyperactivated spermatozoa should promote vigorous stirring of the fluid surrounding the spermatozoon and this could aid the diffusion of ions and metabolites (Katz et al., 1978; Katz and Yanagimachi, 1980; Dresdner and Katz, 1981). This may promote the acrosome reaction (see Section II.D.4).

e. Gamete Contact

The spatial volumes swept out by hyperactivated spermatozoa are much larger than those of non-hyperactivated cells and this may improve their chances of achieving contact with the cumulus mass (Katz et al., 1978), although the freshly ovulated egg and its investments are known to occupy the lumen of the oviduct (Shalgi and Kraicer, 1978).

f. Sperm-Egg Fusion

Hyperactivated motility may be required for penetration of the vitellus (see Section II.E.7).

10. Development in the Epididymis of the Ability of Sperm to Display Hyperactivated Motility

Immature rabbit spermatozoa recovered from the oviducts 16 h after insemination swim in straight lines in-vitro (Orgebin-Crist, 1967a) but it is not known if hyperactivated motility were displayed nearer to ovulation. Caput and corpus sperm from the hamster do not exhibit hyperactivation when removed from the female tract after insemination (Cummins, 1976). However, in-vitro, motility patterns of some spermatozoa freshly removed from the rabbit caput epididymidis (Gaddum et al., 1968) (termed "darting": Gaddum, 1968)) resemble hyperactivated spermatozoa in the flexion of the neck and spermatozoa from the mouse corpus epididymidis undergo hyperactivation in-vitro somewhat sooner than cauda sperm in low ionic strength mdium (Table 9), probably because they are less inhibited by DF than more mature cells. The inability of testicular, caput or corpus sperm to reach the perivitelline space of rabbit eggs in-vitro (Brackett et al., 1978) may be a demonstration of their inadequate motility for penetration of the zona pellucida.

11. Involvement of the Epididymis in Permitting Hyperactivation

Immature sperm do not develop the ability to undergo hyperactivation in-vivo if they are first aged in the epididymis (Cummins, 1976), suggesting an epididymal- rather than age-dependent maturation of this function. The "circling" of mature spermatozoa induced in-vitro (Singh et al., 1983) is associated with large amplitude bends that resemble the flexuous tails of immature bovine caput sperm cells (Acott et al., 1983; Acott and Hoskins, 1983) but it remains to be demonstrated that the epididymal secretion product FMP is really a DF and that conversion of "circling" (Hyperactivation?) to forward motility upon maturation in the epididymis is reversed during capacitation (Suarez et al., 1983). As the "circling" of immature sperm is rather obvious, but "circling" of hyperactivated cells requires detailed micrographic analysis, the two forms of motions may be more dissimilar than the simple analysis above implies.

Intrinsic changes in the flagellum must also occur upon maturation of the hamster axoneme, because demembranated testicular spermatozoa do not display the vigorous motility resembling that of hyperactivation shown by demembranated mature cells in response to ATP and cAMP (Mohri and Yanagimachi, 1980), and demembranated caput sperm are less able than mature cells to bend into small arcs (Yeung, 1984).

12. Summary

The exaggerated form of hyperacitvated motility is necessary for spermatozoa to penetrate the zona pellucida in all species and fusion with the vitellus in some. While the characteristics of hyperactivated motility are becoming understood as a result of advances in high-speed cinematography and video technology, less progress has been made in understanding why the flagellum responds in this way. Once the ability to flagellate has been acquired, the change to hyperactivation may be merely a result of loss, at an appropriate time, of epididymal coating proteins that prevent this expression.

13. References

Acott TS and Hoskins DD (1983) Cinematographic analysis of bovine epididymal sperm motility: epididymal maturation and forward motility protein. J Submicroscop Cytol 15:77–82

Acott TS, Katz DF and Hoskins DD (1983) Movement characteristics of bovine epididymal spermatozoa. Effects of forward motility protein and epididymal maturation. Biol Reprod 29:389–399

Ahuja KK (1984) Lectin-coated agarose beads in the investigation of sperm capacitation in the hamster. Develop Biol 104:131–142
Ahuja KK (1985) Inhibitors of glycoprotein biosynthesis block fertilization in the hamster. Gamete Research 11:179–189
Aitken RJ, Ross A and Lees ML (1983) Analysis of sperm function in Kartagener's Syndrome. Fert Steril 40:696–698
Aonuma S, Kabe M, Kawaguchi M and Kishi Y (1980) Studies of sperm capacitation. IX. Movement characteristics of spermatozoa in relation to capacitation. Chem Pharm Bull 25:1497–1502
Austin CR (1951) Observations on the penetration of the sperm into the mammalian egg. Aust J Sci Res 34:581–596
Barros C and Berrios M (1977) Is the activated spermatozoon really capacitated? J exp Zool 201:65–72
Battalia DE and Yanagimachi R (1979) Enhanced and co-ordinated movement of the hamster oviduct during the periovulatory period. J Reprod Fert 56:515–520
Bavister BD (1980) Recent progress in the study of early events in mammalian fertilization. Develop Growth Diff 22:385–402
Bavister BD and Yanagimachi R (1977) The effects of sperm extracts and energy sources on the motility and acrosome reaction of hamster spermatozoa in vitro. Biol Reprod 16:228–237
Bedford JM (1983) Significance of the need for sperm capacitation before fertilization in Eutherian mammals. Biol Reprod 28:108–120
Bedford JM and Calvin HI (1974) The occurrence and possible significance of -S-S- cross links in sperm heads, with particular reference to Eutherian mammals. J exp Zool 188:137–156
Bedford JM, Cooper GW and Calvin HI (1972) Post-meiotic changes in the nucleus and membranes of mammalian spermatozoa. In: The Genetics of the Spermatozoon. Eds RA Beatty and S Glucksohn-Waelsh, Bogtrykkeriet Forum, Copenhagen. pp 69–89
Blandau RJ (1978) Gamete transport in oviducts of rats. Anat Rec 190:593 abst
Brackett BG (1970) In vitro fertilization of rabbit ova: time sequence of events. Fert Steril 21:169–176
Brackett BG and Oliphant G (1975) Capacitation of rabbit spermatozoa in vitro. Biol Reprod 12:260–274
Brackett BG, Hall JL and Oh YK (1978) In-vitro fertilizing ability of testicular, epididymal, and ejaculated rabbit spermatozoa. Fert Steril 29:571–582
Bradley MP and Garbers DL (1983) The stimulation of bovine caudal epididymal sperm forward motility by bovine cumulus-egg complexes in vitro. Biochem Biophys Res Commun 115:777–787
Bronson R and Hamada Y (1977) Gamete interactions in vitro. Fert Steril 28:570–576
Burkman LK (1984) Characterization of hyperactivated motiity by human spermatozoa during capacitation: comparison of fertile and oligospermic sperm populations. Archs Androl 13:153–165
Calvin HI and Bedford JM (1974) Stimulation of actinomycin D-binding to Eutherian sperm chromatin by reduction of disulphide bonds. J Reprod Fert 36:225–229
Cooper TG (1984) The onset and maintenance of hyperactivated motility of spermatozoa from the mouse. Gamete Res 9:55–74
Cooper TG and Woolley DM (1982) Stroboscopic illumination for the assessment of hyperactivated motility of mouse spermatozoa. J exp Zool 223:291–294
Cooper GW, Overstreet JW and Katz DF (1979) The motility of rabbit spermatozoa recovered from the female reproductive tract. Gamete Research 2:35–42
Corselli J and Talbot P (1986) An in vitro technique to study penetration of hamster oocyte-cumulus complexes by using physiological numbers of sperm. Gamete Res 13:293–308
Cummins JM (1976) Effects of epididymal occlusion on sperm maturation in the hamster. J exp Zool 197:187–190
Cummins JM (1982) Hyperactivated motility patterns of ram spermatozoa recovered from the oviducts of mated ewes. Gamete Res 6:53–63
Cummins JM and Yanagimachi R (1982) Sperm-egg ratios and the site of the acrosome reaction during in vitro fertilization in the hamster. Gamete Res 5:239–256

Dresdner RD and Katz DF (1981) Relationship of mammalian sperm motility and morphology to hydrodynamic aspects of cell function. Biol Reprod 25:920–930

Fleming AD and Yanagimachi R (1981) Effects of various lipids on the acrosome reaction and fertilizing capacity of guinea pig spermatozoa with special reference to the possible involvement of lysophospholipids in the acrosome reaction. Gamete Res 4:253–273

Fleming AD and Yanagimachi R (1982) Fertile life of acrosome-reacted guinea pig spermatozoa. J exp Zool 220:109–115

Fraser LR (1977) Motility patterns in mouse spermatozoa before and after capacitation. J exp Zool 202:439–444

Fraser LR (1979) Accelerated mouse sperm penetration in vitro in the presence of caffeine. J Reprod Fert 57:377–384

Fraser LR (1981) Dibutyryl cyclic AMP decreases capacitation time in vitro in mouse spermatozoa. J Reprod Fert 62:63–72

Fraser LR (1982) Potential contraception by interference with capacitation. In: Progress towards a male contraceptive. Eds SL Jeffcoate and M Sandler, John Wiley and Sons, Chichester. pp 185–208

Fraser LR (1984a) Mechanisms controlling mammalian fertilization. In: Oxford Reviews of Reproductive Biology, Vol.6. Ed JR Clarke, Clarendon Press, Oxford. pp 174–225

Fraser LR (1984b) Mouse sperm capacitation in vitro involves loss of a surface-associated inhibitory component. J Reprod Fert 72:373–384

Fraser LR and Quinn PJ (1981) A glycolytic product is obligatory for initiation of the sperm acrosome reaction and whiplash motility required for fertilization in the mouse. J Reprod Fert 61:25–35

Friend DS (1977) The organization of the spermatozoal membrane. In: Immunobiology of Gametes. Eds M Edidin and MH Johnson, University Press Cambridge. pp 5–30

Friend DS and Heuser JE (1981) Orderly particle arrays on the mitochondrial outer membrane in rapidly frozen sperm. Anat Rec 199:159–175

Friend DS and Rudolf I (1974) Acrosomal disruption in sperm. Freeze-fracture of altered membranes. J Cell Biol 63:466–479

Gaddum P (1968) Sperm maturation in the male reproductive tract. Development of motility. Anat Rec 161:471–482

Gaddum P, Blandau RJ and Hayashi R (1968) Sperm maturation in the male reproductive tract. (Film). University of Washington Press

Gwatkin RBL and Anderson CF (1969) Capacitation of hamster spermatozoa by bovine follicular fluid. Nature (Lond) 224:111–1112

Green DPL and Purves RD (1984) Mechanical hypothesis of sperm penetration. Biophys J 45:659–662

Hunter RHF (1981) Sperm transport and reservoirs in the pig oviduct in relation to the time of ovulation. J Reprod Fert 63:109–117

Hunter RFH (1984) Pre-ovulatory arrest and peri-ovulatory redistribution of competent spermatozoa in the isthmus of the pig oviduct. J Reprod Fert 72:203–211

Hunter RHF and Nichol R (1983) Transport of spermatozoa in the sheep oviduct: preovulatory sequestration of cells in the caudal isthmus. J exp Zool 228:121–128

Hunter RHF and Wilmut I (1983) The rate of functional sperm transport into the oviducts of mated cows. Anim Reprod Sci 5:167–173

Hunter RHF and Wilmut I (1984) Sperm transport in the cow: peri-ovulatory redistribution of viable cells within the oviduct. Reprod Nutr Develop 24:597–608

Hunter RHF, Nichol R and Crabtree SM (1980) Transport of spermatozoa in the ewe: timing of the establishment of a functional population in the oviduct. Reprod Nutr Develop 20:1869–1875

Hunter RHF, Barwise L and King R (1982) Sperm transport, storage and release in the sheep oviduct in relation to the time of ovulation. Brit Vet J 138:225–232

Hunter RHF, Cook B and Poyser NL (1983) Regulation of oviduct function in pigs by local transfer of ovarian steroids and prostaglandins: a mechanism to influence sperm ascent. Europ J Obstet Gynaecol Reprod Biol 14:225–232

Ishijima S and Mohri H (1985) A quantitative description of flagellar movement in golden hamster spermatozoa. J exp Biol 114:463–475

Johnson LL, Katz DF and Overstreet JW (1981) The movement characteristics of rabbit spermatozoa before and after activation. Gamete Res 4:275–282

Katz DF and Overstreet JW (1980) Mammalian sperm movement in the secretions of the male and female genital tracts. In: Testicular Development, Structure, and Function. Eds A Steinberger and E Steinberger, Raven Press, New York. pp 481–489

Katz DF and Yanagimachi R (1980) Movement characteristics of hamster spermatozoa within the oviduct. Biol Reprod 22:759–764

Katz DF and Yanagimachi R (1981) Movement characteristics of hamster and guinea pig spermatozoa upon attachment to the zona pellucida. Biol Reprod 25:785–791

Katz DF, Yanagimachi R and Dresdner RD (1978) Movement characteristics and power output of guinea-pig and hamster spermatozoa in relation to activation. J Reprod Fert 52:167–172

Kinsey WH and Koehler JK (1978) Cell surface changes associated with in vitro capacitation of hamster sperm. J Ultrastruct Res 64:1–13

Koehler JK and Gaddum-Rosse P (1975) Media induced alterations of the membrane associated particles of the guinea pig sperm tail. J Ultrastruct Res 51:106–118

Lambert H (1981) Temperature dependence of capacitation in bat sperm monitored by zona-free hamster ova. Gamete Res 4:525–533

Lambert H (1984) Role of sperm-surface glycoproteins in gamete recognition in two mouse species. J Reprod Fert 40:281–284

Llanos MN and Meizel S (1983) Phospholipid methylation increases during capacitation of golden hamster sperm in vitro. Biol Reprod 28:1043–1051

Mahi CA and Yanagimachi R (1973) The effects of temperature, osmolality and hydrogen ion concentration on the activation and acrosome reaction of golden hamster spermatozoa. J Reprod Fert 35:55–66

Meizel S, Lui CW, Working PK and Mrsny RJ (1980) Taurine and hypotaurine: their effects on motility, capacitation and the acrosome reaction of hamster sperm in vitro and their presence in sperm and reproductive tract fluids of several mammals. Develop Growth Diff 22:483–494

Mohri H and Yanagimachi R (1980) Characteristics of motor apparatus in testicular, epididymal and ejaculated spermatozoa. A study using demembranated sperm models. Exp Cell Res 127:191–196

Mrsny RJ and Meizel S (1980) Evidence suggesting a role for cyclic nucleotides in acrosome reactions of hamster sperm in vitro. J exp Zool 211:153–157

Mrsny RJ, Waxman L and Meizel S (1979) Taurine maintains and stimulates motility of hamster sperm during capacitation in vitro. J exp Zool 210:123–128

Orgebin-Crist MC (1967) Maturation of spermatozoa in the rabbit epididymis: fertilizing ability and embryonic mortality in does inseminated with epididymal spermatozoa. Ann biol Anim Biochem Biophys 7:373–389

Overstreet JW and Cooper GW (1979) Effect of ovulation and sperm motility on the migration of rabbit spermatozoa to the site of fertilization. J Reprod Fert 55:53–59

Overstreet JW, Cooper GW and Katz DF (1978) Sperm transport in the reproductive rtract of the female rabbit: II. The sustained phase of transport. Biol Reprod 19:115–132

Overstreet JW, Katz DF and Johnson LL (1980) Motility of rabbit spermatozoa in the secretions of the oviduct. Biol Reprod 22:1083–1088

Phillips DM (1972) Comparative analysis of mammalian sperm motility. J Cell Biol 53:561–573

Phillips DM and Olson G (1975) Mammalian sperm motility-structure in relation to function. In: The Functional Anatomy of the Spermatozoon. Ed BA Afzelius, Pergamon Press, Oxford. pp 117–126

Santos-Sacchi J and Gordon M (1980) Induction of the acrosome reaction in guinea pig spermatozoa by cGMP analogues. J Cell Biol 85:798–803

Satir P (1979) Basis of flagellar motility in spermatozoa: current status. In: The Spermatozoon. Eds DW Fawcett and JM Bedford, Urban and Schwarzenberg, Baltimore. pp 81–90

Shalgi R and Kraicer PF (1978) Timing of sperm transport, sperm penetration and cleavage in the rat. J exp Zool 204:353–360

Singh JP, Babcock DF and Lardy HA (1983) Motility activation, respiratory stimulation and alteration of Ca^{2+} transport in bovine sperm treated with amine local anaesthetics and calcium transport antagonists. Archs Biochem Biophys 221:291–303

Suarez SS, Katz DF and Overstreet JW (1983) Movement characteristics and acrosomal status of rabbit spermatozoa recovered at the site and time of fertilization. Biol Reprod 29:1277–1287

Suarez SS, Katz DF and Meizel S (1984) Changes in motility that accompany the acrosome reaction in hyperactivated hamster spermatozoa. Gamete Res 10:253–265

Talbot P and Chacon R (1981) Detection of modificiations in the tail of capacitated guinea pig sperm using lectins. J exp Zool 216:435–444

Talbot P and Franklin LE (1978a) Surface modification of guinea pig sperm during in vitro capacitation: an assessment using lectin-induced agglutination of living sperm. J exp Zool 203:1–14

Talbot P and Franklin LE (1978b) Trypsinization increases lectin-induced agglutinability of uncapacitated guinea pig sperm. J exp Zool 204:291–297

Thibault C (1973) Sperm transport and storage in vertebrates. J Reprod Fert Suppl 18:39–53

Williamson RA, Koehler JK, DianneSmith W and Karp LE (1984) Entry of immotile spermatozoa into zona-free hamster ova. Gamete Res 10:319–325

Wilmut I and Hunter RHF (1984) Sperm transport into the oviduct of heifers mated early in oestrus. Reprod Nutr Develop 24:461–468

Wolf DE, Hagopian SS and Ishijima S (1986) Changes in sperm plasma membrane lipid diffusibility after hyperactivation during in vitro capacitation in the mouse. J Cell Biol 102:1372–1377

Yanagimachi R (1966) Time and progress of sperm penetration into hamster ova in vivo and in vitro. J Reprod Fert 11:359–370

Yanagimachi R (1969) In vitro capacitation of hamster spermatozoa by follicular fluid. J Reprod Fert 18:275–286

Yanagimachi R (1970) The movement of golden hamster spermatozoa before and after capacitation. J Reprod Fert 23:193–196

Yanagimachi R (1972) Fertilization of guinea pig sperm in vitro. Anat Rec 174:9–20

Yanagimachi R (1975) Acceleration of the acrosome reaction and activation of guinea pig spermatozoa by detergents and other reagents. Biol Reprod 13:519–526

Yanagimachi R (1978) Calcium requirement for sperm-egg fusion in mammals. Biol Reprod 19:949–958

Yanagimachi R (1981) Mechanisms of Fertilization in Mammals. In: Fertilization and Embryonic Development In Vitro. Eds L Mastroianni and JD Biggers, Plenum Press, London. pp 81–182

Yanagimachi R and Chang MC (1963) Sperm ascent through the oviduct of the hamster and rabbit in relation to the time of ovulation. J Reprod Fert 6:413–420

Yanagimachi R and Usui N (1974) Calcium dependence of the acrosome reaction and activation of guinea pig spermatozoa. Exp Cell Res 89:161–174

Yanagimachi R, Kamiguchi Y, Sugawaras S and Mikamo K (1983) Gametes and fertilization in the chinese hamster. Gamete Res 8:97–117

Yeung CH (1984) Effects of cyclic AMP on the motility of mature and immature hamster epididymal spermatozoa studied by reactivation of the demembranated cells. Gamete Res 9:99–114

Yeung CH and Woolley D M(1983a) Localized reactivation of the principal piece of demembranated hamster sperm by iontophoretic application of ATP. J Submicroscop Cytol 15:327–331

Yeung CH and Woolley DM (1983b) A study of bend formation in locally reactivated hamster sperm flagella. J Muscle Res and Cell Motil 4:625–645

Zamboni L (1972) Fertilization in the mouse. In: Biology of Mammalian Fertilization and Implantation. Eds KS Moghissi and ESE Hafez, Charles C Thomas, Springfield, Illinois. pp 213–262

F. Sperm-Egg Fusion

1. Binding to the Vitellus

Proteinaceous "Sperm Receptors" on the egg plasma membrane facilitate binding of sperm (Wolf et al., 1976). The inability of trypsin inhibitors to prevent binding of mouse sperm to eggs (Saling, 1981) suggests that acrosin activity is not required for gamete contact, but Ca^{2+} is necessary for sperm-egg contact in both hamsters (Yanagimachi, 1982) and mice (Saling, 1981).

2. Fusion with the Vitellus

Cross-species fertilisation occurs more readily with zona-free eggs from the hamster than with those from others species which have a zona-block to polyspermy (see Hanada and Chang, 1972, 1978; Quinn, 1979). This characteristic of the zona-free hamster egg has proved useful in the study of sperm-egg fusion where this would otherwise be difficult in homologous inseminations, *e.g.* for human spermatozoa (see Paulsen, 1983; Prasad, 1984; Yanagimachi, 1984; Rogers, 1985), however, guinea pig sperm do not readily bind to zona-free eggs from other species (Hanada and Chang, 1978).

A clear demonstration that sperm enter eggs by fusion and not by phagocytosis is the mosaic membrane of the fertilised egg formed by the intermingling of labelled sperm and egg membranes (Yanagimachi et al., 1973; Gabel et al., 1979; Gaunt, 1983; Jones et al., 1985). Fusion requires K^+ in the guinea pig (Rogers et al., 1981); temperature (Hirao and Yanagimachi, 1978b) and pH (Yanagimachi et al., 1980) are critical in the hamster, and Ca^{2+} is required in many species (Yanagimachi, 1978, 1982). Simple sugars that inhibit sperm binding to the zona (see Section II.C.4.a) do not inhibit fertilisation of zona-free eggs (Ahuja, 1985). While inhibitors of glycoprotein biosynthesis prevent sperm-egg fusion in hamsters, these effects may be related to the membrane-active properties of these compounds.

The role of the acrosomal enyzmes in fusion is unclear: early studies failed to show an effect of trypsin inhibitors on binding (Hartmann and Gwatkin, 1971); later inhibition was demonstrated (Wolf, 1977) but could have reflected failure of sperm to undergo the acrosome reaction (see Section II.D.1). However, trypsin inhibitors that do not affect the acrosome reaction still can prevent sperm-egg fusion (Dravland and Meizel, 1982). Certain membrane phospholipids are necessary for fusion because phospholipase C treatment of the egg (Hirao and Yanagimachi, 1978a) prevents fusion, whereas inhibitors of phospholipase A do not (Yanagimachi, 1981; Dravland and Meizel, 1982). As would be anticipated, factors affecting membrane fluidity, low temperature (Hirao and Yanagimachi, 1978b) or thiol-reducing agents (Yanagimachi et al., 1983), prevent fusion of sperm with egg.

3. Relationship to Capacitation

It was shown in 1970 that capacitated sperm fuse readily with the vitellus in homologous inseminations (Yanagimachi and Noda, 1970b) it is now accepted that capacitation is a prerequisite for fusion of the sperm with the egg membrane (see Yanagimachi, 1981). In fact, the fertilisation of zona-free eggs is now taken as evidence that capacitation has taken place (Yanagimachi, 1972b; Lorton and First, 1979). In-vivo all sperm arriving in the perivitelline space must be capacitated to have penetrated the zona pellucida (see Sections II.D,E).

4. Relationship to the Acrosome Reaction

Only acrosome-reacted spermatozoa bind to the vitellus in homologous inseminations (see Yanagimachi, 1981). In heterologous inseminations of human spermatozoa with the zona-free hamster egg acrosome reactions are necessary for their attachment (Talbot and Chacon, 1982; Koehler et al., 1983) and abnormal human spermatozoa lacking acrosomes do not penetrate zona-free eggs even though they are motile (Weissenberg et al., 1983; Jeyendran et al., 1985).

Fusion of porcine spermatozoa to zona-free hamster eggs can be promoted by Ca^{2+} in the presence of ionophores (Smith et al., 1983), conditions that promote the acrosome reaction (see Section II.D.3.a). However, the acrosome reaction by itself is not sufficient for fusion, for acrosome-reacted sperm bind to, but do not fuse with, the vitellus if conditions of temperature (Hirao and Yanagimachi, 1978b) or Ca^{2+} concentration (Yanagimachi, 1982) are inappropriate.

Physically removing the acrosome can promote the fusion of sperm with somatic cells (Siroky et al., 1979).

5. Relationship to Sperm Motility

Whereas early reports did not consider motility essential for fusion (Austin, 1951; Austin and Bishop, 1958), observations of motile spermatozoa in the perivitelline space (see Section II.E.8) supported the alternative view (Yanagimachi, 1966; Brackett et al., 1970; Wolf and Armstrong, 1978). Although binding of sperm to the egg membrane is reduced when the motility of spermatozoa is impaired by lanthanum (Saling, 1982), the ion itself may have interfered with the membrane surface involved in gamete recognition. Requirements for hyperactivation for fusion vary between species (see Section II.E.7). Immotile human spermatozoa can enter the zona-free hamster egg (Aitken et al., 1983; Williamson et al., 1984; Moryan et al., 1986), but these abnormal spermatozoa may also display other abnormalities, *e.g.* have a predisposition towards abnormal fusing ability.

6. Regions of the Sperm Head Involved in Fusion

The areas of the sperm head which contact the egg depend on whether insemination is in-vivo or in-vitro (Shalgi and Phillips, 1980a,b). Although acrosome-intact sperm may bind avidly to the vitellus in a manner resembling that of acrosome-reacted cells (Koehler et al., 1984) they do not fuse with them and different areas of the sperm head are involved in direct contact (Phillips and Yanagimachi, 1982).

During the course of neormal fertilisation the acrosome reaction occurs before fusion takes place, as the sperm has had to penetrate the zona, and three areas of the sperm head have been considered as the sites of initial contact with the egg.

a. The Inner Acrosomal Membrane

This membrane, protected in the cell interior until the acrosome reaction, suffers no visible change on approach to the egg in the hamster (Yanagimachi et al., 1973) and appears to be the site of attachment of human sperm to the zona-free hamster egg (Talbot and Chacon, 1982; Koehler et al., 1983). Where more sperm can be observed (in the scanning electron microscope), the inner acrosomal membrane is the first to contact the ovum in-vitro (Shalgi and Phillips, 1980b). However, this membrane appears to be too rigid to partake in the fusion process (Huang and Yanagimachi, 1985) and in addition, the presence of bound hydrolytic enzymes on this membrane (see Section II.D.9) and the trauma of passage through the zona pellucida make it unlikely that this membrane arrives at the vitellus unchanged. It has recently been implied that the sperm head membranes that are "protected" by not undergoing vesiculation at the time of the acrosome reaction are involved in sperm-egg contact (Bedford et al., 1979).

b. The Post-Acrosomal Membrane

Anti-sperm antibodies, lost from the acrosome during the acrosome reaction, are retained on post-acrosomal membranes (Koehler, 1976) and this site is considererd to be the only protected site where the equatorial segment of the acrosome is obliterated (Friend et al., 1977) or vesiculated (Barros and Herrera, 1977) after the acrosome reaction. Early electron microscopical observations also indicated this region as the site of fusion (Yanagimachi and Noda, 1970b; see Yanagimachi, 1981). The clearance of intramembranous particles from the post-acrosomal membrane during capacitation, by analogy with the clearance of these particles from the acrosomal membranes preceding the acrosome reaction (see Section II.D.2), supports the view that this membrane is involved with subsequent fusion events, and protein-free liposomes attach and fuse with this membrane in-vitro (Friend et al., 1977).

c. The Equatorial Segment of the Acrosome

This region appears to act to prevent or limit the extent of vesiculation of the anterior acrosome. The region is not vesiculated during the acrosome reaction of rabbit sperm (Bedford, 1967b, 1968, Bedford et al., 1979) and in the boar, vesicles formed from the equatorial segment remain close to the subacrosomal membrane (Russell et al., 1980). Because it persists intact during passage through the zona pellucida (Moore and Bedford, 1978; Bedford et al., 1979), this region may be preserved as a site of fusion. The scanning electron microscope has confirmed this segment as the initial site of contact of sperm with eggs in-vivo (Shalgi and Phillips, 1980a).

No change is seen in this membrane during the acrosome reaction (Moore and Bedford, 1979; Bedford et al., 1979) but decreases in lectin acceptors (Schwarz and Koehler, 1979) may indicate that changes in the anterior acrosome are transmitted to the equatorial segment at this time.

7. Basis of Fusion

Sperm-egg interactions can be studied by mixing spermatozoa with other cell types and observing the production of binucleate cells as evidence of fusion; however, phagocytosis has not been ruled out where no fusing agent was used (Bendich et al., 1974; Higgins et al., 1975; Phillips et al., 1976). Fusion induced by Sendai virus (Podoplepov et al., 1979), which attaches to, but does not penetrate the sperm membrane (Brackett et al., 1971), can be inhibited by phospholipase B (Kohn, 1965; Barbanti-Brodena et al., 1971). This enzyme degrades lysolecithin, and lysolecithin can promote fusion of sperm to other cells (Croce et al., 1972; Gledhill et al., 1972; Koprowski and Croce, 1973), liberating acrosomal contents and mimicking the acrosome reaction in-vitro (Croce et al., 1972). A variery of other agents can induce sperm-cell fusion (Goehringer and Dilliplane, 1976; Holt and Dott, 1980) and it is the mid-region of the bovine sperm head (Garbera et al., 1973) and the equatorial segment of ovine sperm (Holt and Dott, 1980) that is the site of fusion.

Material with affinity for malachite-green (identified as choline plasmalogen) within the head of the mature spermatozoon (Teichmann et al., 1972, 1974) may be a natural fusinogen (Soupart et al., 1979).

8. Development in the Epididymis of the Ability of Sperm to Fuse with Eggs

Epididymal spermatozoa are more able to bind with zona-free eggs the more distally they are taken from the epididymis in both homologous and heterologous inseminations (Table 9). The ability of sperm to bind to eggs precedes their ability to fuse with them (Table 2), except for the mouse, but the conditions of assay may account for the discrepancy. Where fusion does not occur (the boar) there is still an increasing tendency for sperm to bind to eggs as they mature.

Lanthanum acts to depress binding to eggs of poorly motile, immature sperm (Saling, 1982), suggesting that the ion acts on the binding process and not *via* decreases in motility, as for mature cells (see Section II.F.5). Poor binding or fusion of immature sperm could reflect inadequate capacitation, and the failure of these sperm to fuse when bound may reflect suboptimal conditions for the acrosome reaction (see Section II.D.7). However, even where morphologically normal acrosome reactions are induced by fusinogens in-vitro, ovine testicular spermatozoa do not fuse with chicken erythrocytes (Holt and Dott, 1980). In this case either the acrosome reaction induced did not transmit the necessary changes to the equatorial segment (see Section II.F.6.c) or other changes must occur to the equatorial segment on maturation in the epididymis.

Even more immature germ cells (spermatids) can fuse with other cell types (Nyormoi et al., 1973; Elsevier and Ruddle, 1976) but they represent morphologically underdeveloped cells and the site of fusion cannot be determined. The conntent of choline plasmalogen in the sperm head increases upon maturation in the epididymis (Teichman et al., 1972, 1974) and may be involved with the increasing ability of sperm to fuse with eggs if it acts as a natural fusinogen (Soupart et al., 1979).

9. Involvement of the Epididymis in Permitting Sperm-Egg Fusion

This has not been examined systematically. The loss of membrane lipids (see Section II.B.9.b) would facilitate membrane fusion events, but it is not known if there are changes in lipids in the relevant membranes. The malachite-green affinity staining material in rabbit sperm heads increases upon ageing them in a ligated epididymis, indicating an age- rather than epididymis-dependent development (Cummins and Teichmann, 1974).

10. Summary

Changes in membrane lipids that permit fusion may underly the increased ability of mature sperm over immature to fuse with eggs. The nature of the changes and the involvement of the epididymis are not established.

11. References

Ahuja KK (1985) Inhibitors of glycoprotein biosynthesis block fertilization in the hamster. Gamete Research 11:179–189

Aitken RJ, Ross A and Lees ML (1983) Analysis of sperm function in Kartagener's Syndrome. Fert Steril 40:696–698

Austin CR (1951) Observations on the penetration of the sperm into the mammalian egg. Aust J Biol Sci B4:589–596

Austin CR and Bishop MWH (1958) Role of the rodent acrosome and perfatorium in fertilization. Proc Roy Soc (Lond) 149B:241–248

Barbanti-Brodano G, Possati L and Laplaca M (1971) Inactivation of polykaryocytogenic and hemolytic activities of sendai virus by phospholipase B (lysolecithinase). J Virol 8:796–800

Barros C and Herrera E (1977) Ultrastructural observations of the incorporation of guinea-pig spermatozoa into zona-free hamster oocytes. J Reprod Fert 49:47–50

Bedford JM (1967) Experimental requirement for capacitation and observations on ultrastructural changes in rabbit spermatozoa during fertilization. J Reprod Fert Suppl 2:35–48

Bedford JM (1968) Ultrastructural changes in the sperm head during fertilization in the rabbit. Am J Anat 123:329–358

Bedford JM, Moore HDM and Franklin LE (1979) Significance of the equatorial segment of the acrosome of the spermatozoon in Eutherian mammals. Exp Cell Res 119:119–126

Bendich A , Borenfreund E and Sternberg SS (1974) Penetration of somatic mammalian cells by sperm. Science 183:857–859

Brackett BG, Baranska W, Sawicki W and Koprowski H (1971) Uptake of heterologous genome by mammalian spermatozoa through its transfer to ova through fertilization. Proc Nat Acad Sci US 68:353–357

Croce CM, Gledhill BL, Garbara B, Sawicki W and Koprowski H (1972) Lysolecithin-induced fusion of rabbit spermatozoa with hamster somatic cells. Adv Biosci 8:187–200

Cummins JM and Teichmann RJ (1974) The accumulation of malachite green stainable phospholipid in rabbit spermatozoa during maturation in the epididymis, and its possible role in capacitation. Biol Reprod 10:555–564

Dravland JE and Meizel S (1982) The effect of inhibitors of trypsin and phospholipase A2 on the penetration of zona pellucida-free hamster eggs by acrosome reacted hamster sperm. J Androl 3:388–395

Elsevier SM and Ruddle FH (1976) Haploid genome reactivation and recovery by cell hybridization. Induction of DNA synthesis in spermatid nuclei. Chromosoma 56:227–241

Friend DS, Orci L, Perrelet A and Yanagimachi R (1977) Membrane particle changes attending the acrosome reaction in guinea pig spermatozoa. J Cell Biol 74:561–577

Gabel CA, Eddy EM and Shapiro BM (1979) After fertilisation, sperm surface components remain as a patch in sea urchin and mouse embryos. Cell 18:207–215

Garbara B, Gledhill BL, Croce CM, Cesarini JP and Koprowski H (1973) Ultrastructure of rabbit spermatozoa after treatment with lysolecithin and in the presence of hamster somatic cells. Proc Soc exp Biol Med 143:1120–1124

Gaunt SJ (1983) Spreading of a sperm surface antigen within the plasma membrane of the egg after fertilization in the rat. J Embryol exp Morphol 75:259–270

Gledhill BL, Sawicki W, Croce CM and Koprowski H (1972) DNA synthesis in rabbit spermatozoa after treatment with lysolecithin and fusion with somatic cells. Exp Cell Res 73:33–40

Goeringer GC and Dilliplane D (1976) Some aspects of the fusion of avian erythrocytes and bovine spermatozoa in the presence of chemical fusing agents. Anat Rec 184:412

Hanada A and Chang MC (1972) Penetration of zona-free eggs by spermatozoa of different species. Biol Reprod 6:300–309

Hanada A and Chang MC (1978) Penetration of the zona-free or intact eggs by foreign spermatozoa and the fertilization of deer mouse eggs in vitro. J exp Zool 203:277–286

Hartmann JF and Gwatkin RBL (1971) Alteration of sites on the mammalian sperm surface following capacitation. Nature (Lond) 234:479–481

Higgins PJ, Borenfreund E and Bendich A (1975) Appearance of foetal antigens in somatic cells after interaction with heterologous sperm. Nature (Lond) 257:488–489

Hirao Y and Yanagimachi R (1978a) Effects of various enzymes on the ability of hamster egg plasma membranes to fuse with spermatozoa. Gamete Res 1:3–12

Hirao Y and Yanagimachi R (1978b) Temperature dependence of sperm-egg fusion and post-fusion events in hamster fertilization. J exp Zool 205:433–438

Holt WV and Dott HM (1980) Chemically induced fusion between ram spermatozoa and avian erythrocytes: an ultrastructural study. J Ultrastruct Res 71:311–320

Huang TTF and Yanagimachi R (1985) Inner acrosomal membrane of mammalian spermatozoa: its properties and possible function in fertilization. Am J Anat 174:249–268

Jeyendran RS, van der Ven HH, Kennedy WP, Heath E, Perez-Pelaez M, Sobrero AJ and Zaneveld LJD (1985) Acrosomeless sperm. A cause of primary male infertility. Andrologia 17:31–36

Jones R, Brown CR, von Glos KI and Gaunt J (1985) Development of a maturation antigen on the plasma membrane of rat spermatozoa in the epididymis and its fate during fertilization. Exp Cell Res 156:31–44

Koehler JK (1976) Changes in antigenic site distribution on rabbit spermatozoa after incubation in "capacitating" media. Biol Reprod 15:444–456

Koehler JK, de Curtis I, Stenchever MA and Smith D (1983) Freeze-fracture observations on the interaction of human spermatozoa with zona-free hamster oocytes. In: The Sperm Cell, Ed J Andre, Martinus-Nijhoff, The Hague. pp 94–98

Koehler JK, Smith D and Karp LE (1984) The attachment of acrosome-intact sperm to the surface of zona-free hamster oocytes. Gamete Res 9:197–205

Kohn A (1965) Polykaryocytosis induced by Newcastle disease virus in monolayers of animal cells. Virology 26:228–245

Koprowski H and Croce CM (1973) Fusion of somatic and gametic cells with lysolecithin. In: Methods in Cell Biology, Volume VII. Ed DM Prestcott, Acad Press, London, pp 251–260

Lorton SP and First NL (1979) Hyaluronidase does not disperse the cumulus oophorus surrounding bovine ova. Biol Reprod 21:301–308

Moore HDM and Bedford JM (1978) Ultrastructure of the equatorial segment of hamster spermatozoa during penetration of oocytes. J Ultrastruct Res 62:110–117

Moryan AI, Guay AT and Tulchinsky D (1986) Normal penetration of hamster ova by human spermatozoa with dyskinetic motility. Fert Steril 45:735–736

Nyormoi O, Coon HG and Sinclair JH (1973) Proliferating hybrid cells formed between rat spermatids and an established line of mouse fibroblasts. J Cell Sci 13:863–878

Paulsen CA (1983) Another look at the sperm penetration assay. Fert Steril 40:302–304

Phillips DM and Yanagimachi R (1982) Difference in the manner of association of acrosome-intact and acrosome-reacted hamster spermatozoa with egg microvilli as revealed by scanning electron microscopy. Develop Growth Diff 24:543–552

Phillips SG, Phillips DM, Dev VG, Miller DA, van Diggelen OP and Miller OJ (1976) Spontaneous cell hybridization of somatic cells present in sperm suspensions. Exp Cell Res 98:429–443

Podoplelov II, Popivanov RP, Vulchanov VH, Glynsky IA, Kirov KI, Kryukov VG, Zhivkov SM and Nakov LS (1979) The interaction of human spermatozoa with in vitro cultured somatic cells and the potential of this system for the study of fertilizing capacity of human spermatozoa. Biol et Immunologia Reproductionis 1:12–21

Prasad MRN (1984) The in vitro sperm penetration test: a review. Int J Androl 7:5–22

Quinn P (1979) Failure of human spermatozoa to penetrate zona free mouse and rat ova in vitro. J exp Zool 210:497–506

Rogers BJ (1985) The sperm penetration assay: its usefulness reevaluated. Fert Steril 43:821–840

Rogers BJ, Ueno M and Yanagimachi R (1981) Fertilization by guinea pig spermatozoa requires potassium ions. Biol Reprod 25:639–648

Russell L, Peterson RN and Freund M (1980) On the presence of bridges linking the inner and outer acrosomal membranes of boar spermatozoa. Anat Rec 198:449–459

Saling PM (1981) Involvement of trypsin-like activity in binding of mouse spermatozoa to zonae pellucidae. Proc Natn Acad Sci US 78:6231–6235

Saling PM (1982) Development of the ability to bind to zonae pellucidae during epididymal maturation: reversible immobilization of mouse spermatozoa by lanthanum. Biol Reprod 26:429–436

Schwartz MA and Koehler JK (1979) Alterations in lectin binding to guinea pig spermatoozoa accompanying in utero capacitation and the acrosome reaction. Biol Reprod 21:1295–1307

Shalgi R and Phillips DM (1980a) Mechanics of in vitro fertilization in the hamster. Biol Reprod 23:433–444

Shalgi R and Phillips D (1980b) Mechanics of sperm entry in cycling hamsters. J Ultrastruct Res 71:154–161

Siroky J, Spurva V, Kopency V and Tkadlecok L (1979) Acrosomal removal induces features analogous to fertilization during mouse sperm-fusion with somatic cells. J exp Zool 208:245–254

Smith M, Peterson RN and Russell LD (1983) Penetration of zona-free hamster eggs by boar sperm treated with the ionophore A23187 and inhibition of penetration by antiplasma membrane antibodies. J exp Zool 225:157–160

Soupart P, Anderson ML, Albert DH, Coniglio JG and Repp JE (1979) Accumulation, nature, and possible function of the malachite green affinity material in ejaculated human spermatozoa. Fert Steril 32:450–454

Talbot P and Chacon RS (1982) Ultrastructural observations of binding and membrane fusion between human sperm and zona pellucida-free hamster oocytes. Fert Steril 37:240–248

Teichmann RJ, Fujimoto M and Yanagimachi R (1972) A previously unrecognized material in mammalian spermatozoa as revealed by malachite green and pyronine. Biol Reprod 7:73–81

Teichmann RJ, Cummins JM and Takei GH (1974a) The characterization of a malachite green stainable, glutaraldehyde extractable phospholipid in rabbit spermatozoa. Biol Reprod 10:565–577

Weissenberg R, Eshkol A, Rudak A and Lunenfield B (1983) Inability of round acrosomeless human spermatozoa to penetrate zona-free hamter ova. Archs Androl 11:167–169
Williamson RA, Koehler JK, DianneSmith W and Karp LE (1984) Entry of immotile spermatozoa into zona-free hamster ova. Gamete Res 10:319–325
Wolf DP (1977) Involvement of a trypsin-like activity in sperm penetration of zona-free mouse ova. J exp Zool 199:149–156
Wolf DP and Armstrong PB (1978) Penetration of the zona-free mouse egg by capacitated epididymal sperm: cinematographic observations. Gamete Res 1:39–46
Wolf DP, Inoue M and Stark RA (1976) Penetration of zona-free mouse ova. Biol Reprod 15:213–221
Yanagimachi R (1966) Time and process of sperm penetration into hamster ova in vivo and in vitro. J Reprod Fert 11:359–370
Yanagimachi R (1972) Penetration of guinea-pig spermatozoa into hamster eggs in vitro. J Reprod Fert 28:477–480
Yanagimachi R (1978) Calcium requirement for sperm-egg fusion in mammals. Biol Reprod 19:949–958
Yanagimachi R (1981) Mechanisms of Fertilization in Mammals. In: Fertilization and Embryonic Development In Vitro. Eds L Mastroianni and JD Biggers, Plenum Press, London. pp 81–182
Yanagimachi R (1982) Requirement of extracellular calcium ions for various stages of fertilization and fertilization-related phenomena in the hamster. Gamete Res 5:323–344
Yanagimachi R (1984) Zona-free hamster eggs: their use in assessing fertilizing capacity and examining chromosomes of human spermatozoa. Gamete Res 10:187–232
Yanagimachi R and Noda YD (1970) Physiological changes in the postnuclear cap region of mammalian spermatozoa: a necessary preliminary to the membrane fusion between sperm and egg cells. J Ultrastruct Res 31:486–493
Yanagimachi R, Nicolson GL, Noda YF and Fujimoto M (1973) Electron microscopic observations of the distribution of acidic and anionic residues on hamster spermatozoa and eggs before and during fertilization. J Ultrastruct Res 43:344–353
Yanagimachi R, Miyashiro LH and Yanagimachi H (1980) Reversible inhibition of sperm-egg fusion in the hamster by low pH. Develop Growth Diff 22:281–288
Yanagimachi R, Huang TTF, Fleming AD, Kosower NS and Nicolson GL (1983) Dithiothreitol, a disulphide-reducing agent, inhibits capacitation, acrosome reaction and interaction with eggs by guinea pig spermatozoa. Gamete Res 7:145–154

G. Post-Fusion Events

1. Activation of the Eggs

Fusion of spermatozoa with eggs not only permits provision of the genetic material from the male but also activates various processes in the egg that are essential for zygote development, *e.g.* initiating the second meiotic division, extruding the second polar body and synthesising DNA, as well as those that prevent aneuploidy, *e.g.* release cortical granule contents facilitating vitelline or zona blocks to polyspermy.

Activation of eggs can be brought about in the absence of sperm cells by a variety of stresses (see Gwatkin, 1977), including hyaluronidase (see Section II.D.10.b) and calcium. Activation of eggs induced by the ionophore A23187 occurs in the absence of exogenous calcium ions and may be related to the displacement of Ca^{2+} from egg membranes by the drug (Steinhart et al., 1974).

2. Decondensation of Sperm Chromatin

After fusion the sperm head undergoes swelling and the chromatin is dispersed, a process that requires Ca^{2+} in the hamster (Yanagimachi, 1982). Related events occurring at this time include an early dispersion of arginine-rich nuclear proteins (Kopecny and Pavlok, 1975) and increased nuclear uptake of Giemsa and Toluidine blue stains (Miller and Masui, 1982; Krzanowski, 1982).

Exposure of sulphydryl groups is implicated in these changes in chromatin structure because there is uptake of the sulphydryl group blocker *N*-ethylmaleimide (NEM: Miller and Masui, 1982) and similar effects can be brought about in-vitro by disulphide reducing agents: dispersion of mature sperm chromatin by detergents only occurs after reduction of disulphide bonds (see Mahi and Yanagimachi, 1975). Conversely, when disulphide bonds in sperm are increased in number by ageing (Calvin and Bedford, 1971), chromatin takes longer to decondense (Beil and Graves, 1977). Sperm nuclear proteins are known to be arginine- and cysteine-rich and the sulphydryl groups are oxidised in mature sperm cells (Berry and Meyer, 1960; Coelingh et al., 1972; Marushige and Marushige, 1974). Reducing the disulphide bonds in-vitro increases the uptake of Giemsa and Toluidine blue dyes and labelling with ^{3}H-NEM (Miller and Masui, 1982; Krzanowski, 1982); increases the uptake of actinomycin D (which competes with DNA for nucleoprotein binding sites: Calvin and Bedford, 1974), promotes the Feulgen staining of DNA (Esnault, 1973) and releases template restrictions on DNA (Witkin et al., 1975; Heston et al., 1975; Zirkin et al., 1976).

Although sperm can fuse with eggs at all stages of the egg's development, decondensation of mature sperm chromatin in the ooplasm only occurs after breakdown of the germinal vesicle, whether sperm fuse naturally (Yanagimachi and Usui, 1972; Usui and Yanaginachi, 1976) or sperm heads are injected into the eggs (Thadani, 1979; Perreault et al., 1984). Injected sperm nuclei do not decondense if the eggs are killed (Markert, 1983) or if the reducing power of the egg cytoplasm is decreased (Perreault et al., 1984), demonstrating the importance of the metabolic generation of reducing equivalents by the egg. Eggs with lowered reducing ability, however, can support the decondensation of sperm heads pretreated with disulphide reducing agents before injection (Perreault et al., 1984).

Decondensation of chromatin from sperm incorporated into other cell types is rare (Sawicki and Koprowski, 1971; van Meel and Pearson, 1979), although changes in morpohology (Johnson et al., 1970) and staining (Zelenin et al., 1974) have been observed. DNA synthesis occurs if lysolecithin is the fusinogen (Croce et al., 1972; Gledhill et al., 1972; Koprowski and Croce, 1973) and the expression of foetal antigens within the somatic cells of heterokaryons has been claimed (Higgins et al., 1975).

Hydrolytic enzymes within the chromatin can facilitate the decondensation and remove restrictions to DNA template activity (Zirkin and Chang, 1977), but are not obligatory for these processes (see Zirkin et al., 1985).

3. Formation of the Male Pronucleus

The hamster egg fosters decondensation of sperm from the many different species that can fuse with it (Uehora and Yanagimachi, 1976; see Yanagimachi, 1981) but the form of the pronuceus is dictated by the egg (Imai et al., 1977, 1980). Male pronucleus formation is dependent upon decondensation and is thus prevented if the decondensation of chromatin is inhibited by blocking the sulphydryl groups of the egg with iodoacetate (IAA: Perreault et al., 1984). Decondensation alone, however, is not sufficient stimulus to pronucleus formation, since DTT-treated sperm, which decondense in IAA-treated eggs, in germinal vesicle eggs or in fertilised eggs at the pronuclear stage (Perreault et al., 1984) or the second of two sperm, which decondenses if injected immediately after the first one (Markert, 1983), do not form pronuclei. Clearly, some other factors are required and exogenous calcium is necessary in the hamster (Yanagimachi, 1982).

4. Chromosome Condensation

Fusion between eggs and sperm occur in heterologous inseminations and this has permitted the analysis of human sperm chromosomes (Rudak et al., 1978; Martin et al., 1982, 1983; see Martin, 1984; Yanagmachi, 1984; Chaudhury and Yanagimachi, 1984). Little is known of the requirements for sperm chromatin to condense into chromosomes, but more may be learnt by injecting spermatozoa into Xenopus eggs (Ohsumi et al., 1986).

5. Development in the Epididymis of the Ability of Sperm to Interact with the Vitellus

a. Egg Activation

Orgebin-Crist (1967, 1969) suggested that the reduced fertilising ability of immature rabbit spermatozoa in-vivo could be explained by inadequate activation of eggs. There is also abnormal activation of eggs in-vitro by testicular, caput and corpus spermatozoa from the rabbit (Brackett et al., 1978) and corpus spermatozoa from the human (Mahadevan and Trounson, 1985). Although an activation process occurred after in-vitro insemination, rabbit spermatozoa were absent from the perivitelline space of the activated eggs. If not due to the caffeine that was used in an attempt to improve the motility of the testicular spermatozoa, these observations may suggest that some component emanating from the sperm could activate the eggs parthogenetically, possibely hyaluronidase (see Section II.D.10.b).

b. Chromatin Decondensation

That the chromatin of immature spermatozoa is less condensed than that of mature cells is shown by the nuclear swelling and decondensation of the nuclear material following in-vitro treatment with detergents or heparin alone, in the absence of disulphide reducing agents (Johnson et al., 1970; Calvin and Bedford, 1971; Bedford et al., 1973; Bedford and Calvin, 1974; Marushige and Marushige, 1975a; Meistrich et al., 1975, 1976; Gall and Ohsumi, 1976; Beil and Graves, 1977; Pellicari et al., 1983; Carrenco et al.,1983).

There is biochemical confirmation of the increasing oxidation of sulphydryl groups in maturing spermatozoa: the reduction in binding of ^{3}H-NEM (Miller and Masui, 1982), and ^{14}C-iodoacetamide in rats and rabbits (Calvin et al., 1973; Marushige and Marushige, 1975a; Huang and Nieschlag, 1984) and boars (Tobita et al., 1984) and the characteristics of sulphydryl-containing proteins from mature and immature sperm (Sattasaya and Panyim, 1982).

Changes in Feulgen stainability for DNA occur as sperm mature in the epididymis of rabbits (Bouters, 1967a,b; Porcelli et al., 1982), rams (Esnault, 1973; Esnault and Nicolle, 1976; Nicolle et al., 1985), voles (Clarke et al., 1980), stallions (Bustos-Obregon, 1980) and bulls (Gledhill, 1966; Gledhill et al., 1966a) and are thought to reflect changes in the accessibility of the dye to the chromatin rather than a change in the total amount of DNA, secondary to folding consequent on disulphide bond oxidation (Esnault, 1973).

Conformational changes in the structure of chromatin in maturing sperm indicated by changes in thermal stability (Ringertz et al., 1970) and thought to occur in the caput epididymidis (Gledhill, 1972) were not confirmed when pure suspensions of testicular and epididymal spermatozoa were studied (Gledhill and Amann, 1973). Nevertheless, conformational changes in chromatin structure are demonstrated by the inability of testicular sperm to recognise antibodies raised to non-histone protein antigens exposed in mature spermatozoa (Redi et al., 1979); the greater binding of Actinomycin D (Calvin and Bedford, 1974; Daryzynkiewisz et al., 1969) and dyes (Miller and Masui, 1982; Krzanowski, 1982) to immature spermatozoa; greater resistance of mature sperm nuclei to nuclease (Tobita et al., 1984) or incubation in uterine fluid (Nicolle et al., 1985) and cytofluorimetric measurements of nucleoproteins (Pellicari et al., 1983). The conformational stability may prevent aneuploidy (Redi et al., 1982) or prevent possible damage to the genetic material caused by deformation of the sperm head that occurs as it passes through the zona pellucida (Bedford et al., 1972; Bedford and Calvin, 1974).

c. Pronucleus Formation

When sperm are injected directly into eggs, chromatin from spermatids and from spermatozoa from the testis, caput, corpus and cauda epididymidis decondense (Uehora and Yanagimachi, 1977; Perreault et al., 1984). Male pronuclei are also formed with all cells except those from the caput epididymidis (Uehora and Yanagimachi, 1977.

6. Involvement of the Epididymis in Permitting Post-Fusion Events

Disulphide bonds in immature spermatozoa increase upon ageing them in the caput epididymis or in-vitro to an extent even greater than that normally achieved during maturation in the epididymis (Calvin and Bedford, 1971), suggesting that while this is an age-dependent event, the extent of disulphide bond bridging is controlled by the epididymis: the free sulphydryl content of sperm heads in the epididymis is influenced by the testicular androgens (Huang and Nieschlag, 1984). Sulphydryl oxidase in the male reproductive tract and epididymal fluids, which has increasing activity distally, may act directly or indirectly on sperm in a protective way (Chang and Morton, 1973, 1975; Zirkin and Chang, 1978). Glutathione in the epididymis and sperm, together with glutathione oxidase, glutathione transferase and catalase (see Section III.G.13) may also control the extent of sperm disulphydryl oxidation in-vivo.

7. Summary

The increased oxidation state of sulphydryl groups in sperm chromatin that are found on maturation in the epididymis can be achieved in-vitro by ageing alone. The presence of epididymal enzymes which can control the extent of sulphydryl bond oxidation, however, suggests that the epididymis may control the extent of oxidation.

8. References

Bedford JM and Calvin HI (1974) The occurrence and possible significance of -S-S- cross links in sperm heads, with particular reference to Eutherian mammals. J exp Zool 188:137–156

Bedford JM, Cooper GW and Calvin HI (1972) Post-meiotic changes in the nucleus and membranes of mammalian spermatozoa. In: The Genetics of the Spermatozoon. Eds RA Beatty and S Glucksohn-Waelsh, Bogtrykkeriet Forum, Copenhagen. pp 69–89

Bedford JM, Calvin H and Cooper GW (1973) The maturation of spermatozoa in the human epididymis. J Reprod Fert Suppl 18:199–213

Beil RE and Graves GN (1977) Nuclear decondensation of mammalian spermatozoa changes during maturation and in vitro storage. J exp Zool 202:235–240

Berry RE and Mayer DT (1960) The histone-like basic protein of bovine spermatozoa. Exp Cell Res 20:116–126

Bouters R, Esnault C, Ortavant R and Salisbury GW (1967a) Comparison of DNA revealed by Feulgen and by ultra-violet light in rabbit spermatozoa after storage in the male efferent ducts. Nature (Lond) 213:181–182

Bouters R, Esnault C, Salisbury GW and Ortavant R (1967b) Discrepancies in analyses of deoxyribonucleic acid in rabbit spermatozoa, involving Feulgen staining (Feulgen-DNA) and ultraviolet light absorption (UV-DNA) measurements. J Reprod Fert 14:355–363

Brackett BG, Hall JL and Oh YK (1978) In-vitro fertilizing ability of testicular, epididymal, and ejaculated rabbit spermatozoa. Fert Steril 29:571–582

Bustos-Obregon E (1980) Cytochemical characterization of sperm nuclei during epididymal maturation in mammals. Arch Biol Med Exp 13:335–341

Calvin H and Bedford JM (1971) Formation of disulphide binds in the nucleus and accessory structures of mammalian spermatozoa during maturation in the epididymis. J Reprod Fert Suppl 13:65–75

Calvin HI and Bedford JM (1974) Stimulation of actinomycin D-binding to Eutherian sperm chromatin by reduction of disulphide bonds. J Reprod Fert 36:225–229

Calvin HI, Yu CC and Bedford JM (1973) Effects of epididymal maturation, zinc (II) and copper (II) on the reactive sulphydryl content of structural elements in rat spermatozoa. Exp Cell Res 81:333–341

Carranco A, Reyes R, Magdaleno VM, Huacuja L, Hernandez O, Rosado A, Merchant H and Delgado NM (1983) Heparin-induced decondensation of mammalian epididymal spermatozoa. Archs Androl 10:213–218

Chang TSK and Morton B (1975) Epididymal sulfhydryl oxidase: a sperm-protective enzyme from the male reproductive tract. Biochem Biophys Res Commun 66:309–315

Chang TSK and Morton B (1978) Distribution of sulfhydryl oxidase activity in the rat and hamster male reproductive tract. Biol Reprod 18:745–748

Chaudhury JP and Yanagimachi R (1984) An improved method to visualize human sperm chromosomes using zona-free hamster eggs. Gamete Res 10:233–239

Clarke JR, Esnault C and Nicolle JC (1980) Feulgen-DNA changes in the germ cells of the male vole (Microtus agrestis) during their development. Reprod Nutr Develop 20:183–190

Coelingh JP, Monfoort CH, Rozijn TH, Gevers Leuven JA, Schiphof R, Steyn-Parve EP, Braunitzer G, Schrank B and Ruhfus A (1972) The complete amino acid sequence of the basic nuclear protein of bull spermatozoa. Biochem Biophys Acta 285:1–14

Croce CM, Gledhill BL, Garbara B, Sawicki W and Koprowski H (1972) Lysolecithin-induced fusion of rabbit spermatozoa with hamster somatic cells. Adv Biosci 8:187–200

Darzynkiewicz Z, Gledhill BL and Ringertz NR (1969) Changes in deoxyribonucleoprotein during spermiogenesis in the bull ^{3}H-actinomycin binding capacity. Exp Cell Res 58:435–438

Esnault C (1973) Reactivation of the Feulgen reaction of ram spermatozoa by dithiothreitol. J Reprod Fert 32:153–157

Esnault C and Nicolle JC (1976) Evolution de l'ADN et des proteins nucleaires basiques au cours de la maturation des cellules germinales du bélier. Etude microspectrophotometrique. Ann Histochem 21:189–197

Gall WE and Ohsumi Y (1976) Decondensation of sperm nuclei in vitro. Exp Cell Res 102:349–358

Gledhill BL (1966) Studies on the DNA content, dry mass and optical area of bull spermatozoal heads during epididymal maturation. Acta Vet Scand 7:131–142

Gledhill BL (1972) Further studies on the nuclear chromatin of morphologically abnormal bull spermatozoa. J Reprod Fert 29:431–472

Gledhill BL and Amann RP (1973) Stability of spermatozoal deoxyribonucleoprotein during passage through the bovine epididymis. J Reprod Fert 33:531–534

Gledhill BL, Gledhill MP, Rigler R and Ringertz NR (1966a) Changes in deoxyribonucleoprotein during spermatozogenesis in the bull. Exp Cell Res 41:652–665

Gledhill BL, Sawicki W, Croce CM and Koprowski H (1972) DNA synthesis in rabbit spermatozoa after treatment with lysolecithin and fusion with somatic cells. Exp Cell Res 73:33–40

Gwatkin RBL (1977) Fertilization mechanisms in man and mammals. Plenum Press, New York

Heston WDW, Zirkin BR and Coffey DS (1975) Release of chromatin template restriction in rabbit spermatozoa. Biochem Biophys Res Commun 64:162–168

Higgins PJ, Borenfreund E and Bendich A (1975) Appearance of foetal antigens in somatic cells after interaction with heterologous sperm. Nature (Lond) 257:488–489

Huang HFS and Nieschlag E (1984) Alteration of free sulphydryl content of rat sperm heads by suppression of intratesticular testosterone. J Reprod Fert 70:31–38

Imai H, Niwa K and Iritani A (1977) Penetration in vitro of zona-free hamster eggs by ejaculated boar spermatozoa. J Reprod Fert 51:495–497

Imai H, Niwa K and Iritani A (1980) Ultrastructural observations of boar spermatozoa penetrating zona-free hamster eggs. Biol Reprod 23:481–486

Johnson RT, Rao PN and Hughes HD (1970) Mammalian cell fusion. III. A HeLa cell inducer of premature chromosome condensation active in cells from a variety of animal species. J Cell Physiol 76:151–158

Kopecny V and Pavlok A (1975) Autoradiographic study of mouse spermatozoan arginine-rich nuclear protein in fertilization. J exp Zool 191:85–95

Koprowski H and Croce CM (1973) Fusion of somatic and gametic cells with lysolecithin. In: Methods in Cell Biology. Ed DM Prestcott, Acad Press, London, Volume VII, pp 251–260

Krzanowska H (1982) Toluidene blue staining reveals changes in chromatin stabilization of mouse spermatozoa during epididymal maturation and penetration of ova. J Reprod Fert 64:97–101

Mahadevan MM and Trounson AO (1985) Removal of the cumulus oophorus from the human oocyte for in vitro fertilization. Fert Steril 43:263–267

Mahi CA and Yanagimachi R (1975) Induction of nuclear decondensation of mammalian spermatozoa in vitro. J Reprod Fert 44:293–296

Markert CL (1983) Fertilization of mammalian eggs by sperm injection. J Exp Zool 228:195–201

Martin RH (1984) Comparison of chromosomal abnormalities in hamster egg and human sperm pronuclei. Biol Reprod 31:819–825

Martin RH, Lin CC, Balkan W and Burns K (1982) Direct chromosomal analysis of human spermatozoa: preliminary results from 18 normal men. Am J Human Genet 34:459–468

Martin RH, Balkan W, Burns K, Rademaker AW, Lin CC and Rudd NL (1983) The chromosome constitution of 1000 human spermatozoa. Human Genetics 63:305–309

Marushige Y and Marushige K (1974) Properties of chromatin isolated from bull spermatozoa. Biochem Biophys Acta 340:498–508

Marushige Y and Marushige K (1975) Transformation of sperm histone during formation and maturation of rat spermatozoa. J Biol Chem 250:39–45

Meistrich ML, Hughes TJ and Bruce WRT (1975) Alteration of epididymal sperm transaport and maturation in mice by oestrogen and testosterone. Nature (Lond) 258:145–147

Meistrich ML, Reid BO and Barcellona WJ (1976) Changes in sperm nuclei during spermiogenesis and epididymal maturation. Exp Cell Res 99:72–78s

Miller MA and Masui Y (1982) Changes in the stainability and sulphydryl level in the sperm nucleus during sperm-oocyte interaction in mice. Gamete Res 5:167–179

Nicolle JC, Fournier-Delpech S and Courot M (1985) Influence of uterine secretions on the chromatin of ram spermatozoa at different stages of maturation: cytofluorimetric study of Feulgen-DNA after in vitro incubation. Gamete Res 11:321–328

Ohsumi K, Katagiri C and Yanangimachi R (1986) Development of pronuclei from human spermatozoa injected microsurgically into frog (Xenopus) eggs. J exp Zool 237:319–325

Orgebin-Crist MC (1967) Maturation of spermatozoa in the rabbit epididymis: fertilizing ability and embryonic mortality in does inseminated with epididymal spermatozoa. Ann biol Anim Biochem Biophys 7:373–389

Orgebin-Crist MC (1969) Studies on the function of the epididymis. Biol Reprod Suppl 1:155–175

Pellicciari C, Hosokawa Y, Fukuda M and Manfredu Romani MG (1983) Cytofluorimetric study of nuclear sulphydryl and disulphide groups during sperm maturation in the mouse. J Reprod Fert 68:371–376

Perreault SD, Wolff RA and Zirkin BR (1984) The role of disulfide bond reduction during mammalian nuclear decondensation in vivo. Develop Biol 101:160–167

Porcelli F, Cozzi B and Ferrandi B (1982) Cytochemical behaviour of rabbit spermatozoa chromatin in male and female genital tracts. Bas Appl Histochem 26:289–302

Redi CA, Garagna S, Merani MS, Capanna E, Bianchi NO and Romanini M (1982) Microdensitometric evaluation of the DNA content, as ploidy parameters of spermatozoa in the polymorphic chromosomal system of akodon molinae cabrera (Rodentia, Cricetidao). Gamete Res 5:345–354

Ringertz NR, Gledhill BL and Darzynkiewicz Z (1970) Changes in deoxyribonucleoprotein during spermiogenesis in the bull. Sensitivity to heat denaturation. Exp Cell Res 62:204–218

Rudak E, Jacons PA and Yanagimachi R (1978) Direct analysis of the chromosome constitution on human spermatozoa. Nature (Lond) 274:911–913

Sattayasai N and Panyim S (1982) Nature of the proteins which form disulfide bonds during the maturation of rat spermatozoa. Int J Androl 5:337–344

Sawicki W and Koprowski H (1971) Fusion of rabbit spermatozoa with somatic cells cultivated in vitro. Exp Cell Res 66:145–151

Steinhardt RA, Epel D, Carroll EJ and Yanagimachi R (1974) Is calcium ionophore a universal activator for unfertilized eggs? Nature (Lond) 252:41–43

Thadani YM (1979) Injection of sperm heads into immature rat oocytes. J exp Zool 210:161–168

Tobita T, Tanaka H, Tanaka K, Tanaka T, Kojim S and Nakano M (1984) Characteristics of cysteinyl sulfhydryls and nuclease sensitiviy of boar sperm nuclei during epididymal maturation. Biochem Int 9:161–168

Tung KSK, Yanagimachi H and Yanagimachi R (1982) Sperm autoantigens and fertilization. III. Ultrastructural localization of guinea pig autoantigens. Anat Rec 202:241–253

Uehora T and Yanagimachi R (1976) Microsurgical injection of spermatozoa into hamster eggs with subsequent transformation of sperm nuclei into male pronuclei. Biol Reprod 15:467–470

Uehora T and Yanagimachi R (1977) Behaviour of nuclei of testicular, caput and cauda epididymal spermatozoa injected into hamster eggs. Biol Reprod 16:315–321

Usui N and Yanagimachi R (1976) Behaviour of hamster sperm nuclei incorporated into eggs at various stages of maturation, fertilization, and early development. The appearance and disappearance of factors involved in sperm chromatin decondensation in egg cytoplasm. J Ultrastruct Res 57:276–288

van Meel FCM and Pearson PL (1979) Do human spermatozoa reactivate in the cytoplasm of somatic cells? J Cell Sci 35:105–122

Witkin SS, Korngold GC and Bendich A (1975) Ribonuclease-sensitive DNA-synthesising complex in human sperm heads and seminal fluid. Proc Natn Acad Sci US 72:3295–3299

Yanagimachi R (1981) Mechanisms of Fertilization in Mammals. In: Fertilization and Embryonic Development In Vitro. Eds L Mastroianni and JD Biggers, Plenum Press, London. pp 81–182

Yanagimachi R (1982) Requirement of extracellular calcium ions for various stages of fertilization and fertilization-related phenomena in the hamster. Gamete Res 5:323–344

Yanagimachi R (1984) Zona-free hamster eggs: their use in assessing fertilizing capacity and examining chromosomes of human spermatozoa. Gamete Res 10:187–232

Yanagimachi R and Usui N (1972) The appearance and disappearance of factors involved in sperm chromatin decondensation in the hamster egg. J Cell Biol 55:293a

Zelenin AV, Shapiro IM, Kolesnkov VA and Senin VM (1974) Physico-chemical properties of chromatin of mouse sperm nuclei in heterkaryons with chinese hamster cells. Cell Diff 3:95–101

Zirkin BR and Chang TSK (1977) Involvement of endogenous proteolytic activity in thiol-induced release of DNA template restrictions in rabbit sperm nuclei. Biol Reprod 17:131–137

Zirkin BR, Boisin A, Heston WDW and Coffey DS (1976) Release of DNA template restrictions in rabbit spermatozoa and rat liver nuclei. J exp Zool 197:283–288

Zirkin BR, Soucek DA, Chang TSK and Perreault S (1985) In vitro and in vivo studies of mammalian sperm nuclear decondensation. Gamete Res 11:349–365

Part III. Function of the Epididymis and Its Secretory Products

A. Epididymal Structure and Function

1. Introduction

The involvement of the epididymis in the process of sperm maturation appears to be concerned with enabling gamete contact to occur (see Parts I,II) and the more terminal events of fertilisation may not require epididymal input (Glover and Nicander, 1971). The organ has evolved along with sperm maturation (Glover and Nicander, 1971) and the need for capacitation (Bedford, 1983), although spermatozoa at all stages of their maturation require conditions compatible with their survival.

The actions of the epididymis in influencing spermatozoa is thought to be mediated by the fluid surrounding the cells (see Jones and Glover, 1975), direct contact with the epithelial cell microvilli observed in micrographs (Hamilton, 1975) being perhaps coincidental but not essential to their maturation. The composition of the fluid varies along the length of the duct as a consequence of modification of the prevaling distal current of testicular exocrine secretion, by the processes of secretion into, and resorption from, the lumen across, and mediated by, the epithelium. Fluid in the tail of the epididymis (see Jones and Glover, 1975; Jones, 1978) is more likely to influence the maintenance of sperm fertilising ability (sperm storage) than sperm maturation, which occurs more proximally (see Part I).

Sperm maturation and sperm storage are clearly different processes in that the latter shows a greater susceptibility to inhibition by progestins in hamsters (Lubicz-Nawrocki, 1973b) and the deleterious effects of abdominal temperature in the cryptepididymal state in rats (Bedford, 1978b; Wong et al., 1982; Foldesy and Bedford, 1982) and rabbits (Bedford, 1977; Warner and Nienenhuis, 1983). Indeed, evidence has been presented for the scrotal position being an adaptation to epididymal, rather than testicular, requirements (Glover and Nicander, 1971; Bedford, 1977, 1978a), although this has been challenged (Stowe, 1981). The epididymis may assist in the descent of the testis into the scrotum (Minniberg and Schlossberg, 1983) and lower scrotal temperatures (see Carrick and Setchell, 1977) may reduce damage to sperm cells caused by lipid peroxidation (Alvarez and Storey, 1985).

The androgen-dependence of epididymal structure has long been recognised (see Orgebin-Crist et al., 1975) and explains the androgen requirements of sperm maturation. Certain specific epididymal secretion products known to influence events in fertilisation are known to be androgen-dependent, *e.g.* AEG and other sperm-coating proteins (see Section III.G.8) and carnitine (see Section III.E), whereas non-specific contributions may provide the conditions for long-term storage of sperm cells. The following section deals with the ways in which the epididymis elaborates and accumulates these secretions.

2. Epididymal Structure

Differences in the structure of the epididymis between species have been known for a long time (see Benoit, 1926; Hamilton, 1972) but three anatomical areas are grossly discernible in most species: a major "head" (caput) receiving spermatozoa and fluid from the testicular efferent ducts – and incorporating them in some species (Cooper and Jackson 1974) – and a narrower "body" (corpus or isthmus) which connects the head to a globose "tail" (cauda) which stores sperm before ejaculation, (see Hinton et al., 1979, for gross structure). The primate epididymis differs from this pattern (see Amann et al., 1976) and the epididymis of a sub-human primate is the best model for that of man (Moore and Pryor, 1981).

Many different epididymal regions can be discerned by teasing tubules apart along natural planes of cleavage which separate regions of the tubule with different diameters. In this way up to 18 regions can be recognised in the whole guinea pig epididymis (Fawcett and Hollenberg, 1963; Hoffer and Greenberg, 1978), 5 in the proximal part of the mouse epididymis (Abou-Haila and Fain-Maurel, 1984) and at least 3 within the "initial segment" of rats (Hoffer and Fawcett, 1979; Nicander et al., 1983). Histological appearance has suggested fewer regions: 3 in the dog (Chandler et al., 1981); 5 in the mouse (Takano et al., 1980; Sorenzo et al., 1982); 6 in the rat (Reid and Cleland, 1957; Zunke and Gossrau, 1981); horse, ram and bull (Nicander, 1957b), 7 in the guinea pig (Hoffer and Greenberg, 1978) and 8 in rabbits (Nicander, 1957a) and man (Holstein, 1969). Ultrastructural observations of the main cell types (see Section III.A.6), however, differentiate only 3 regions in the rat (Hamilton, 1975), rabbit (Nicander and Ploen, 1979) and hamster (Nicander and Glover, 1973; Flickinger et al., 1978), and no distinct regions were discernible in the monkey epididymis by Ramos and Dym (1977) or Moore and Pryor (1981).

A simpler terminology introduced by Glover and Nicander (1971), based on a likely evolutionary development of function, has not received universal acceptance. Three regions only were envisaged – an initial, a middle and a terminal segment – each distinct histologically but not synonymous with caput, corpus and cauda. The initial and middle segments were considered to be involved with sperm maturation and the terminal segment with sperm storage. While this relationship with histology holds well for the hamster (Nicander and Glover, 1973) and testicondi (Jones and Djakiew, 1978) and the initial segment of the epididymis is accepted as structurally unique in many species, more than one region within it can be made out (Fawcett and Hoffer, 1979; Nicander et al., 1983). Clearly, the terminology can not be applied to epididymides for which the "fertility profile" (see Part I) is not established, and the rabbit corpus, known to be the region that supports sperm maturation (and therefore by definition the "middle" segment) was later classified as terminal segment (Nicander and Glover, 1973), because under certain experimental conditions the region was capable of maintaining the fertilising capacity of sperm precociously matured there. It is not applicable to the guinea pig epididymis (Hoffer and Greenberg, 1978) and there is no relationship whatever with protein synthetic activity (Brooks, 1981).

Variation in tubule and epithelial dimensions for all mammalian species include a decrease in cell and microvillus height and an increase in tubule diameter and muscle layer as the storage region of the organ is approached, but this is less pronounced in primates (Moore and Pryor, 1981).

3. Blood Supply

The epididymis receives a dual blood supply, from the internal spermatic artery supplying the head and body, and from the deferential artery that supplies mainly the epididymal tail (see Gunn and Gould, 1975). The capital and vasal arteries anastomose in rats, mice, rabbits (see Clavert et al., 1981; Chubb and Desjardins, 1982) and man (Harrison and Barclay, 1948; Harrison, 1949; MacMillan, 1954a) and as a consequence the vasal artery can maintain the structure (MacMillan, 1953, 1954b) and function (Richard, 1929) of the cauda if the capital artery is occluded. Similarly, occluding the deferential artery causes focal damage to the cauda but proximal regions are only affected when stagnation of the luminal contents in the affected region causes rupture of the proximal canal (MacMillan, 1953).

The composition of the blood supplying the head and body of the epididymis may differ from that arriving from the systemic circulation because of the possibility of counter-current movement of substances from the spermatic veins to the spermatic artery within the pampiniform plexus (see Free, 1976, 1981; Free and Tillson, 1976). Steroids found in high concentration in the testicular venous effluent (see Section III.B.2.b), are most likely to be transported in this fashion, although the significance for the epididymal circulation has been questioned (Free and Jaffe, 1978). Anastomoses between veins and artery, established in bulls (Hees et al., 1984; Noordhuizen-Stassen et al., 1985) and rats (Weerasooriya and Yamamoto, 1985), occur near the branching of the epididymal artery and would enable even large molecules to be refluxed into the arterial circulation. An anastomosis between veins draining the cauda epididymidis and the testis was only observed in the rabbit (Chubb and Desjardins, 1982), a species in which drainage in the abdomen is not uniform (Cameron et al., 1983).

The microvasculature of the epididymis has been studied in rats (Kormano, 1968), man (Kormano and Reijonen, 1976) and mice (Suzuki, 1982; Abe et al., 1984). The capillary density is highest in the initial segment of the epididymis in mice (Takano, 1980; Abe et al., 1984) and rabbits (Clavert et al., 1981), the regions where blood flow is highest in rats, rams (Setchell et al., 1964; Waites et al., 1973; Jaffe and Free, 1979) and rabbits (Brown and Waites, 1972). In the rabbit the epididymal epithelial surface area and tubule perimeter are also greater in the area of high capillary density so that the ratio between the two is constant along the length of the duct (Clavert et al., 1981). In the mouse the capillary arrangement in this region of the epididymis resembles that in the cauda and vas deferens, with centripetal vessels stemming from a peritubular vascularisation, whereas in the efferent ducts, distal caput and corpus, a testicular arrangement is present (Suzuki, 1982). Fenestrated capillaries are found solely in the initial segment of the mouse epididymis (Suzuki, 1982; Abe et al., 1984) but in the rat they are found in the caput (Hamilton, 1975) as well as in other regions (Hoffer and Hinton, 1984).

The extent of the capillary network in the initial segment is lost on androgen deprivation caused by bilateral castration, but unilateral castration reveals changes only on the ipsilateral side (Clavert et al., 1981). Loss of capillary fenestrations (Abe et al., 1984) and blood flow (Brown and Waites, 1972) follow ligation of the efferent ducts, suggesting a local control of vascularisation.

4. Lymphatics

Major lymphatics drain the caput, corpus and cauda epididymidis of rats (Pérez-Clavier et al., 1982) and rabbits (Clavert et al. 1981). There are wide intertubular lymphatic sinusoids in the mouse (Suzuki, 1982) and rat (Hamilton, 1975; Kazeem, 1983) and lymphocytes are found in the epididymis (Dym and Romrell, 1975; Hoffer et al., 1973; Miller and Killian, 1983). In man these are more numerous in the cauda than more proximal regions (Wang and Holstein, 1983) and they have been shown to be T-lymphocytes; those in the epithelium ("suppressor/cytotoxic") differing from those in the interstitium ("helper/inducer": Ritchie et al., 1984). The differentiation of these cells may be related to supressing influences from luminal sperm antigens; after vasectomy spermatozoa escaped from the duct are found in regional lymph nodes in rams and boars (Ball and Setchell, 1983) and man (Ball et al., 1982).

5. Methods of Study

The techniques of renal physiology has been successfully applied to examination of epididymal function with micropuncture (Howards et al., 1975; Hinton and Howards, 1982b) providing information on the composition of fluid in various regions (see Turner, 1979; Hinton, 1980). Obtaining luminal fluid from the cauda epididymidis of domestic animals at slaughter by cannulation of the vas and application of pressure (Czarnetsky and Henle, 1938) has been applied to smaller animals (Henle, 1938; Brooks et al., 1974). The collection of fluid from living animals *via* fistulae or cannulae has been accomplished in boars (Wierzbowski and Wierschos, 1969; Johnson and Pursell, 1975; Einarsson and Funqvist, 1979), bulls (Bennett and Rowson, 1963; Amann et al., 1963), rams (White et al., 1959; Voglmayr et al., 1977) and rats (Gilmore et al., 1971) and fluid has been collected from the vas and cauda of anaesthetised rabbits (Jones and Glover, 1973a), rats, guinea pigs and hamsters (Back et al., 1974; Back and Shenton, 1975).

Less is known about the composition of human epididymal fluid. Most material is obtained from a small segment of the vas deferens removed at vasectomy and this contains lower concentrations of organic compounds and higher concentration of ions to support osmolality (Brooks and Howards, 1979; Setchell and Hinton 1981; Hinton and Setchell, 1981; Hinton et al., 1981).

These fluid collection techniqes have proved invaluable in studying steady-state conditions existing inside the tubules at the time of fluid colection. More dynamic studies can be done if tracers are injected into the bloodstream of the animals before collection of fluid and if epididymal luminal contents are replaced by solutions of known composition perfused continuously through the lumen or under conditions of stop-flow. The luminally perfused preparation is a sensitive one and detects changes in the resorption of sodium and water after treatment with the anti-androgen cyproterone acetate (Wong et al., 1978, 1979), the antifertility agents α-chlorohydrin (Wong et al., 1977) and 6-chloro-6-deoxyglucose (Wong et al., 1980) and adrenalectomy (Au et al., 1978), when no differences in the composition of epididymal fluid could be discerned with cyproterone acetate (Back et al., 1977; Jones, 1977), α-chlorohydrin (Back et al., 1975) or aldosterone antagonists (Wong and Lee, 1982).

To the familiar techniques of incubation of tissue slices, minces and homogenates can now be added the study of isolated cells (see III.A.6.e), explants (Öfner et al., 1979), tubule culture (Orgebin-Crist and Tichenor, 1972) and perifusion (Orgebin-Crist and Ménézo, 1980; Klinefelter and Hamilton, 1984), the latter techniques permitting study of the interrelationships between epididymal secretions and spermatozoa in-situ.

6. Epithelial Cells

a. Cell Types

The epithelial cell types of the pseudostratified epithelium lining the epididymal lumen comprise (with percentages in the rat epididymis (Yeung and Cooper, 1982) in parentheses): principal (chief) cells (59%), clear (light) cells (10%), basal cells (27%). Wandering (halo) cells or leucocytes (Hoffer et al., 1973) are now thought to be lymphocytes (Dym and Romrell, 1975; Miller and Killian, 1983) and in man have been identified as T lymphocytes (Ritchie et al., 1984; see Section III.A.4). Macrophages are present in the epididymal epithelium of man (Wang and Holstein, 1983) and bull (Goyal, 1985). Cells with apically situated nuclei include apical cells in the caput (see below) and narrow cells (Sun and Flickinger, 1980)

b. Principal Cells

Principal cells are the major transporting cells, mediating secretion and resorption across the epithelium and can remove proteins from the lumen into multivesicular bodies (Friend et al., 1969; Moore and Bedford, 1979a,b; Yeung and Cooper, 1982). Evidence for secretory processes is slim (Hoffer et al., 1973; Nicander and Malmqvist, 1977; Nicander and Ploen, 1979), although the long time that sperm spend in the duct may obviate the need for high rates of secretion and their attendant morphological expression. However, experimental radioautography at the electron micrograph level has shown that the normal route of transfer of amino acids and sugars from the endoplasmic reticulum to the Golgi occurs in the epididymis (see Flickinger et al., 1984; Flickinger, 1985). An endocrine role for principal cells was postulated to account for the presence of granules in the basal part of these cells close to the blood vessels (Ramos and Dym, 1977; Ramos, 1980; Prakash et al., 1980; Moore and Pryor, 1981; Abe et al., 1983b; Goyal, 1985) but the only strong evidence for an endocrine role is for steroids (see Section III.B.4) and secretion of lipid soluble products would not necessarily be evident from micrographs.

c. Clear Cells

Clear cells are no longer thought to be holocrine in nature (Martan, 1969) and while they have been considered to be secretory cells (Prakash et al., 1979; Anand Kumar et al., 1980) and are capable of synthesisng protein (Dadoune et al., 1985), they are demonstrably resorptive (Moore and Bedford, 1979a; Cooper and Yeung, 1980; Yeung and Cooper, 1982). They lack true microvilli (Hamilton et al., 1977) and their membranes differ from those of principal cells (Brown and Montesano, 1981). The nature of the cell's vacuolar contents, lipid in the caput and glycoprotein in the cauda (Kreth, 1965; Anand Kumar et al., 1980), may reflect the composition of luminal fluid in these regions. The clear cells of the rat epididymis accumulate lipid-like inclusions when luminal spermatozoa degenerate under the influence of a variety of experimental conditions (Flickinger and Loving, 1976; Flickinger, 1977a,b, 1978; Flickinger and Charlton, 1978). Apical cells in the caput serve the same absorptive function as clear cells in that region (Moore and Bedford, 1979a) and both cell types contain carbonic anhydrase (Cohen et al., 1976). In species lacking clear cells (rabbits) some specialised principal cells contain vacuoles resembling those of clear cells (Jones et al., 1979a) and this may indicate a resorptive role for these cells.

d. Ultrastructure

The ultrastructure of cells in the epididymal epithelium has been presented in man (Holstein, 1969), monkey (Ramos and Dym, 1977, Ramos, 1980; Moore and Pryor, 1981), rat (Hamilton, 1975; Moore and Bedford, 1979a,b), rabbit (Nicander 1965; Nicander and Ploen, 1979; Jones et al., 1979a,b), guinea pig (Hoffer and Karnovsky, 1981; Greenberg and Forssman, 1983a,b), hamster (Nicander and Glover, 1973; Flickinger et al., 1978), mouse (Abe et al., 1983), ram (Nicander, 1979) and bull (Goyal, 1985).

e. Epididymal Cell Culture

Unit gravity sedimentation (Killian et al., 1976, 1977; Killian, 1981) and cell elutriation (Klinefelter and Amann, 1980; Killian and Chapman, 1980) have been employed for isolation of epithelial cells from the epididymis. They have been examined morphologically (Olson et al., 1982, 1983; Wagley et al., 1984; Byers et al., 1985a) and used for the study of cellular enzymes (Kemp and Killian, 1978; Chapman and Killian, 1984) and oxygen uptake (Brooks, 1975) as well as synthesis of DNA (Kierzenbaum et al., 1981: Olson et al., 1983; Byers et al., 1985a), synthesis and uptake of protein (Byers et al., 1985b; see Section III.G), synthesis of glycerophosphocholine (see Section III.D), the uptake of carnitine (see Section III.E) and conversion of androgens (see Section III.B).

7. Testicular Control of Epididymal Function

The androgen-dependence of epididymal weight, histological appearance and biochemical content is well known (see Maneely, 1959) and the steroid influence on synthetic and transporting capacities of the epididymis is documented in the pertinent sections of Part III. The ability of injected androgens to maintain most epididymal functions indicates that circulating androgens from the testis mediate this control.

Greater amounts of injected androgens are required to return the epididymal weight towards pre-castration values than those required for other accessory organs (Cavazos and Melampy, 1956), in part because of the influence of with drawal of testicular fluid on the initial segment (see below). Rajalakshmi et al. (1976) used this as a rationale basis for the design of a contraceptive for males, arguing that depletion of androgen would first affect those organs with a high threshold for androgens without affecting other accessory organs. However, the reversible fertility caused by cyproterone acetate in rats observed by Prasad et al. (1970) could not be repeated (Elger and von-Berswordt-Wallrabe, 1973; Neumann and Schenk, 1976; Chatterjee et al., 1977) and other workers found the accessory organs to be more sensitive than the epididymis to androgen depletion (Back et al., 1977; Rastogi, 1979). Despite claimed changes in the epididymal epithelial cells after this treatment (Rajalakshmi et al., 1971) no difference in the composition of epididymal fluid from the cauda was observed in rat and rabbits (Back et al., 1977; Jones, 1977), although the luminally perfused rat preparation detected a reduction in sodium and water resorption after cyproterone acetate administration (Wong et al., 1978, 1979).

The initial segment of the epididymis receives testicular fluid directly into its lumen and its structure is dependent on it, for this region involutes ipsilaterally after unilateral castration or ligation of the efferent ducts in mice (Elliot, 1965), goats (Goyal, 1983), hamsters (Moniem et al., 1978), bulls (Gustafsson, 1966), rabbits (Danzo et al., 1977) and rats (Reijonen et al., 1975; Fawcett and Hoffer, 1979; Nicander et al., 1983). The reduced tissue mass after this operation may explain the subsequent decrease in blood flow (Brown and Waites, 1972) and metabolism (Elliot, 1965), but reduced protein synthetic biosynthetic capacity in rats (Jones et al., 1980a), rabbits (Jones et al., 1981), rams (Jones et al., 1982), biochemical content (Brooks, 1979) and loss of fenestrated capillaries (Abe et al., 1984) may reflect loss of specific components of testicular fluid.

Because the structure of the initial segment is maintained when flow of sperm-free testicular fluid is maintained in X-irradiated mice (Abe et al., 1983), the fluid rather than the sperm in it were considered important for maintaining the tissue.

Further, because caput epididymal weight is partially restored when progesterone maintains the testosterone concentration of rete testis fluid of hypophysectomised rats (Harris and Bartke, 1975), luminal steroids in particular may be important. The failure of supraphysiological concentrations of circulating testosterone to maintain the initial segment of efferent duct-ligated rats (Fawcett and Hoffer, 1979) may reflect the loss of ABP (see Section III.B) and the concomitant failure of the steroid to reach the lumen in adequate concentration, or decreased conversion to an active metabolite, *e.g.* DHT (Aafjes and Vreeburg, 1972; see Section III.B.3.c).

Ligating the testicular efferent ducts increases the uptake of labelled macroglobulin by cells of the proximal caput epididymidis by lowering the amount of endogenous macroglobulin arriving from the testis which normally saturate the receptors mediating intracellular uptake (Djakiew et al., 1985).

The mitotic rate of epithelial cells in the adult epididymis is low in rats (Messier and Leblond, 1960; Clermont and Flannery, 1970; Tuohimaa and Niemi, 1974) and rabbits (Pabst and Schick, 1979) but is higher in the initial segment of hamsters (Nagy and Edmunds, 1975), rabbits, boars and rams (Nicander, 1957a,b, 1965, 1967). Cell proliferation in the epididymis is under androgenic control (Niemi and Tuohimaa, 1971; Tuohimaa and Niemi, 1974), but the physiological relevance of other factors such as growth factor in rete testis fluid (Brown et al., 1982a,b) or mitogens produced by Sertoli cells (Feig et al., 1980, 1983) has not been elucidated.

8. Permeability of the Epididymal Epithelium

a. Anatomical Evidence

As in other exocrine glands, the composition of fluid in the epididymal lumen differs from that of blood plasma (see Tables 11, 12, 13, 15, 16, 17 and 21), and this is maintained not only by active secretion and resorption (see Section III.B-G), but also by a barrier residing in the epithelium at the level of tight junctions between the cells. These effectively block the transfer of lanthanum in the capillaries from passing between the cells (Friend and Gilula, 1972; Moore and Bedford, 1979b; Hoffer and Hinton, 1984). Freeze-fracture studies have revealed the tight junctions to consist of rows of intramembranous particles parallel to the surface in the rat (Friend and Gilula, 1972) and these are more extensive in the cauda than the corpus or caput in the rat (Suzuki and Nagano, 1978). In the guinea pig epididymis tight junctional strands are more extensive in the proximal regions (Greenberg and Forssman, 1983b) and in the monkey epididymis, the strands of particles are not parallel and they do not exclude lanthanum from the junction (Cavicchia 1979). Gap junctions in the epididymal epithelium of the rat (Friend and Gilula, 1972) and mouse (Revel et al., 1971) may provide electrotonic coupling between adjacent cells.

b. Physiological Evidence

Tight junctions are presumably effective in-situ against the entry of large molecules gaining access to the lumen *via* an intercellular route since the evidence from short-term experiments indicates that macromolecules (usually inulin, Mol Wt 5200) injected into the bloodstream or placed outside epididymal tubules in-vitro are excluded from the tubule lumen (see Hinton, 1985; Turner and Howards, 1985: Table 10). Androgen binding protein (see Section III.B.3.c), although present in the circulation (Gunsalas et al., 1978a,b, 1980) is not specifically accumulated by the epididymis (Danzo and Eller, 1984). There seems to be rather greater movement of labelled inulin across monolayers of polarised caput epididymal epithelial cells than occurs in-vivo (Byers et al., 1986).

Slow transport of molecules across the epithelium could occur over the period of time that sperm spend in the epididymal duct; this would not be detected in acute experiments and some recent circumstantial evidence suggests that proteins can move across the epithelium: (a) although there was no evidence from radioautographic studies that albumin entered the epididymis to a greater extent than that anticipated to arrive *via* the efferent ducts (Mancini et al., 1965), Aragon et al. (1972) reported that horse radish peroxidase (HRP) could enter the lumen of the mouse epididymis within 15 minutes; however, reports by the same workers of the movement of HRP to the seminiferous tubules could not be substantiated (Dym, 1973); (b) specific proteins from the hamster epididymis that promote binding to eggs when added to corpus sperm in-vitro, also promote sperm-egg binding by immature sperm contained in corpus tubules in culture when added to culture medium (González Echevería et al., 1984); (c) antibodies to specific rat epididymal secretory proteins can inhibit the fertilising capacity of mature rat spermatozoa contained in cultured caudal tubules when added to the medium (Cuasnicú et al., 1984c); (d) active immunisation of male rats against a specific epididymal prealbumin involved in sperm-egg binding leads to a reduced fertility (Fournier-Delpech et al., 1985); (e) IgG molecules enter the rabbit cauda by 6 weeks after active immunisation (Bedford et al., 1982; Weininger et al., 1982). While the permeability of the cultured tubule epithelium remains unchecked with established markers and it is known that unravelling the rat corpus tubule is responsible for gross leakage (Cooper et al., 1985) the suggestion of protein transfer across the epithelium must remain unproven. Once this is clarified the nature of the epididymal extract that conferred fertilising ability upon immature spermatozoa contained in rabbit corpus tubules in culture (see Part I) will become clearer.

The epididymal epithelium appears to be freely permeable to small molecules, and transports against a concentration gradient compounds found in high concentration within the lumen (Table 10). Lysine can also enter the perfused lumen rapidly (Fig. 4; Dacheux, 1980) and to a greater extent than non-metabolised amino acids or sugars. Saturable carriers appear to mediate the transport of inositol, amino acids and glucose (Hinton and Howards, 1982b; Cooper et al., 1985) and there is stereo-specific transport of glucose (Hinton and Howards, 1982b) and carnitine (Yeung et al., 1980). Of lipid soluble molecules cholesterol never entered the lumen, but no evidence that it ever left the bloodstream was presented. Other steroids were influenced by the nature and concentration of protein in the epididymal lumen (Cooper and Waites, 1979b, 1981; Cooper, 1980).

Regional transport of molecules into the caput epididymal lumen is greater than that into the corpus or cauda for inositol, carnitine and α-aminoisobutyrate (Brooks et al., 1973; Hinton et al., 1979; Yeung et al., 1980; Setchell and Hinton, 1981; Hinton and Howards, 1982b). This is unlikely to be due to the higher blood flow in that region (see Section III.A.3) because there is no greater uptake of 3-*O*-methylglucose in the caput in-vivo (Hinton and Howards, 1982b) and there is higher uptake of carnitine into epididymal tubules from the caput in-vitro (Johansen and Bøhmer, 1979), revealing inherent differences in transport by the organ.

c. Anti-luminal Transport

This has not received much attention: pyroantimonate in the lumen is visibly excluded from the tight junctions in the rat (Friend and Gilula, 1972) and inulin introduced into the epididymal lumen is not found in epididymal tissue or deferential vein blood (Wong and Yeung, 1978) or removed from the cauda of rats (Wong and Yeung, 1978; Cooper, 1982) or hamsters (Turner et al., 1983) after perfusion through the lumen. Carnitine is removed from the lumen in proximal parts of the epididymis (Hinton and Hernandez, 1985), although the significance of this in view of the inward-transporting activities of the epithelium (see Section III.E) remains to be established.

The large cellular traffic of recycling membranes expected to mediate the extensive resorption of proteins from the lumen between caput and cauda (see Steinman et al., 1983; Turner, 1984) would ensure a continuous flux of vesicles that could transport molecules from the blood stream in the other direction. The HRP placed in the lumen which becomes associated with the Golgi apparatus of principal cells (Yeung and Cooper, 1982) could be evidence for this recycling. One might expect transfer of proteins from the epididymal lumen to the bloodstream to be a normal function of the epithelium if inhibin in testicular fluid is resorbed from this site (LeLannou and Chambon, 1977a,b,c; LeLannou et al., 1979; Lipner and Rush, 1981), although this premise has been challenged (Walton et al., 1978; Blanc et al., 1978; Davies et al., 1979).

There is little loss of small water-soluble molecules placed in the caudal lumen (inositol, 3-*O*-methylglucose: Cooper, 1982; Turner et al., 1983) but testosterone perfused through the lumen of the rat cauda in-vivo does enter deferential vein blood as dihydrotestosterone (Fig. 5. see Section III.B.4) and steroid infused into the lumen of the canine vas deferens also enters the hind leg circulation (Boulanger et al., 1983).

d. Androgen Dependence

The effectiveness of tight junctions in sealing the intercellular space between epithelial cells is lost on castration (Moore and Bedford, 1979b) and this may explain the greater amount of albumin in rat epididymal tissue (Cameo and

Blaquier, 1976; Brooks and Higgins, 1980) and rabbit epididymal fluid (Jones and Dott, 1980) in androgen-deficient animals.

9. Resorptive Activity of the Epididymis

The absorptive nature of the epididymis is at once evident from the increased packed cell volume or sperm cell concentration of luminal fluid in various parts of the tract (Table 14), since sperm cells are an ideal volume marker, not being metabolised, being too large to move between epithelial cells and – apart from small regions at the termination of the vas deferens in rats (Cooper and Hamilton, 1977a), cats (Murakami et al., 1984a,b) and rabbits (Murakami et al., 1985) – not being phagocytosed to any extent by the epithelium (see Bedford, 1975; Turner, 1984). Calculations reveal that over 90% of fluid from the testis is resorbed in the efferent ducts and caput in the bull and boar (Crabo, 1965) and rat (Turner, 1984). That fluid resorption occurs is also supported by the observation that the time taken for the testis to swell after ligation of the caput epididymidis is longer than when the ligatures are placed on the efferent ducts (Young, 1933; van Wagenen, 1924). Such resorptive activity will also influence the concentrations of compounds within the epididymal lumen (Setchell and Hinton, 1981; Turner, 1984).

Early studies revealed the resorptive nature of the epididymis by the removal of luminal compounds (*e.g.* dyes) entering the epididymal lumen after passing through the seminiferous tubules (Wagenseil, 1928; Young, 1933; Mason and Shaver, 1952) or of particulate matter injected into the rete testis or efferent ducts and subsequently found within epithelial cells of the efferent ducts or caput of rats (Shaver, 1954; Macmillan, 1957; English, 1979), rabbits (Nicander, 1965), hamsters (Burgos et al., 1959; Burgos, 1960, 1964; Sedar, 1966; Montiorzi and Burgos, 1967) and rams (Gunn, 1936).

The efferent ducts and proximal regions of the rat caput display different activities towards dyes and India ink (Mason and Shaver, 1952) and whereas non-specific uptake into the cells of the proximal caput epididymidis of rabbits was noted, cells of the initial segment and distal cauda were selective in their uptake (Nicander and Ploen, 1979). Recent work has indicated that non-ciliated cells of the efferent ducts display both absorptive and fluid-phase endocytosis (Morales and Hermo, 1983; Hermo and Morales, 1984; Byers et al., 1985b) and fluid-phase and receptor-mediated mechanisms have been identified in the rat caput (Djakiew et al., 1984, 1985). In both cases the markers passed from apical vesicles and tubules to vacuoles to multivesicular bodies and eventually to lysozomes in the basal part of the cells. Labelled albumin placed in the lumen is also taken up into lysosomes of epididymal by cells (Mayorga and Bertini, 1983).

Whether the transfer of particulate material to macrophages in the epididymal interstitium of the rat (Shaver, 1954) and rabbit (Richard, 1929, 1930) after injection into the lumen was caused by damage done to the tissue at the time of injection is not clear, but water soluble compounds placed in the lumen of the tail of the epididymis were later found in the urine of anaesthetised rabbits in the latter study and transepithelial transport of steroids from the lumen of the epididymis has been confirmed (see Section III.B.4).

An intercellular route of chanelling of water from the lumen *via* the formation of iso- or hypo-tonic resorbates has been postulated from the rat caput and cauda (Wong et al., 1978, 1979) and distended intercellular spaces have been observed in caudal tubules transporting water out of the lumen in-vitro (YC Wong et al., 1978) and in-vivo (Cooper and Yeung, 1980; Yeung and Cooper, 1982). Few intercellular spaces are distended in-situ (Hamilton, 1975) but here, rates of resorption may be lower than those stimulated experimentally by perfusion of unusually high concentrations of sodium through the lumen. An intracellular vacuolar route of fluid transport has been proposed for the hamster (Nicander and Glover, 1973; Flickinger et al., 1978).

10. Secretory Activity of the Epididymis

The composition of epididymal fluid reflects not only the resorptive activity of the epididymis but the elaboration of specific substances in it. The maturation of spermatozoa may be dependent on their sequential interaction with epididymal secretions accumulated in different regions of the epididymis. The changing composition of epididymal fluid are documented in the remaining parts of the review: steroids (Section III.B), ions (Section III.C), glycerophosphocholine (Section III.D), carnitine (Section III.E), *myo*-inositol (Section III.F) and proteins (Section III.G).

11. Summary

The epithelium lining the lumen through which the sperm cells pass consists of a variety of cell types, the main functions of which are the resorption and modification of testicular fluid entering the epididymis and the elaboration of epididymal secretions. Resorption of fluid increases the concentration of sperm cells and that of secreted impermeant compounds. The interaction of the secretions (proteins, carnitine) with spermatozoa are thought to modulate the function of spermatozoa.

12. References

Aafjes JH and Vreeburg JTM (1972) Distribution of 5α-dihydrotestosterone in the epididymis of bull and boar, and its concentration in rat epididymis after ligation of efferent testicular ducts, castration and unilateral gonadectomy. J Endocr 53:85–93

Abe K, Takano H and Ito K (1983a) Response of epididymal duct to the temporary depletion of spermatozoa induced by testicular irradiation in mice. Anat Record 207:17–24

Abe K, Takano H and Ito T (1983b) Ultrastructure of the mouse epididymal duct with special reference to the regional differences of the principal cells. Arch Histol Japonicum 46:51–68

Abe K, Takano H and Ito T (1984) Microvasculature of the mouse epididymis, with special reference to fenestrated capillaries localized in the initial segment. Anat Record 209:209–218

Abou-Haila A, Fain-Maurel MA (1984) Regional differences of the proximal part of mouse epididymis: morphological and histochemical characteristics. Anat Record 209:197–200

Alvarez JG and Storey BT (1985) Spontaneous lipid peroxidation in rabbit and mouse epididymal spermatozoa: dependence of rate on temperature and oxygen concentration. Biol Reprod 32:342–351

Amann RP, Hokanson JF and Almquist JO (1963) Cannulation of the bovine ductus deferens for quantitative recovery of epididymal spermatozoa. J Reprod Fert 6:65–69

Amann RP, Johnson L, Thompson DL and Pickett BW (1976) Daily spermatozoal production, epididymal spermatozoal reserves and transit times of spermatozoa through the epididymis of the rhesus monkey. Biol Reprod 15:586–592

Anand Kumar TC, Prakash A and Prasad MRN (1980) Ultrastructure of the clear cell lining the epididymal lumen of the rat. Anat Anz (Jena) 147:220–230

Aragon JA, Lustig L and Mancini RE (1972) Uptake of horseradish peroxidase by the testis and epididymis of mice. I. Light-microscopical study. J Reprod Fert 28:299–302

Au CL, Ngai HK, Yeung CH and Wong PYD (1978) Effect of adrenalectomy and hormone replacement on sodium and water transport in the perfused rat cauda epididymis. J Endocr 77:265–266

Back DJ and Shenton JC (1975) A comparison of the composition of epididymal plasma from the cauda epididymidis of the rat, hamster and guinea-pig. Experentia 31:464–465

Back DJ, Shenton JC and Glover TD (1974) The composition of epididymal plasma from the cauda epididymidis of the rat. J Reprod Fert 40:211–214

Back DJ, Glover TD, Shenton JC and Boyd GP (1975) The effect of α-chlorohydrin on the composition of rat and rabbit epididymal plasma. A possible explanation of species differences. J Reprod Fert 45:117–128

Back DJ, Glover TD, Shenton JC and Boyd CP (1977) Some effects of cyproterone acetate on the reproductive physiology of the male rat. J Reprod Fert 49:237–243

Ball RY and Setchell BP (1983) The passage of spermatozoa to regional lymph nodes in testicular lymph following vasectomy in rams and boars. J Reprod Fert 68:145–153

Ball RY, Naylor CPE and Mitchinson MJ (1982) Spermatozoa in an abdominal lymph node after vasectomy in a man. J Reprod Fert 66:715–716

Bedford JM (1975) Maturation, transport, and fate of spermatozoa in the epididymis. In: Handbook of Physiology, Section VII Endocrinology, Volume V, Male Reproductive System. Eds RO Greep and DW Hamilton, American Physiol Soc, Washington DC. pp 303–317

Bedford JM (1977) Evolution of the scrotum: the epididymis as the prime mover? In: Reproduction and Evolution. Eds JH Calaby and CH Tyndale-Biscoe, Aust Acad Sci. pp 171–182

Bedford JM (1978a) Anatomical evidence for the epididymis as the prime mover in the evolution of the scrotum. Am J Anat 152:483–508

Bedford JM (1978b) Influence of abdominal temperature on epididymal function in the rat and rabbit. Am J Anat 152:509–522

Bedford JM (1983) Significance of the need for sperm capacitation before fertilization in Eutherian mammals. Biol Reprod 28:108–120

Bedford JM, Weininger RB, Fisher S and Rifkin J (1982) Specific IgG levels appearing in the cauda epididymidis of male rabbits. Int J Androl Suppl 5:48–52

Bennett JP and Rowson LEA (1963) A fistula for the collection of epididymal semen from the bull. J Reprod Fert 6:61–64

Benoit MM (1926) Recherches anatomiques, cytologiques et histophysiologiques sur les vois excretices du testicle chez les mammifères. Contribution a l'étude de quelques problemes de cytologie génerale relatifs à la cellule glandulaire. Arch Anat Histol Embryol (Strasbourg) 5:173–412

Blanc MR, Dacheux JL, Cahoreau C, Courot M, Hochereau de Reviers MT, Lacroix A and Pisselet C (1978) The role of testicular fluid on blood plasma levels of FSH and LH in the ram. Ann Biol Anim Biochem Biophys 19:1027–1032

Boulanger P, de Saulniers M, Bleau G, Roberts KD and Chapdelaine A (1983) Sex steroid concentrations in plasma from the canine deferential vein. J Endocr 96:223–228

Brooks DE (1975) The isolation of epithelial cells from the rat epididymis. Andrologia 7:241–253

Brooks DE (1979a) Biochemical Environment of maturing spermatozoa. In: The Spermatozoon. Eds DW Fawcett and JM Bedford, Urban and Schwarzenberg, Baltimore. pp 23–34
Brooks DE (1979b) Influence of testicular secretion on tissue weight and on metabolic and enzyme activities in the epididymis of the rat. J Endocr 82:305–313
Brooks DE (1981) Secretion of proteins and glycoproteins by the rat epididymis: regional differences, androgen-dependence, and effects of protease inhibitors, procaine, and tunicamycin. Biol Reprod 25:1099–1117
Brooks DE and Higgins SJ (1980) Characterization and androgen-dependence of proteins associated wih luminal fluid and spermatozoa in the rat epididymis. J Reprod Fert 59:363–375
Brooks DE and Howards SS (1979) cited in Brooks DE (1979a)
Brooks DE, Hamilton DW and Mallek AH (1973) The uptake of L-(*methyl*-^{3}H)carnitine by the rat epididymis. Biochem Biophys Res Commun 52:1354–1360
Brooks DE, Hamilton DW and Mallek AH (1974) Carnitine and glycerylphosphorylcholine in the reproductive tract of the rat. J Reprod Fert 36:141–146
Brown PDC and Waites GMH (1972) Regional blood flow in the epididymis of the rat and rabbit: effect of efferent duct ligation and orchidectomy. J Reprod Fert 28:221–223
Brown D and Montesano R (1981) Membrane specialization in the rat epididymis. II. The clear cell. Anat Rec 201:477–483
Brown KD, Blakeley DM, Henville A and Setchell BP (1982a) Rete testis fluid contains a growth factor for cultures of fibroblasts. Biochem Biophys Res Commun 105:391–397
Brown KD, Henville A and Setchell BP (1982b) A growth factor in rete testis fluid. Ann New York Acad Sci US 383:428–429
Burgos M (1960) The role of amorphous cellular coatings in active transport. Anat Rec 137: 171a
Burgos M (1964) Uptake of colloidal particles by cells of the caput epididymidis. Anat Rec 148:517–523
Burgos MH, de Oca HM and Montorzi NM (1959) La incorporcion de Fe^{59} en los conductos efferentes del hamster. Rev Soc Argent Biol 35:36–39
Byers SW, Djakiew D and Dym M (1985a) Structural features of rat epididymal epithelial cells in vitro. J Reprod Fert 75:401–411
Byers SW, Musto NA and Dym M (1985b) Culture of ciliated and nonciliated cells from rat ductuli efferentes. J Androl 6:271–178
Byers SW, Hadley MA, Djakiew D and Dym M (1986) Growth and characterization of polarized monolayers of epididymal epithelial cells and Sertoli cells in dual environment culture chambers. J Androl 7:59–68
Cameo MS and Blaquier JA (1976) Androgen-controlled specific proteins in rat epididymis. J Endocr 69:47–55
Cameron DF, Snydle FE and Smith LC (1983) Three patterns of extra-testicular venous drainage in the rabbit. Anat Embryol 167:103–112.
Carrick FW and Setchell BP (1977) The evolution of the scrotum. In: Reproduction and evolution. Eds JH Calaky and CH Tyndale-Biscoe, Aust Acad Sci. pp 165–170
Cavazos LF and Melampy RM (1956) Effects of differential testosterone propionate levels on rat accessory gland activity. Iowa State College Journal of Science 31:19–24
Cavicchia JC (1979) Fine structure of the monkey epididymis: a correlated thin-section and freeze-fracture study. Cell Tiss Res 201:451–458
Chandler JA, Sinowatz F and Pierrepoint CG (1981) The ultrastructure of dog epididymis. Urol Res 9:33–44
Chapman DA and Killian GK (1984) Glycosidase activities in principal cells, basal cells, fibroblasts and spermatozoa isolated from the rat epididymis. Biol Reprod 31:627–636
Chatterjee A, Pal AK, Ray P and Basu R (1977) Cyproterone acetate. II. Failure to reduce sperm fertility in the surgically separated epididymis of rat. Contraception 15:429–434
Chubb C and Desjardins C (1982) Vasculature of the mouse, rat, and rabbit testis-epididymis. Am J Anat 165:357–372
Clavert A, Cranz C and Brun B (1981) Epididymal vascularization and microvascularization. In: Epididymis and Fertility: Biology and Pathology. Eds C Bollack and A Clavert, S Karger, London. Progr Reprod Biol 8:48–57

Clermont Y and Flannery J (1970) Mitotic activity in the epithelium of the epididymis in young and old adult rats. Biol Reprod 3:283–292

Cohen JP, Hoffer AP and Rosen S (1976) Carbonic anhydrase localization in the epididymis and testis of the rat: histochemical and biochemical analysis. Biol Reprod 14:339–346

Cooper ERA and Jackson H (1977) A histological study of the post-testicular duct system in a variety of mammals. Private Publication, Manchester University Press.

Cooper TG (1982) Secretion of inositol and glucose by the perfused rat cauda epididymidis. J Reprod Fert 64:373–379

Cooper TG and Hamilton DW (1977) Phagocytosis of spermatozoa in the terminal region and gland of the vas deferens of the rat. Am J Anat 150:247–268

Cooper TG and Waites GMH (1979) Factors affecting the entry of testosterone into the lumen of the cauda epididymidis of the anaesthetized rat. J Reprod Fert 56:165–174

Cooper TG and Waites GMH (1980) Role of proteins in the transport of steroids into the epididymis. In: Testicular Development, Structure, and Function. Eds A Steinberger and E Steinberger, Raven Press, New York. pp 299–304

Cooper TG and Yeung CH (1980) Epithelial structure of the rat cauda epididymidis after luminal perfusion. Int J Androl 3:361–374

Cooper TG, Yeung CH, Lui W and Yang CZ (1985) Luminal secretion of myo-inositol by the rat epididymis perfused in vitro. J Reprod Fert 74:135–144

Crabo B (1965) Studies on the composition of epididymal content in bulls and boars. Acta Vet Scand Vol 6, Suppl 5

Cuasnicú PS, Gonzáles Echeverría F, Piazza AD, Cameo MS and Blaquier JA (1984) Antibodies against epididymal glycoproteins block fertilizing ability in rat. J Reprod Fert 72:467–471

Czarnetzky EJ and Henle W (1938) A pressure device for the separation of mammalian spermatozoa from the isolated epididymis. Proc Soc exp Biol Med 38:63–64

Dacheux J (1980) An in vitro luminal perfusion technique to study epididymal secretion. IRCS Med Sci 8:137

Danzo BJ and Eller BC (1984) Clearance, metabolic fate and tissue distribution of an injected bolus of photoaffinity-labeled rat androgen binding protein. Biol Reprod 31:259–270

Danzo BJ, Cooper TG and Orgebin-Crist MC (1977) Androgen binding protein (ABP) in fluid colected from the rete testis and cauda epididymidis of sexually mature and immature rabbits and observations on morphological changes in the epididymis following ligation of the ductuli efferentes. Biol Reprod 17:64–77

Dadoune JP, Fain-Maurel MA and Baures M (1985) Autoradiographic study of labeled amino-acid incorporation into clear cells of mouse epididymis. Archs Androl 14:199–207

Davies RV, Main SJ and Setchell BP (1979) Inhibin in ram rete testis fluid. J Reprod Fert Suppl 26:87–95

Djakiew D and Jones RC (1982) Stereological analysis of the epididymis of the echidna, Tachyglossus acilleatus and Wistar rat. Aust J Zool 30:865–875

Djakiew D, Byers SW and Dym M (1984) Receptor-mediated endocytosis of alpha2-macroglobulin and transferrin in rat caput epithelial cells in vitro. Biol Reprod 31:1073–1085

Djakiew D, Byers SW, Lewin DM and Dym M (1985) Receptor-mediated endocytosis of alpha2-macroglobulin by principal cells in the proximal caput epididymidis in vivo. J Androl 6:190–196

Dym M (1973) The fine structure of the monkey (Macaca) Sertoli cell and its role in establishing the blood-testis barrier. Anat Rec 175:639–65675

Dym M and Romrell LJ (1975) Intraepithelial lymphocytes in the male reproductive tract of rats and rhesus monkeys. J Reprod Fert 42:1–7

Einarsson S and Funkquist B (1979) Surgical technique for establishment of ductus deferens fistula in the boar. Int J Androl 2:581–584

Elger W and von Berswordt-Wallrabe R (1973) Failure to induce sterility in male rats with continously released microquantities of cyproterone acetate and norgesterol. Acta Endocrinol 173:120 (abst)

Elliott PR (1965) The effect of efferentiectomy and orchidectomy on the metabolism of the epididymis of the mouse. J Cell Comp Physiol 66:293–302

English HF (1979) Absorption in the ductuli efferentes testis of the rat. Anat Rec 193:531 (abst)

Fawcett DW and Hoffer AP (1979) Failure of exogenous androgen to prevent regression of the initial segments of the rat epididymis after efferent duct ligation or orchidectomy. Biol Reprod 20:162–181

Fawcett DW and Hollenberg RD (1963) Changes in the acrosome of guinea-pig spermatozoa during passage through the epididymis. Z Zellforsh Mikr Anat 60:276–292

Feig LA, Bellve AR, Erickson NH and Klagsbrun (1980) Sertoli cells contain a mitogenic polypeptide. Proc Natn Acad Sci US 77:4774–4778

Feig LA, Klagsbrun and Bellve AR (1983) Mitogenic poplypeptide of the mammalian seminiferous epithelium: biochemical characterization and partial purification. J Cell Biol 97:1435–1443

Flickinger CJ (1977a) Effects of clomiphene on the structure of the testis, epididymis and sex accessory glands of the rat. Am J Anat 149:533–562

Flickinger CJ (1977b) The influence of progestin and androgen on the fine structure of the male reproductive tract of the rat. II. Epididymis and sex accessory organs. Anat Rec 187:431–462

Flickinger CJ (1978) Effects of testosterone enanthate on the structure of the male reproductive tract of the rat. Anat Rec 192:555–584

Flickinger CJ (1985) Radioautographic analysis of the secretory pathway for glycoproteins in principal cells of the mouse epididymis exposed to (^{3}H)fucose. Biol Reprod 32:377–389

Flickinger CJ and Charlton JS (1978) Ultrastructure of the testis and different regions of the epididymis in normal and vasectomised hamsters. In: Vasectomy, Immunologic and Pathophysiologic Effects in Animals and Man. Eds IH Lepow and R Crozier, Acad Press, New York. pp 25–51

Flickinger CJ and Loving CK (1976) Fine structure of the testis and epididymis of rats treated with cyproterone acetate. Am J Anat 146:359–384

Flickinger CJ, Howards SS and English HF (1978) Ultrastrucural differences in efferent ducts and several regions of the epididymis of the hamster. Am J Anat 152:557–586

Flickinger CJ, Wilson KM and Gray HA (1984) The secretory pathway in the mouse epididymis as shown by electron microscope radioautography of principal cells exposed to monensin. Anat Rec 210:435–448

Foldesy RG and Bedford JM (1982) Biology of the scrotum.I. Temperature and androgen as determinants of the sperm storage capacity of the rat cauda epididymidis. Biol Reprod 26:673–682

Fournier-Delpech S, Courot M and Dubois MP (1985) Decreased fertility and motility of spermatozoa from rats immunized with a prealbumin epididymal specific glycoprotein. J Androl 6:246–250

Free MJ (1976) Blood supply to the testis and its role in local exchange and transport of hormones in the male reproductive tract. In: The Testis, Volume 4. Eds AD Johnson and WR Gomes, Acad Press, New York. pp 39–90

Free MJ and Tillson SA (1976) Local increase in concentration of steroids by venous arterial transfer in the pampiniform plexus. In: The Hormonal Regulation of Spermatogenesis. Eds FS French, V Hansson EM Ritzen and SN Nayfeh, Plenum Press, New York. pp 181–194

Free MJ and Jaffe RA (1978) Target organs for testosterone transferred from vein to artery in the pampiniform plexus: the epididymis. Biol Reprod 18:639–642

Friend DS (1969) Cytochemical staining of multivesicular bodies and Golgi vesicles. J Cell Biol 41:269–279

Friend DS and Gilula, NB (1972) Variations in tight and gap junctions in mammalian tissues. J Cell Biol 53:758–776

Gilmore DP, Hooker RH and Chang MC (1971) A technique for the collection by fistula of epididymal spermatozoa from the rat. J Reprod Fert 25:137–140

Glover TD and Nicander L (1971) Some aspects of structure and function in the mammalian epididymis. J Reprod Fert Suppl 13:39–50

González Echeverría F, Cuasnicú PS, Piazza A, Piñeiro L and Blaquier JA (1984) Addition of an androgen-free epididymal protein extract increases the ability of immature hamster spermatozoa to fertilize in vivo and in vitro. J Reprod Fert 71:432–437

Goyal HO (1983) Histoquantitative effets of orchiectomy with and without testosterone enanthate treatment on the bovine epididymis. Am J Vet Res 44:1085–1090

Goyal HO (1985) Morphology of the bovine epididymis. Am J Anat 172:155–172
Greenberg J and Forssman WG (1983a) Studies of the guinea pig epididymis. I. Ultrastructure and quantitative morphology of the prinicipal cells. Anat and Embryol 168:173–194
Greenberg J and Forssmann WG (1983b) Studies of the guinea pig epididymis. II. Intercellular junctions of principal cells. Anat and Embryol 168:195–209
Gunn RMC (1936) Fertility in sheep. Artificial production of seminal ejaculation and the characteristics of the spermatozoa contained therein. CSIRO Bulletin 94, Commonwealth of Australia.
Gunn SA and Gould TC (1975) Vasculature of the testes and adnexia. In: Handbook of Physiology, Section VII Endocrinology, Volume 5 Male Reproductive System, Eds DW Hamilton and RO Greep, American Physiological Society, Washington DC. pp 117–142
Gunsalas GL, Musto NA and Bardin CW (1978a) Immunoassay of androgen binding protein in blood: a new approach for study of the seminiferous tubule. Science 200:65–66
Gunsalas GL, Musto NA and Bardin CW (1978b) Factors affecting blood levels of androgen binding protein in the rat. Int J Androl Suppl 2:482–493
Gunsalas GL, Musto NA and Bardin CW (1980) Bidirectional releases of a Sertoli cell product, androgen binding protein, into the blood and seminiferous tubule. In: Testicular Development, Structure, and Function. Eds A Steinberger and E Steinberger, Raven Press, New York. pp 291–297
Gustafsson B (1966) Luminal contents of the bovine epididymis under conditions of reduced spermatogenesis, luminal blockage and certain sperm abnormalities. Acta Vet Scand Suppl 17.
Hamilton DW (1972) The Mammalian Epididymis. In: Reproductive Biology. Eds H Balin and S Glasser, Excerpta Medica, Amsterdam. pp 268–337
Hamilton DW (1975) Structure and function of the epithelium lining the ductuli efferentes, ductus epididymidis, and ductus deferens in the rat. In: Handbook of Physiology, Section VII Endocrinology, Volume 5 Male Reproductive System. Eds DW Hamilton and RO Greep, Amer Physiol Soc, Washington DC. pp 259–301
Hamilton DW, Olson GE and Cooper TG (1977) Regional variation in the surface morphology of the epithelium of the rat ductuli efferentes, ductus epididymidis and vas deferens. Anat Rec 188:13–28
Harris ME and Bartke A (1975) Maintenance of rete testis fluid testosterone and dihydrotestosterone levels by pregnenolone and other C_{21} steroids in hypophysectomised rats. Endocrinology 96:1396–1402
Harrison RG (1949) The distribution of the vasal and cremasteric arteries to the testis and their functional importance. J Anat 83:267–280
Harrison RG and Barclay AE (1948) The distribution of the testicular artery (internal spermatic artery) to the human testis. Brit J Urol 20:57–66
Hees H, Leiser R, Kohler T and Wrobel KH (1984) Vascular morphology of the bovine spermatic cord and testis. I. Light- and scanning electron miroscopic studies on the testicular artery and pampiniform plexus. Cell Tiss Res 237:31–38
Henle W (1938) The specificity of some mammalian spermatozoa. J Immunol 34:325–336
Hermo L and Morales C (1984) Endocytosis in non-ciliated epithelial cells of the ductuli efferentes in the rat. Am J Anat 171:59–74
Hinton BT (1980) The epididymal microenvironment. A site of attack for a male contraceptive? Invest Urol 18:1–10
Hinton BT (1985) The Blood-Epididymis Barrier. In: Male Fertility and its Regulation. Eds TJ Lobl and ESE Hafez, MTP, Press, Lancaster. pp 371–382
Hinton BT and Hernandez H (1985) Selective luminal absorption of L-carnitine from the proximal regions of the rat epididymis. Possible relationship to development of sperm motility. J Androl 6:300–305
Hinton BT and Howards SS (1982a) Rat testis and epididymis can transport (^{3}H)3-*O*-methyl-D-glucose, (^{3}H)inositol and (^{3}H)α-aminoisobutyric acid across its epithelia in vivo. Biol Reprod 27:1181–1189
Hinton BT and Howards SS (1982b) Micropuncture and microperfusion techniques for the study of testicular physiology. Ann NY Acad Sci US 383:29–43

Hinton BT and Setchell BP (1981) Micropuncture and microanalytical studies of Rhesus monkey and baboon epididymis and the human ductus deferens. Am J Primatol 1:251–256

Hinton BT, Snoswell AM and Setchell BP (1979) The concentration of carnitine in the luminal fluid of the testis and epididymis of the rat and some other mammals. J Reprod Fert 56:105–111

Hinton BT, Pryor JP, Hirsch AV and Setchell BP (1981) The concentration of some inorganic ions and organic compounds in luminal fluid of the human ductus deferens. Int J Androl 4:457–461

Hoffer AP and Greenberg JA (1978) The structure of the epididymis, efferent ductules and ductus deferens of the guinea pig: a light microscopical study. Anat Rec 190:659–678

Hoffer AP and Hinton BT (1984) Morphological evidence for a blood-epididymis barrier and the effects of gossypol on its integrity. Biol Reprod 30:991–1004

Hoffer AP and Karnovsky MC (1981) Studies on zonation in the epididymis of the guinea pig. I. Ultrastructural and biochemical analysis of the zone rich in large lipid droplets (Zone II). Anat Rec 201:623–633

Hoffer AP, Hamilton DW and Fawcett DW (1973) The ultrastructure of the principal cells and intraepithelial leucocytes in the initial segment of the rat epididymis. Anat Rec 175:169–202

Holstein AF (1969) Morphologische Studien am Nebenhoden des Menschen. In: Zwanglose Abhandlungen aus dem Gebiet der normalen und pathologischen Anatomie, Eds W Bargman and W Doerr, Georg Thieme Verlag, Stuttgart. pp 1–91

Howards SS, Johnson A and Jessee S (1975) Micropuncture and microanalytical studies of the rat testis and epididymis. Fert Steril 26:13–19

Jaffe RA and Free MJ (1979) The determination of blood flow in reproductive organs. In: Animal Models for Research on Contraception and Fertility. Ed NJ Alexander, Harper and Row, Hagerstown. pp 496–509

Johansen L and Bøhmer T (1979) Uptake of ^{3}H-L-carnitine by isolated rat epididymal tubules. Archs Androl 2:117–121

Johnson LA and Pursell VG (1975) Cannulation of ductus deferens of the boar: a surgical technique. Am J Vet Res 36:315–317

Jones R (1977) Effects of testosterone, testosterone metabolites and anti-androgens on the function of the male accessory glands in the rabbit and rat. J Endocr 74:75–78

Jones R (1978) Comparative biochemistry of mammalian epididymal plasma. Comp Biochem Biophys 61B:365–370

Jones RC and Djakiew D (1978) The role of the excurrent ducts from the testes of testicond mammals. Austral Zool 20:201–210

Jones R and Dott HM (1980) Changes in luminal plasma and disappearance of spermatozoa from the ligated cauda epididymidis of the androgen-deficient rabbit. J Reprod Fert 60:65–72

Jones R and Glover TD (1973) The collection and composition of epididymal plasma from the cauda epididymidis of the rabbit. J Reprod Fert 34:395–403

Jones R and Glover TD (1975) Interrelationships between spermatozoa, the epididymis and epididymal plasma. In: The Biology of the Male Gamete. Eds JG Duckett and PM Racey, Suppl No 1 to the Biol J Linnaen Soc 7:367–384

Jones R, Hamilton DW and Fawcett DW (1979a) Morphology of the epithelium of the extra-testicular rete testis, ductuli efferentes and ductus epididymidis of the adult nmale rabbit. Am J Anat 156:373–400

Jones R, Hamilton DW and Fawcett DW (1979b) Some effects of testosterone, 5α-dihydrotestosterone, and 5α-androstane-3α,17β-diol on the morphology of the rabbit epididymis and prostate. J Ultrastruct Res 67:194–208

Jones R, von Glos KI and Brown CR (1980) The synthesis of a sperm-coating protein in the initial segment of the rat epididymis is stimulated by factors in testicular fluid. IRCS Med Sci 8:56

Jones R, von Glos KI and Brown CR (1981) Characterization of hormonally regulated secretory proteins from the caput epididymidis of the rabbit. Biochem J 196:105–114

Jones R, Fournier-Delpech S and Willadson SA (1982) Identification of androgen-dependent proteins synthesised in vitro by the ram epididymis. Reprod Nutr Develop 22:495–504

Kazeem AA (1983) The assessment of epididymal lymphatics within the concept of immunologically privileged sites. Lymphology 16:168–173

Kemp WR and Killian GJ (1978) Glycosidase activity in epididymal epithelial cells isolated from normal and α-chlorohydrin treated male rats. Contraception 17:93–101

Kierszenbaum AL, Lea O, Petrusz P, French FS and Tres LL (1981) Isolation, culture, and immunocytochemical characterization of epididymal epithelial cells from pubertal and adult rats. Proc Natn Acad Sci US 78:1675–1679

Killian GJ (1981) A rapid and simple method for isolating principal cells from the rat caput epididymidis. J Androl 2:255–258

Killian GJ and Chapman DA (1980) Glycerylphosphorylcholine, sialic acid and protein in epithelial cells isolated from the rat caput epididymidis by elutriation. Biol Reprod 22:846–850

Killian GJ, Amann RP and Snyder J (1976) Isolation of principal and basal cells from the epithelium of the hamster caput epididymidis by unit gravity sedimentation. Biol Reprod 15:266–279

Killian GJ, Snyder J and Amann RP (1977) Unit gravity sedimentation separation of cells comprising the caput epididymidis of the rat. Cell Tiss Res 183:371–378

Klinefelter GR and Amann RP (1980) Metabolism of testosterone by principal cells and basal cells isolated from the rat epididymal epithelium. Biol Reprod 22:1149–1154

Klinefelter GR and Hamilton DW (1984) Organ culture of rat caput epididymal tubules in a perifusion chamber. J Androl 5:243–258

Kormano M (1968) Microvasculature structure of the rat epididymis. Annls Med exp Biol Fenn 46:113–118

Kormano M and Reijonen K (1976) Microvasculature structure of the human epididymis. Am J Anat 145:23–32

Kreth HW (1965) Ueber die hellen Zellen im Ductus Epididymidis der Ratte. Z Zellforsch 68:28–42

le Lannou D and Chambon Y (1977a) Effets de la ligature des canaux efferentes ou de l'épididdidyme, sur la fonction testiculaire et la secretion des gonadotrophines chez le rat. Comptes Rendus Soc de Biol 171:197–200

le Lannou D and Chambon Y (1977b) Presence dans l'épididyme d'un facteur abaissant fortement le taux sanguin de FSH chez le rat. Comptes Rendus Soc de Biol 171:636–638

le Lannou D and Chambon Y (1977c) Presence dans l'épididyme d'un facteur inhibant la synthese hypothalamique de FSH-RH chez le rat. Comptes Rendus Soc de Biol 171:1064–1067

le Lannou D, Chambon Y and le Calve M (1979) Role of the epididymis in reabsorption of inhibin in the rat. J Reprod Fert Suppl 26:117–121

Lipner H and Rush ME (1981) Site of absorption of testicular inhibin. Fed Proc 40: 473 (abst)

Lubicz-Nawrocki CM (1973) Anomalous effects of progesterone on the maturation and survival of spermatozoa in the epididymis of the golden hamster. J Endocr 58:199–205

MacMillan EW (1953) The effects of interruption of the vasal and inferior epididymal arteries on the cauda epididymidis and testis. Proc Soc Fert 5:12–19

MacMillan EW (1954a) The blood supply of the epididymis in man. Brit J Urol 26:60–71

MacMillan EW (1954b) Observations on the isolated vaso-epididymal loop and on the effects of experimental sub-capital epididymal obstructions. Studies on Fertility 6:57–64

MacMillan EW (1957) The caput epididymidis and trypan blue. J Anat 91:616

Mancini RE, Villar O, Alvarez B and Seiguer AC (1965) Extravascular and intratubular diffusion of labelled serum proteins in the rat testis. J Histochem Cytochem 13:376–385

Maneely RB (1959) Epididymal structure and function: a historical and critical review. Acta Zool 40:1–21

Martan J (1969) Epididymal histochemistry and physiology. Biol Reprod 1:134–154

Mason KE and Shaver SL (1952) Some functions of the caput epididymidis. Ann New York Acad Sci 55:585–593

Mayorga L and Bertini F (1983) Two populations of hydrolase-containing particles in rat epididymis. J Cell Physiol 117:135–139

Messier B and Leblond CP (1960) Cell proliferation and migration as revealed by radioautography after injection of thymidine-^{3}H into male rats and mice. Am J Anat 106:247–285

Miller RJ and Killian GJ (1983) Rat epididymal halo cells have lymphocytic characteristics. Biol Reprod 28:130 (abst)

Minniberg DT and Schlossberg S (1983) The role of the epididymis in testicular descent. J Urol 129:1207–1208

Moniem KA, Glover TD and Lubicz-Nawrocki CM (1978) Effect of duct ligation and orchidectomy on histochemical reactions in the hamster epididymis. J Reprod Fert 54:173–176

Montorzi NM and Burgos MH (1967) Uptake of colloidal particles by cells of the ductuli efferentes of the hamster. Z für Zellforsch 83:58–69

Moore HDM and Bedford JM (1979a) The differential absorptive activity of epithelial cells of the rat epididymus before and after castration. Anat Rec 193:313–328

Moore HDM and Bedford JM (1979b) Short-term effects of androgen withdrawal on the structure of different epithelial cells in the rat epididymis. Anat Rec 193:293–312

Moore HDM and Pryor JP (1981) The comparative ultrastructure of the epididymis in monkeys and man: a search for a suitable animal model for studying epididymal physiology in Primates. Am J Primatol 1:241–250

Morales C and Hermo L (1983) Demonstration of fluid-phase endocytosis in epithelial cells of the male reproductive system by means of horseradish peroxidase-colloidal gold complex. Cell Tiss Res 230:503–510

Murakami M, Nishida T, Iwanaga S and Shiromoto M (1984a) Scanning and transmission electron microscopic evidence of epithelial phagocytosis of spermatozoa in the terminal regions of the vas deferens of the cat. Experientia 40:958–960

Murakami M, Iwanaga S, Nishida T and Aiba T (1984b) Phagocytosis of latex beads by the epithelial cells in the terminal region of the vas deferens of the cat: SEM and TEM study. Andrologia 16:548–553

Murakami M and Nishida T (1985) Phagocytosis of spermatozoa and latex beads by epithelial cells of the ampulla vasis deferentis of the rabbit – a combined SEM and TEM study. Arch Histol Japon 48:269–278

Nagy F and Edmonds RH (1975) Cellular proliferation and renewal in the various zones of the hamster epididymis after colchicine administration. Fert Steril 26:460–468

Neumann F and Schenk B (1976) New antiandrogens and their mode of action. J Reprod Fert Suppl 24:129–145

Nicander L (1957a) On the regional histology and cytochemistry of the ductus epididymidis in rabbits. Acta Morph Neerl Scand 1:99–118

Nicander L (1957b) Studies on the regional histology and cytochemistry of the ductus epididymis in stallions, rams and bulls. Acta Morph Neerl Scand 1:337–362

Nicander L (1965) An electron microscopical study of absorbing cells in the posterior caput epididymis of rabbits. Z Zellforsch 66:829–847

Nicander L (1979) Fine structure of principal cells in the initial segment of the epididymis of the ram. Zb Vet Med C Anat Histol Embryol 8:318–330

Nicander L and Glover TD (1973) Regional histology and fine structure of the epididymal duct in the golden hamster (*Mesocricetus auratus*). J Anat 114:347–364

Nicander L and Malmqvist M (1977) Ultrastructural observations suggesting merocrine secretion in the initial segment of the mammalian epididymis. Cell Tiss Res 184:487–490

Nicander L and Ploen L (1979) Studies on regional fine structure and function in the rabbit epididymis. Int J Androl 2:463–481

Nicander L, Osman DI, Ploen L, Briggs HP and Knisgaard KN (1983) Early effects of efferent ductule ligation on the proximal segment of the rat epididymis. Int J Androl 6:91–102

Niemi M and Tuohimaa P (1971) The mitogenic activity of testosterone in the accessory glands of the rat in relation to its conversion to dihydrotestosterone. In: Basic Action of Sex Steroids on Target Organs. Eds PO Hubinot, F Leroy and P Galand, Karger, Basel. pp 258–264

Noordhuizen-stassen EN, Charbon GA, de Jong FH and Wensing CIG (1985) Functional arteriovenous anastomosis between the testicular artery and the pampiniform plexus in the spermatic cord of rams. J Reprod Fert 75:193–201

Öfner P, Vena RL, Leav I and Hamilton DW (1979) Metabolism of C_{19}-radiosteroids by explants of canine prostate and epididymis with disposition as hydroxylated products: a possible mechanism for androgen inactivation. J Steroid Biochem 11:1367–1379

Olson GE, Jonas-Davies J, Hoffmann LH and Orgebin-Crist MC (1982) Structural characterization of isolated rat epididymal epithelial cells. Gamete Res 6:161–178

Olson GE, Jonas-Davies J, Hoffman LH and Orgebin-Crist MC (1983) Structural features of cultured epithelial cells from the adult rat epididymis. J Androl 4:347–360

Orgebin-Crist MC and Ménézo Y (1980) A continuous flow method for organ cultrue of rabbit epididymis: morphology, amino acid utilization, glucose uptake, RNA, and protein synthesis. J Androl 1:289–298

Orgebin-Crist MC and Tichenor P (1972) Technique for studying sperm maturation in vitro. Nature (Lond.) 239:227–228

Orgebin-Crist MC, Danzo BJ and Davies J (1975) Endocrine control of the development and maintenance of sperm fertilizing ability in the epididymis. In: Handbook of Physiology, Section VII Endocrinology, vol 5, Male Reproductive System, Eds. RO Greep and DW Hamilton, Am Physiol Soc, Washington DC, pp 319–338.

Pabst R and Schick P (1979) Proliferation of epithelial cells of vas deferens and epididymis of young adult rabbits. Archs Androl 2:183–186

Pérez-Clavier R, Harrison RG and MacMillan EW (1982) The pattern of lymphatic drainage of the rat epididymis. J Anat 134:667–675

Prakash A, Anand Kumar TC and Prasad MRN (1980) Ultrastructure of the epididymis in the rhesus monkey. In: Non-human Primate models for study of Human Reproduction. Ed TC Anand Kumar, S Karger, Basel. pp 129–135)

Prasad MRN, Singh SP and Rajalakshmi M (1970) Fertility control in male rats by continuous release of microquantities of cyproterone acetate from subcutaneous silastic capsules. Contraception 2:165–178

Rajalakshmi, Singh SP and Prasad MRN (1971) Effects of microquantitites of cyproterone acetate released through silastic capsules on the histology of the epididymis of the rat. Contraception 3:335–346

Rajalakshmi M, Arora T, Bose TK, Dinakar N, Gupta G, Thampan TNRV and Prasad MRN (1976) Physiology of the epididymis and induction of functional sterility in the male. J Reprod Fert Suppl 24:71–94

Ramos A (1980) Ultrastructural and histochemical observations on the principal cells of monkey epididymis. Archs Androl 5:159–168

Ramos AS and Dym M (1977) Fine structure of the monkey epididymis. Am J Anat 149:501–532

Rastogi RK (1979) A critical appraisal of the effects of the antiandrogen, cyproterone acetate, on the epididymis in connection with fertility control in the mouse. J steroid Biochem 11:101–105

Reid BL and Cleland KW (1957) The structure and function of the epididymis. I. The histology of the rat epididymis. Aust J Zool 5:223–246

Reijonen K, Kormano M and Ericsson RJ (1975) Studies on the rat epididymal blood vessels following alpha-chlorohydrin administration. Biol Reprod 12:483–490

Revel JP, Yee AG and Hudspeth AJ (1971) Gap junctions between electrotonically coupled cells in tissue culture and in brown fat. Proc Natn Acad Sci US 68:2924–2927

Richard M (1929) Über Resorption und Sekretion des Nebenhodens. Schweiz Med Wochenschr 10:1269–1271

Richard M (1930) Über Sekretion und Resorption des Nebenhodens. Deutsch Zeitschrift Chirurg 226:98–109

Ritchie AWS, Hargreave TB, James K and Chisholm GD (1984) Intra-epithelial lymphocytes in the normal epididymis. A mechanism for tolerance to sperm auto-antigens? Brit J Urol 56:79–83

Sedar AW (1966) Transport of exogenous peroxidase across the epithelium of the ductuli efferentes. J Cell Biol 31:102A (abst)

Setchell BP and Hinton BT (1981) The effects on the spermatozoa of changes in the composition of luminal fluid as it passes along the epididymis. In: Epididymis and Fertility: Biology and Pathology. Eds C Bollack and A Clavert, S Karger, London. Progr Reprod Biol 8:58–66

Setchell BP, Waites GMH and Till AR (1964) Variations in flow of blood within the epididymis and testis of the sheep and rat. Nature (Lond) 203:317–318

Shaver SL (1954) The role of stereocilia in removing India ink particles from the lumen of the rat epididymis. Anat Rec 119:177–185

Soranzo L, Dadoune JP and Fain-Maurel MA (1982) Segmentation of the epididymal duct in mouse: an ultrastructural study. Reprod Nutr Develop 22:999–1012

Steinman RM, Millman IS, Muller WA and Cohn ZA (1983) Endocytosis and the recycling of plasma membrane. J Cell Biol 96:1–27

Suzuki F (1982) Microvasculature of the mouse testis and excurrent duct system. Am J Anat 163:309–325

Stowe BA (1981) Thermal characteristics of the testis and epididymis of the boar. J Reprod Fert. 63:551–557

Sun EL and Flickinger CJ (1979) Development of cell types and of regional differences in the post natal rat epididymis. Am J Anat 154:27–56

Suzuki F and Nagano T (1978) Regional differences of cell junctions in the excurrent duct epithelium of the rat testis as revealed by freeze fracture. Anat Rec 191:503–520

Takano H (1980) Qualitative and quantitative histology and histogenesis of the mouse epididymis, with special reference on the regional difference. Acta Anat Nippon 55:573–587

Tuohimaa P and Niemi M (1974) Cell renewal and mitotic activity of testosterone in male acccessory organs. In: Male Accessory Organs. Structure and Function in Mammals. Ed DE Brandes, Acad Press, New York. pp 329–343

Turner TT (1979) On the epididymis and its function. Invest Urol 16:311–321

Turner TT (1984) Resorption versus secretion in the rat epididymis. J Reprod Fert 72:509–514

Turner TT and Howards SS (1985) The tenacity of the blood-testis and blood-epididymis barriers. In: Male Fertility and its Regulation. Eds TJ Lobl and ESE Hafez, MTP Press, Lancaster. pp 383–393

Turner TT, D'Addario DA and Howards SS (1983) The transepithelial movement of ^{3}H-3-*O*-methyl-D-glucose in the hamster seminiferous and cauda epididymal tubules. Fert Steril 40:530–535

Van Wagenen G (1924a) Changes in the testis of the rat following ligation of the ductuli efferenti. Anat rec 29:399 (abst)

Van Wagenen G (1924b) Degeneration of germinal epithelium in the testis of the rat as a result of efferent duct ligation. Anat Rec 27:189–190 (abst)

Voglmayr JK, Musto NA, Saksena SK, Brown-Woodman PDC, Marley PB and White IG (1977) Characteristics of semen collected from the cauda epididymidis of conscious rams. J Reprod Fert 49:245–251

Wagenseil F (1928) Experimentakuntersuchungen am Nebenhoden der Maus. Z Zellforsch 7:141–176

Wagley LM, Versluis TD, Brown DV and Amann RP (1984) Culture of principal cells from the ram epididymis. A comparison of the morphology of principal cells in culture and in situ. J Androl 5:389–408

Waites GMH, Setchell BP and Quinlan D (1973) Effect of local heating of the scrotum, testes and epididmides of rats on cardiac output and regional blood flow. J Reprod Fert 34:41–49

Walton JS, Evins JD and Waites GMH (1978) Effect of chronic removal of testicular fluid on the release of follicle stimulating hormone in hemicastrated rams. J Endocr 77:421–422

Wang YF and Holstein AF (1983) Intraepithelial lymphocytes and macrophages in the human epididymis. Cell Tiss Res 233:517–521

Warner CW and Nienenhuis RJ (1983) Influence of abdominal temperature and luminal contents on the cauda epididymidis of the rat. Biol Reprod 29:1011–1017

Weerasooriya TR and Yamamoto T (1985) Three-dimensional organisation of the vasculature of the rat spermatic cord and testis. A scanning electron-microsopic study of vascular corrosion casts. Cell Tiss Res 241:317–323

Weininger RB, Fisher S, Rifkin J and Bedford JM (1982) Experimental studies on the passage of specific IgG to the lumen of the rabbit epididymis. J Reprod Fert 66:251–258

White IG, Larsen LH and Wales RG (1959) Method for the in vivo collection of epididymal spermatozoa and for their comparison with ejaculated cells. Fert Steril 10:571–577

Wierzbowski S and Wierzchos E (1969) Cannulation of the vas deferens of the boar. J Reprod Fert 19:173–175

Wong PYD and Lee WM (1982) Effects of spironolactone (aldosterone antagonist) on electrolyte and water content of the cauda epididymidis and fertility of male rats. Biol Reprod 27:771–777
Wong PYD and Yeung CH (1978) Absorptive and secretory functions of the perfused rat cauda epididymidis. J Physiol 275:13–26
Wong PYD, Yeung CH and Ngai HK (1977) Effect of α-chlorohydrin on transport processes in perfused rat cauda epididymidis. Contraception 16:637–644
Wong YC, Wong PYD and Yeung CH (1978) Ultrastructural correlation of water resorption in isolated cauda epididymides. Experientia 34:485–487
Wong PYD, Au CL and Ngai HK (1978) Electrolyte and water transport in rat epididymis. Its possible role in sperm maturation. Int J Androl Suppl 5:608–628
Wong PYD, Au CL and Ngai HK (1979) Some characteristics of salt and water transport in the rat epididymis. In: The Spermatozoon. Eds DW Fawcett and JM Bedford, Urban and Schwarzenberg, Baltimore. pp 57–63
Wong PYD, Au CL and Ngai HK (1980) Effects of 6-chloro-6-deoxyglucose on electrolyte and water transport in the epididymis and fertility of male rats. Int J Androl 3:82–86
Wong PYD, Au C and Bedford JM (1982) Biology of the scrotum. II. Suppression by abdominal temperature of transepithelial ion and water transport in the cauda epididymidis. Biol Reprod 26:683–689
Yeung CH and Cooper TG (1982) Ultrastructure of the perfused rat epididymis: effect of luminal sodium ion concentration. Cell Tiss Res 226:407–425
Yeung CH, Cooper TG and Waites GMH (1980) Carnitine transport into the perfused epididymis of the rat: regional differences, stereospecificity, stimulation by choline and effects of other luminal compounds. Biol Reprod 23:294–304
Young WC (1933) Die Resorption in den Ductuli efferentes der Maus und ihre Bedeutung für das Problem der Unterbindung im Hoden-Nebenhoden System. Z Zellforsch Mikroskop Anat Abt Histochem 17:729–759
Zunke VU and Gossrau R (1981) Morphology and zonal division of the ductus epididymidis of rats. I. Investigations of semi-thin sections. Anat Anz (Jena) 149:409–424

B. Secretion of Steroids by the Epididymis

1. Steroids in Epididymal Tissue

The steroid content of the epididymis was first demonstrated by androgen bioassay (Gallagher, 1928). Concentrations of 5α-dihydrotestosterone exceed that of testosterone and the proximal epididymis contains more androgens than the distal regions in rats (Vreeburg, 1975; Vreeburg et al., 1976; Purvis and Hansson, 1978), boars (Aafjes and Vreeburg, 1972) and bulls (Aafjes and Vreeburg, 1972; Ganjam and Amann, 1976). However, in the canine epididymis concentrations of DHT are highest in the corpus (Boulanger et al., 1982) and no regional differences are found in the human epididymis (Purvis et al., 1978; Leinonen et al., 1980). Concentrations of dehydroepiandrosterone and testosterone in the rabbit caput are greater than those in the cauda (Frankel and Eik-Nes, 1970a) and although no pregnenolone is detectable in the rabbit epididymis (Frankel and Eik-Nes, 1970b), progesterone concentrations are greater in the distal epididymis of rats (Kumari et al., 1980) and bulls (Ganjam and Amann, 1976). The concentrations of steroids in epididymal fluid are givien in Table 11. Concentrations of cholesterol decrease as concentrations of desmosterol increase in luminal fluids as distal parts of the hamster epididymis are reached (Legault et al., 1979a).

2. Origin of Epididymal Steroids

a. Testicular Fluid

The reduction of both the dihydrotestosterone and testosterone content in the proximal epididymis of the rat following bilateral castration (Aafjes and Vreeburg, 1972; Podesta et al., 1975) indicates either that the testis is a source of the androgens or their precursors, or that castration alters the activity of enzymes responsible for their synthesis (see Section III.B.2.a). While testosterone is delivered directly into the epididymal lumen, with one exception (Pujol et al., 1976), concentrations of the 5-α steroid in rete testis fluid have been found to be lower than those of testosterone in rats (Harris and Bartke, 1975; Vreeburg, 1975), bulls (Ganjam and Amann, 1973, 1976) and rabbits (Lau and Saksena, 1979). The return of dihydrotestosterone to the epididymis of castrated animals given exogenous testosterone (Aafjes and Vreeburg, 1972) with an increased dihydrotestosterone/testosterone ratio (Podesta et al., 1975) together with the maintenance of dihydrotestosterone in the ipsilateral epididymis of unilaterally castrated animals (Aafjes and Vreeburg, 1972) indicate that epididymal dihydrotestosterone arises from testosterone in the circulation.

Decreases in epididymal dihydroteststerone after long-term ligation of the testicular efferent ducts, however, or in the ipsilateral epididymis of unilaterally ligated animals (Aafjes and Vreeburg, 1972), argue for a control of the 5α-reductase by testicular fluid (see Section III.B.3.a). No changes in the steroid profile of the bovine epididymis were observed after ligation of the efferent ducts (Ganjam and Amann, 1976), although concentrations of testosterone and progesterone in the rat cauda decline after this operation (Kumari et al., 1978).

b. The Bloodstream

Circulating steroids have access to the cauda epididymal lumen of rats (Back, 1975) and hamsters (Turner et al., 1981). They still appear after ligation of the corpus (Back, 1975) and enter the isolated perfused cauda (Cooper and Waites, 1979b, 1980) in-vivo proving that they can cross the epithelium rather than gain entry *via* the testicular exocrine secretion. The extent of their entry depends on the nature and concentration of proteins in the epididymal lumen (Cooper and Waites, 1979b, 1980; Cooper, 1980).

There is a possibility that steroid concentrations in epididymal arterial blood are elevated above those in peripheral blood because of the transfer of steroids in the testicular venous effluent to the superior epididymal arterial branch of the testicular artery within the pampiniform plexus. That steroids move across at different rates in the bull (Amann and Ganjam, 1976) may be explained by their lipid solubility, but the presence of functional anastomoses between the artery and veins of the testis (see Section III.A.3) should make possible movement of many compounds from the testicular vein to artery. The high position of the anastomoses suggest that they are of more importance for the epididymis (Hees et al., 1984) but the significance of this for the rat epidiymis has been questioned (Free and Jaffe, 1978). ^{3}H-testosterone injected into the rat testis preferentially enters the ipsilateral corpus epididymidis (Einer-Jensen, 1974) but the contribution of the lymphatics to this transfer is not known. There was only a marginal increase in the movement of testosterone to the ipsilateral perfused cauda epididymal lumen when the steroid was infused into one testicular vein (Cooper and Waites, 1981).

c. Epididymal Biosynthesis

i. By Epididymal Tissue

Acetate is converted into cholesterol by epididymal tissue from mice (Hamilton et al., 1969) and rats (Hamilton and Fawcett, 1970) and further converted into testosterone and other steroids in the epididymides of rats, hamsters, rams and rabbits (Hamilton and Fawcett, 1970; Hamilton, 1971). Frankel Eik-Nes (1970b), however, could find no conversion of acetate or cholesterol into testosterone by the rabbit epididymis.

The first part of *de novo* synthesis of steroids is the formation of polypisoprene units which can be synthesised from mevalonate in the rat epididymis in-vitro (Wenstrom and Hamilton, 1980). The epithelium and not the muscle is responsible for the conversion in the vas deferens (Milner and Hamilton, 1971). Progesterone and pregnenolone are also converted into androgens by the epididymis of rabbits (Frankel and Eik-Nes, 1970b) and rats (Kumari et al., 1978) and there is interconversion of dehydroepiandrosterone and testosterone to androstenediol and androstenedione (Frankel and Eik-Nes, 1970b).

The major pathway for metabolism of androgens in the epididymis is 5α-reduction which has been demonstrated in rats (Gloyna and Wilson, 1969; Inano et al., 1969; Djøseland et al., 1973, 1974, 1976; Djøseland 1976a,b; Tuohimaa and Niemi, 1974), dogs (Sowell and Eik-Nes, 1972), man (Sulcová and Stárka, 1973; Sommerville, 1971; Kinoshita, 1981; Tezón et al., 1983) and rabbits (Danzo and Eller, 1978, 1979, 1980), but in the latter species the enyzme activity is low (Booth and Jones, 1979).

The reductive activity takes place in the epithelium and not the stroma of the rat epididymis (Djøseland et al., 1983) and vas (Fig. 6) and epididymal epithelial cells can convert T to DHT in culture (Klinefelter et al., 1982; Brown et al., 1983; Brown and Amann, 1984; Robaire and Buchanan, 1985). The principal cells are chiefly responsible for the conversion (Klinefelter and Amann, 1980).

There is interconversion of androgens in the epididymis of rabbits (Booth and Jones, 1979) and rats (Hastings and Djøseland, 1977; Djøseland et al., 1976), conjugation of steroids in rats (Hamilton, 1971; Djøseland, 1976) and hydroxylation, especially in the epididymis of dogs (Öfner et al., 1974a,b,c, 1979; Morfin et al., 1980). Histochemical studies have shown an abundance of steroid hydroxylases in epididymides of hamsters (Baillie, 1966; McGadey et al., 1966), rats

(Moniem, 1972), mice (Rastogi et al., 1976), monkeys (Prakash and Moore, 1982) and sulphotransferases have been characterised from the hamster eididymis (Bouthillier et al., 1981, 1984, 1985a,b).

The increase in the concentration of oestrogen in the rat epididymis after hCG injection (Kumari et al., 1978) suggests that aromatisation of androgens can occur; oestrogen sulphates can be hydrolysed by the porcine epididymis (Booth, 1983).

ii. By Spermatozoa

Although rat epididymal spermatozoa contain dihydroteststerone after injection of testosterone into rats (Blaquier, 1971a; Vreeburg and Aafjes, 1971) this must have arisen from conversion elsewhere followed by uptake of the 5α-reduced steroid by the spermatozoa, since no 5α-reductase has been found in spermatozoa from rats (Vreeburg and Scholte, 1973; Rajalakshmi et al., 1978) or bulls (Hammerstedt and Amann, 1976b). The extent of metabolism of androgens differs between species and substrates, but is mainly oxidative in nature (Rajalakshmi et al., 1978), although bovine sperm can interconvert dihydrotestosterone and androstanediols (Hammerstedt and Amann, 1976b; Djøseland et al., 1978). There are few differences in steroid metabolism between testicular and mature epididymal sperm cells (Voglmayr, 1971; Hammerstedt and Amann, 1976a).

3. Testicular Control of Androgenic Function in the Epididymis

a. Steroid Metabolising Enyzmes

De novo steroid biosynthesis in the mouse epididymis is dependent on testicular androgens (Hamilton et al., 1969); these maintain the smooth endoplasmic reticulum in this species and the rabbit (Jones et al., 1979b). Further conversion to and of dihydroteststerone appear to be controlled separately ba 5α-reductase and 3α-hydroxysteroid dehydrogenase.

i. 5α-Reductase

This enzyme has been characterised from the rat epididymis (Scheer and Robaire, 1982) and in nuclear and microsomal membranes is modified by the phospholipid environment (Cooke and Robaire, 1985). The activity of the enzyme is greater in the epididymal head than the tail in rats (Vreeburg and Scholte, 1973; Djøseland et al., 1974; Saksena et al., 1976; Robaire et al., 1977; Robaire, 1979; Pujol and Bayard, 1978; Kumari et al., 1978; Foldesy and Leatham, 1981; Brown et al., 1983) and man (Kinoshita, 1981) with highest activity in the initial segment (Robaire et al., 1981). This may explain the greater release of dihydroteststerone from the caput than the cauda in-vitro (Lau and Saksena, 1981). In the bovine epididymis more 5α-reductase activity is found in the central caput (Hammerstedt and Amann, 1976).

After castration the activity of the enzyme decreases in the whole epididiymis (Djøseland, 1976; de Larminat et al., 1978) and in the caput-corpus (Robaire et al., 1977), but the return of activity with androgen treatment (Djøseland, 1976; Pujol and Bayard, 1978) was not observed in the caput-corpus (Robaire et al., 1977). In the rabbit 5α-reductase activity in the epididymis is lost on castration and returned with androgens (Danzo and Eller, 1978, 1979, 1980).

In rats the enzyme activity can decrease when circulating testosterone is maintained, *e.g.* after bilateral (Robaire et al., 1977; Pujol and Bayard, 1978) or after unilateral (Robaire et al., 1981) ligation of the efferent ducts or unilateral castration (Robaire, 1979), when decreases in enzyme activity are only observed on the ipsilateral side. This is suggestive of an influence of luminal secretions on 5α-reductase, but in the hamster 5α-reductase activity increases after ligation of the efferent ducts (Tsuji et al., 1984).

ii. 3α-Dehydrogenase

The enzyme from the rat epididymis has been characterised by Hastings and Hansson (1979) and Hastings et al. (1980). The activity of this enzyme is greater than that of the reductase (Robaire et al., 1977; Pujol and Bayard, 1978) and it is localised in more distal regions of the rat epididymis (Robaire et al., 1977, 1981; Robaire, 1979). Like 5α-reductase, its activity decreases after bilateral castration and hypophysectomy (Robaire and Zirkin, 1981) but its activity is restored when androgens are replaced (Robaire et al., 1977; Robaire and Zirkin, 1981) or are available from the contralateral testis of unilateral castrates (Robaire, 1979). The dependence on circulating rather than luminal androgens is supported by observations (Robaire et al., 1978, 1981) that the enzyme activity does not decline on unilateral ligation of the efferent ducts.

A decline in the activity of the enzyme in the rat caput after castration was not responsive to androgens (Pujol and Bayard, 1978) but this may reflect the failure of exogenous androgens to support the proximal part of the epididymis after this operation (see Section III.A.7), which was the region of the epididymis studied by these workers.

b. Intracellular Androgen Binding Proteins (Receptors)

There is ample evidence that androgens are taken up and bound to macromolecules in the epididymis (see Hansson et al., 1974, 1975; Danzo et al., 1975a) and translocation of cytosolic receptors to the nucleus has been demonstrated in the rat epididymis (Tezón et al., 1982; Tezón and Blaquier, 1983). Binding of dihydrotestosterone and 5α-androstanediol also occurs (Calandra and Blaquier, 1973; Blaquier and Calandra, 1973; Wilson and French, 1976).

Androgen receptors have been identified in the epididymides of rats (Blaquier et al., 1970; Ritzén et al., 1971; Hansson et al., 1973b; Tindall et al., 1972), rabbits (Danzo et al., 1973; Danzo and Eller, 1975, 1978, 1979), dogs (Younes et al., 1979; Younes and Pierrepoint, 1981b), bulls (Ganjam and Amann, 1976) and rams (Carreau et al., 1984a,b). In mice radioautography only localised specific binding of dihydroteststerone in the nuclei of epididymal epithelial cells (Stumpf and Sar, 1976; Schleicher et al., 1984).

Studies suggesting that the population of androgen receptors in the rat epididymis declined after castration and were maintained or restored with androgen therapy (Calandra et al., 1975; Podesta et al., 1975; Pujol and Bayard, 1978) are now thought to be artefactual (Calandra et al., 1977) and other reports failed to detect decreases in androgen receptors after castration or hypophsectomy (Tindall et al., 1974). A shift in the relative amount of nuclear and cytoplasmic binding observed after castration is restored with testosterone or dihydrotestosterone and mirrored with antiandrogens that prevent translocation of receptors into the nucleus (Tezón et al., 1982; Tezón and Blaquier, 1983).

c. Extracellular Steroid Binding Proteins

i. Androgen Binding Protein (ABP)

Studies of androgen receptors in the epididymis were initially confused by the presence of extracellular androgen binding proteins (Danzo et al., 1973). ABP has been found in the epididymides of rats (Taylor et al., 1980; Danzo et al., 1980; Schmidt et al., 1981), rabbits (Danzo et al., 1975; Danzo Eller, 1978), sheep (Carreau et al., 1984a,b) and guinea pigs (Danzo et al., 1982) but the binding component in the human epididymis cannot easily be separated from contaminating serum testosterone binding globulin (Vigersky et al., 1976; Lippschutz et al., 1977; Hsu et al., 1977; Purvis et al., 1978). In the rat epididymis ABP displays microheterogeneity (Larreau et al., 1981).

Its loss from the castrated epididymis (Podesta et al., 1975) points to a testicular source of the protein and ligation of the testicular efferent ducts reduces the ABP activity in the epididymis of rats (French and Ritzén, 1973) and rabbits (Ritzén and French, 1974; Danzo et al., 1974; see Hansson et al., 1975; Ritzén et al., 1975).

The ability to distinguish between the receptor (CR) and ABP enables the roles of these proteins to be established. In castrated and hypophysectomised rats, in which ABP but not CR is depleted, the binding of androgens to the nucleus is maintained; conversely, when antiandrogens that bind to CR but not ABP are given to rats, a decrease in nuclear binding follows (Tindall et al., 1974). These results are consistent with nuclear binding being dependent on the cytoplasmic receptor and not ABP.

ii. The Role of ABP

Because the steroid binding affinities of CR and ABP differ (Tindall et al., 1978; Kirchoff et al., 1979; Cunningham et al., 1978; Lobl et al., 1980; Lobl, 1981; Rousseau et al., 1980, 1981) and different faces of the steroids are involved, a mechanism for the tranfer of androgen from ABP to receptors has been postulated (Lobl et al., 1980). Why a specific protein should be necessary for the transfer of androgens to the receptor when the epithelial cells are freely permeable to steroids (see Section III.A.8) is not at all clear.

Rommerts et al. (1976) concluded that ABP would not influence the concentration of the physiologically important, free, steroid concentration in the lumen of the seminiferous tubule because (i) the rate of tubular fluid flow is very low relative to lymph and venous blood, (ii) these fluids contain very high concentrations of steroids and (iii) the tubule wall is freely permeable to them. Although the rate of fluid flow through the epididymis is considerably lower than that in the testis (Brooks, 1979) and the walls equally permeable to steroids (see Section III.B.8), venous plasma steroid levels are much lower than those bathing testicular tubules, so ABP may be more important in controlling steroid concentrations in the epididymis (Turner et al., 1984). In support of this, ABP concentrations decrease along the length of the epididymis of rats (French and Ritzén, 1973; Purvis and Hansson, 1978) and rabbits (Danzo et al., 1973; Weddington et al., 1974) as do T and DHT (see Section III.B.1), and in these species ligating the efferent ducts decreases epididymal ABP (French and Ritzén, 1973; Ritzén and French, 1974) and dihydrotestosterone (Vreeburg, 1975: see Section III.B.a.i). Furthermore, in contrast to fluids from the testis, concentrations of ABP exceed those of testosterone and dihydrotestosterone in the caput and cauda lumen: the protein may maintain high luminal levels but low free steroids (Turner et al., 1984). There is little correlation between the testosterone and ABP content of the ovine epididymis (Carreau et al., 1984b).

By flushing out the guinea pig luminal contents it has been established that ABP is located in the lumen (Danzo et al., 1982) and micropuncture samples have revealed higher concentrations of ABP in the caput than cauda of the rat (Vigersky et al., 1978; Turner et al., 1984). In the rabbit the ABP concentration in epididymal fluid is higher than that in rete testis fluid (Danzo et al., 1976, 1977) but DHT concentration in the lumen of this species is not yet known. In the bull there are higher concentrations of dihydrotestosterone in epididymal fluid from the cauda than in testicular fluid (Ganjam and Amann, 1973, 1976) but ABP concentrations have yet to to estimated. In the ram dihydroteststerone concentrations in epididymal fluid are higher than those in rete testis fluid (Voglmayr et al., 1977) and the concentration of ABP is about equal to that in testicular fluid (Voglmayr et al., 1977; Jégou et al., 1979).

There are hints that non-steroidal compounds which specifically bind to ABP but not the androgen receptor, can reduce the progressive motility and sperm-egg binding ability of sperm taken from the cauda when administered to rats (Sirett et al., 1984). This suggests that modifying the amount of total or free luminal androgen may be important. In two animal models, the genetically deficient, restricted (H^{re}) rat and the pregnenolone-injected hypophysectomised rat, positive correlations have been established between the testicular and epididymal content of ABP and the fertilising capacity of spermatozoa (Anthony et al., 1984a,b). Inhibiting the binding of androgens to ABP without affecting androgen receptors does not render rats infertile, although there is a depression of sperm motility (Rousseau et al., 1985).

iii. Uteroglobin

This progesterone binding protein has been identified in the epithelial cells and lumen of the rabbit epididymis (Noske and Fiegelson, 1976; Fiegelson et al., 1977; Noske and Gooding, 1978; El Eltreby et al., 1983) where it may play a role in regulating the amount of free progesterone present in epididymal fluid (Table 11). Uteroglobin has also immunosupressive properties (see Section III.G.12.a.vii).

4. Endocrine Role of the Epididymis

After the discovery of the androgen content of the epididymis (see Section III.B.1.) several attempts were made to determine if the organ had an endocrine function. In immature rats, Lawless (1931) failed to find evidence for an internal secretion affecting the accessory organs of orchidectomised animals, but Tuohiven and Pohjola (1949) concluded that the post-castration atrophy of the prostate was delayed when the epididymis was left intact after removal of the testes in both mature and immature animals. This could be due to the content of androgens bound to androgen-binding proteins. Further indirect evidence that the epididymis has an, albeit transient, androgenic role includes the observations that stallions may retain some male characteristics after castration if the epididymis is left in the body (O'Connor, 1950) and that there is better growth of antlers in reindeer (Ritzén, 1984) and deer (Lincoln, 1975) after this operation,

An endocrine role for the epididymis has been postulated from electron microscopical observations of granular inclusions close to the basal membrane of epithelial cells (see Section III.A.6) but the nature of the secreted products was not speculated upon: steroids are unlikely to manifest themselves this way.

A local effect of steroids on the vas deferens regulating ampullary function in sheep has been demonstrated by the atrophy on the ipsilateral side after unilateral cryptorchidism (Skinner and Rowson, 1968a) and the local maintenance of its structure and secretory function after injection of androgens into the lumen (Skinner and Rowson, 1968b).

Most evidence of a local influence of the epididymis on the vas deferens or other accessory organs comes from the rat. Vasectomy of rats leads to a decrease in the activity of androgen-dependent RNA-polymerase in the prostate (Pierrepoint and Davies, 1973; Pierrepoint, 1974; Pierrepoint and John, 1974; Pierrepoint et al., 1975a,b) which is not due to a depletion of circulating androgens (Mock et al., 1975). The effect is thought to be direct and local since a decline in activity of the ipsilateral prostatic lobe is observed after unilateral vasectomy (Pierrepoint et al., 1975b). Removing both testes and one epididymis (orchidoepididymectomy) also leads to an ipsilateral decline in prostatic RNA polymerase activity, but when androgen levels are maintained by unilateral orchidectomy or bilateral orchidectomy with testosterone supplementation, RNA polymerase activity is maintained (Pierrepoint et al., 1974, 1975b), suggesting that the androgen-stimulated epididymis is involved in this phenomenon. Removing the caput and corpus epididymidis along with the testis does not reduce the prostatic function in testosterone-supplemented rats, indicating that the cauda epididymidis influences the ipsilateral prostate when both circulating androgens and an intact vas are present (Pierrepoint et al., 1975b).

The venous route may be important for mediating this effect because venotomy alone also decreases RNA polymerase activity in the prostate, unilateral venotomy leads to an ipsilateral decline and these effects are greater than those due to vasoligation without venotomy (Pierrepoint et al., 1975a,b). More recently it has been shown that removing the epididymis in androgen-maintained castrates reduces the number of androgen receptors in the prostate, lowers the proportion of nuclear receptors and increases the number of unoccupied cytoplasmic receptors compared to orchidectomised rats (Pierrepoint and Davies, 1984), and venotomy acts in a similar way.

The venous drainage of the prostate and vas deferens has been studies in rats (Lewis and Moffat, 1975) and dogs (Dhabuwala and Pierrepoint, 1977; Boulanger et al., 1983) and in both species the venous drainage is continuous with the valveless venous "circle" that drains the prostatic complex such that material injected into the vasal vein in the direction of blood flow arrives preferentially at the prostate (Lewis and Moffat, 1975; Dhabuwala et al., 1978; Boulanger et al., 1983).

The concentration of testosterone in deferential vein blood is greater than that in peripheral blood and may be as high as that in spermatic venous blood in androgen-maintained animals (Pierrepoint et al., 1975a,b; Boulanger et al., 1983), but there is no direct connection between the spermatic vein and epididymal vein in the dog, as occurs in the rabbit (Chubb and Desjardins, 1982). In rats, mice and rabbits testosterone concentrations in epididymal vein blood exceed those of the deferential vein (Chubb and Desjardins, 1982).

Labelled testosterone injected into the vein in the direction of blood flow accumulates in the ipsilateral prostatic lobe to a greater extent than that in other tissues (Pierrepoint et al., 1975a,b) and ^{3}H-testosterone injected into the lumen of the vas deferens of rats (Pierrepoint and Davies, 1973; Pierrepoint, 1974; Pierrepoint et al., 1975b) and dogs (Pierrepoint, 1974) also reaches the prostate rapidly and at concentrations greater than can be accounted for by blood activity, judging from the activity arriving in other tissues from ^{14}C-testosterone simultaneously injected into the femoral vein (Pierrepoint, 1974). Radioactivity from ^{3}H-testosterone perfused through the canine vas lumen appears in peripheral blood (Boulanger et al., 1983) but the nature of the metabolites was not established. In-vitro androgens perfused through the lumen of the canine vas deferens are metabolised and secreted into the medium (Dupuy et al., 1979) and the rat cauda epididymidis perfused with ^{3}H-testosterone in-vivo releases ^{3}H-dihydroteststerone into deferential vein blood (Fig. 5). Labelled testosterone infused into the femoral artery appears as dihydrotestosterone in epididymal venous blood (Sowell and Eik-Nes, 1972). It is thus quite feasible that the vas deferens acts as a portal system transporting 5α-reduced androgens from the epididymis to adjacent accessory organs. In support of this, dihydrotestosterone is found in spermatic vein blood, which drains the testis-epididymis (Ewing et al., 1975; Ewing and Brown, 1975) and in spermatic lymph, which drains the epididymis as well as the testis (Haltmeyer and Eik-Nes, 1974).

5. Role of Steroids in Epididymal Function

a. Androgens

The presence of androgen receptors in the epididymis (see Section III.B.3.b) are responsible for the action of circulating steroids in maintaining epididymal structure and function, including sperm maturation (see Section I.2.b), by controlling the synthesis of DNA, RNA and proteins (see Section III.G.5). The failure of high levels of circulating testosterone to maintain the structure of the initial segment of the epididymis after ligation of the efferent ducts (see Section III.A.7) signifies that other mechanisms operate to maintain the initial segment of the epididymis.

b. Oestrogens

The paradoxical biphasic effect of hCG on the fertilising capacity of spermatozoa isolated in the ligated cauda epididymidis of hypophysectomised hamsters suggested to Lubicz-Nawrocki and Glover (1973c) an elaboration of inhibitory oestrogens which could be overcome by androgens (Lubicz-Nawrocki, 1973c). Oestrogens do render male hamsters infertile (Lubicz-Nawrocki et al., 1973) and reduce the fertilising capacity of spermatozoa in the isolated cauda epididymidis of intact hamsters, but there the effects are not overcome by testosterone. As oestrogens can inhibit the beneficial effect of testosterone in castrated and hypophysectomised hamsters (Lubicz-Nawrocki, 1974) an anti-androgen action is suggested.

Oestrogens do inhibit the conversion of testosterone to dihydrotestosterone in-vivo and in-vitro (Djøseland, 1976; Tindall et al., 1981) but, curiously, did not inhibit the partially purified 5α-reductase (Monsalve and Blaquier, 1977). They also inhibit the binding of dihydrotestosterone to both nuclear and cytoplasmic androgen receptors in the epididymis (Tindall et al., 1981).

Oestrogen receptors have been found in the epididymis of mature rats (Mulder et al., 1974), immature rats (Danzo et al., 1977), mature rabbits (Danzo and Eller, 1979), immature rabbits (Danzo et al., 1975, 1978, 1983; Hendry et al., 1985), dogs (Younes and Pierrepoint, 1981b), guinea pigs (Danzo et al., 1981) and man (Murphy et al., 1980). There are more receptors in the cauda than caput and translocation of the cytoplasmic receptor to the nucleus has been demonstrated in the rabbit (Danzo et al., 1978). In the mouse, specific binding of oestrogen occurs mainly in the proximal parts of the tract and only in the nucleus of clear cells in the cauda (Schleicher et al., 1984).

Oestrogens stimulate the sialic acid production of the immature rat epididymis (Peyre and Laporte, 1966) and stimulate some growth of accessory organs in castrated rats (Brooks, 1979) but they do not stimulate protein synthesis in this organ (Jones et al., 1980b) or certain enzyme activities promoted by androgens (Brooks, 1976).

c. Corticosteroids

Cortisol is metabolised by the hamster epididymis (Baillie et al., 1966; McGadey et al., 1966) but it has no effect on the uptake of androgens by the epididymis (Tindall et al., 1981) and corticosterone does not inhibit 5α-reductase (Monsalve and Blaquier, 1977) or stimulate protein synthesis by the rat epididymis (Jones et al., 1980b). Its slight effect on the growth of accessory organs (Brooks, 1979; Pearce et al., 1986) may be due to conversion to androgens. Receptors for aldosterone are present in clear cell nuclei (Hinton and Keefer, 1985) and may control ion transport within the epididymis (see Section III.C.3.b).

d. Progestins

There are no progesterone receptors in the rat epididymis (Pujol and Bayard, 1979) although another progesterone binding protein, uteroglobin, is present in the epididymis of the rabbit (see Section III.B.3.c.iii). Progesterone does not stimulate protein synthesis by the rat epididymis (Jones et al., 1980b) or maintain the activity of glycolytic enzymes (Brooks, 1976) but it maintains the accessory organs of castrates, possibly *via* conversion to androgens (Brooks, 1979). Pregnenolone can maintain fertility in the hypophysectomised rat (Vreeburg et al., 1976; Anthony et al., 1984), whereas progesterone cannot maintain the fertilising capacity of spermatozoa in the hamster (Lubicz-Nawrocki, 1973a). Because the action of progesterone in reducing the fertilising capacity of spermatozoa isolated in the epididymis of intact hamsters is overcome by dihydrotestosterone and androstanediol but not testosterone, and because progesterone can prevent the ameliorative effect of testosterone but not dihydrotestosterone or androstanediol in castrated hamsters (Lubicz-Narocki, 1973a), progesterone has been suggested to inhibit the conversion of testosterone to dihydrotestosterone, and this has been confirmed in-vivo and in-vitro (Djøseland, 1976; Brown and Amann, 1984) and in a partially purified enzyme preparation (Monsalve and Blaquier, 1977). Progesterone can also be reduced to 5α-reduced dihydroprogesterone by this enzyme (Monsalve and Blaquier, 1977).

6. Interactions of Steroids with Spermatozoa

a. Endogenous Steroids

The steroid content of bovine epididymal spermatozoa and the fluid in which they reside in the epididymis has been quantified by Ganjam and Amann (1976). The major steroids in epididymal fluid (dihydrotestosterone and testosterone) are not the main steroids in spermatozoa. This observation suggests that steroids do not enter the sperm cell passively, but that some selective accumulation occurs, reflecting, perhaps lipid solubility or protein binding within the cell.

b. Exogenous Steroids

The uptake of ^{3}H-steroids into charcoal-stripped, steroid-free bovine spermatozoa reflects their lipid solubility (Amann and Hammerstedt, 1976) reflecting the order (progesterone, oestradiol, dihydrotestosterone, testosterone) found in spermatozoa of rabbits (Ericsson et al., 1967), man (Briggs, 1974) and rats (Galena et al., 1974). Steroid uptake is increased if the cells are washed (Ericsson et al., 1967; Hyne and Boettcher, 1977). Uptake into the cell is not active (Hyne and Boettcher, 1977) and there is little evidence that it is mediated by classical receptors (see Section III.B.6.c), so non-specific uptake into the membrane is implied. A membranous environment is also suggested by steroid-induced alterations in the fluorescence of human spermatozoa (Mercado and Reyes, 1974) and the binding of tetracycline (Pedron et al., 1973; Briggs, 1974).

c. Steroid Receptors in Spermatozoa

The inability of unlabelled exogenous steroids to compete with bound labelled steroids for binding sites and the lack of increased labelling with time of incubation (Ericsson et al., 1967) or with steroid concentration (Amann and Hammerstedt, 1976) argue against their being receptors for steroids in spermatozoa. However, some non-specific competitive effects are displayed by spermatozoa, whereby steroids other than the bound steroid can displace labelled hormones (Galena et al., 1974; Briggs, 1976; Terner, 1977; Hyne and Boettcher, 1977, 1978; Hernandez-Perez et al., 1979; Cheng et al., 1981a).

d. Site of Binding

Radioautographic studies at the light microscopical level have indicated that the sperm mid-piece accumulates more steroids than the neck or head with the least being bound to the prinicipal piece (Cheng et al., 1981a). Ericsson et al. (1967) stated that activity was localised over the surface of the rabbit spermatozoon and radioautography at the EM level has permitted grains to be localised within 10 μm of the plasma membrane over all regions of the sperm (Cheng et al., 1981b). Most of the bound steroid is associated with the plasma membrane of human spermatozoa (Hernandez-Perez et al., 1979) and Triton reduces the amount of steroid bound, although not all steroid is removed. More testosterone was lost than progesterone and or oestradiol suggesting the penetration of the latter steroids into the cytoplasm.

e. Nuclear Binding

Sanborn et al. (1975) showed by nuclear exchange assay that a considerable amount of androgens bound to the nucleus of rat epididymal spermatozoa and Frankel and Chapman (1984) showed that this nuclear binding to sperm differed from that in other cells by its resistance to salt extraction.

7. Role of Steroids in Spermatozoal Function

a. Permeability

Human spermatozoa lose components absorbing at 260nm if treated with progesterone or oestradiol (Hyne et al., 1978) and a loss of enzyme cofactors in this way could explain the inhibitory effects of steroids on metabolism (see Section III.B.8).

b. Agglutinability

The sperm-agglutinating ability of human serum is due to steroids within it and not immunoglobulins (see Boettcher, 1974), although steroid-lipoprotein complexes may be involved. Steroids added to charcoal-stripped sera induce agglutination of spermatozoa (Boettcher and Kay (1973): testosterone and progesterone promote their association; dehydroepiandrosterone, pregnenolone, oestradiol and oestrone prevent it, and cholesterol has no effect. An action of steroids in binding proteins to the sperm head was postulated to explain the agglutination (Boettcher and Kay, 1973).

Macromolecules are not essential for steroids to agglutinate sperm, however, for washed bull sperm display head-to-head association in the presence of oestradiol and corticosterone, but here testosterone and dihydrotestosterone are inactive and progesterone dissociates the complexes. The action of the steroid in these conditions is dependent on the calcium concentration (Lindahl, 1978). Cortisol promotes tail-to-tail and head-to-tail association of porcine spermatozoa (Cheng et al., 1980). Dilution induces agglutination of epididymal spermatozoa (Dacheux et al., 1983), but whether this is caused by steroids or the prevention of agglutination by epididymal fluid reflects the removal of steroids by binding, is unknown.

c. Sperm Metabolism

Steroids are mainly inhibitory to glycolytic and oxidative pathways of both epididymal spermatozoa (Voglmayr and Amann, 1973a; Hammerstedt and Amann, 1976) and those from the testis (Voglmayr et al., 1970; Voglmayr, 1971; Voglmayr and Amann, 1973a) in-vitro; reduce the conversion of glucose to amino acids by ovine testicular spermatozoa (Voglmayr et al., 1970c) and lower inositol utilisation by bovine testicular spermatozoa (Voglmayr and Amann, 1973b). In contrast, steroids increase lipid synthesis by bovine epididymal and testicular spermatozoa (Voglmayr, 1971; Voglmayr and Amann, 1973a). All these effects are achieved at concentrations of steroids (10–100 μg/ml), *i.e.* far exceeding those encountered by spermatozoa physiologically (see Voglmayr, 1975; Table 11). At physiological levels no effects on the metabolism of spermatozoa were observed (Hammerstedt and Amann, 1976), although it has been argued that the prolonged contact that spermatozoa have with steroids in the epididymis may be equivalent to conditions of short-term high concentrations provided in-vitro (see Voglmayr, 1975). Progesterone and oestradiol reduce glucose oxidation, glycolysis and the Kreb's cycle by human spermatozoa (Hyne and Boettcher, 1978; Hyne et al., 1978) and this may be due to the loss of coenzymes (see Section III.B.7.a).

d. Sperm Motility

The difficulties in assessing sperm motility quantitatively (see Section II.A.3.a) may explain the inconsistencies in reports of steroid effects on human sperm motility (Hyne and Boettcher, 1978; Hyne et al., 1978; Ridley and Blasco, 1981). The distance migrated by spermatozoa in capillary tubes is a better measure of motility than percentage beating cells, but even here the results depend on the medium used. When added to cervical mucus, for example, progesterone delays migration, but because the action is mediated by the cervical mucus which is rendered viscous (Kesseru et al., 1975), a direct effect of the steroid on the sperm cell is unclear. Progesterone is inhibitory to human sperm motility in albumin-containing Tyrode-Ringer (Beck et al., 1976) and blood serum (Hyne and Boettcher, 1978). In Tyrode's medium oestradiol promotes the motility of human spermatozoa (Beck et al., 1976; Cheng and Boettcher, 1981) and cortisol promotes forward motility in porcine spermatozoa but not those of man (Cheng et al. 1980). Migration of spermatozoa in serum exceed those in Tyrode's suggesting that factors in serum can obscure the effects of the added steroids (Cheng and Boettcher, 1979): for example, testosterone, which promotes motility in serum (Hyne and Boettcher, 1978) is less effective in Tyrode's (Beck et al., 1976. Steroids may act by modulating the ion pump activity of spermatozoa (Wester and Foote, 1972).

e. Sperm Capacitation

Contraceptive steroids block the capacitation of spermatozoa by cumulus cells in-vivo and in-vitro (Gwatkin and Williams, 1970) and in-vivo (Briggs, 1973). Desmosterol sulphate also blocks capacitation by cumulus cells (Bleau et al., 1975).

f. Chromatin Condensation

Androgens control the extent of disulphide bond formation in the nuclei of maturing rat sperm (Huang and Nieschlag, 1984).

8. Summary

The androgenic control of epididymal function provides support for the view that the epididymis influences the spermatozoa independent of any ageing of the sperm. While steroid action *via* nuclear receptors appears similar to that in other tissues, the roles of oestrogens and aldosterone have yet to be demonstrated. By maintaining the structure of the epitheium steroids would influence all resorptive and secretory functions of the epithelial cells. The role of extracellular steroid binding proteins and steroid metabolites requires further examination. Steroid metabolites formed in the epididymis and secreted into the lumen may interact directly with spermatozoa, although short-term in-vitro experiments have failed to demonstrate significant effects of steroids at physiological concentrations on spermatozoa. It is doubtful that steroids secreted by the epididymis would influence the spermatozoa after ejaculation.

9. References

Aafjes JH and Vreeburg JTM (1972) Distribution of 5α-dihydrotestosterone in the epididymis of bull and boar, and its concentration in rat epididymis after ligation of efferent testicular ducts, castration and unilateral gonadectomy. J Endocr 53:85–93

Amann RP and Ganjam VK (1976) Steroid production by the bovine testis and steroid transfer across the pampiniform plexus. Biol Reprod 15:695–703

Amann RP and Hammerstedt RH (1976) Binding of steroids by intact bovine spermatozoa. Biol Reprod 15:670–677

Anthony CT, Danzo BJ and Orgebin-Crist MC (1984a) Investigations of the relationship between sperm fertilizing ability and androgen-binding protein in the hypophysectomized, pregnenlonone injected rat. Endocrinology 114:1419–1425

Anthony CT, Danzo BJ and Orgebin-Crist (1984) Investigations on the relationship between sperm fertilizing ability and androgen-binding protein in the restricted rat. Endocrinology 114:1413–1418

Back DJ (1975) The presence of metabolites of ^{3}H-testosterone in the lumen of the cauda epididymidis of the rat. Steroids 25:413–420

Baillie AH, Ferguson MM and Hart DMcK (1966) Developments in steroid histochemistry. Academic Press, London. pp 60–71

Bardin CW, Musto N, Gunsalas G, Kotite N, Cheng SL, Larrea F and Becker R (1981) Extracellular androgen binding proteins. Ann Rev Physiol 43:189–198

Beck KJ, Herschel S, Hungershofer and Schwinger E (1976) The effect of steroid hormones on motility and selective migration of X- and Y-bearing human spermatozoa. Fert Steril 27:407–412

Blaquier JA (1971) The intracellular binding of androgens in rat epididymis. Acta Physiol Latinoam 21:97–99

Blaquier JA and Calandra RS (1973) Intranuclear receptors for androgens in rat epididymis. Endocrinology 93:51–60

Blaquier JA, Cameo MS and Charreau EH (1970) Comparative uptake in androstenediol, testosterone and dihydrotestosterone by tissues of the male rat. J Steroid Biochem 1:327–334

Bleau G, vandenHeuvel WJA, Anderson CF and Gwatkin RBL (1975) Desmosterol sulphate of hamster spermatozoa, a potent inhibitor of capacitation in vitro. J Reprod Fert 43:175–178

Boettcher B (1974) The molecular nature of spermagglutinins and sperm antibodies in human sera. J Reprod Fert Suppl 21:151–167

Boettcher B and Kay DJ (1973) Agglutination of spermatozoa by human sera with added steroids. Andrologia 5:265–271

Booth WD (1983) In-vitro metabolism of unconjugated androgens, oestrogens and the sulphate conjugates of androgens and oestrone by accessory sex organs of the mature domestic boar. J Endocr 96:457–464

Booth WD and Jones R (1979) Metabolism of testosterone and 5α-reduced androgens by the rabbit prostate and epididymis. Studies in vitro and in vivo. J Endocr 82:207–214

Boujard D, Thomas D and Garnier DH (1982) Regional distribution of steroids in the dog epididymis. Reprod Nutr Develop 22:931–939

Boulanger P, de Saulniers M, Dupuy GM, Bleau G, Roberts KD and Chapdelaine A (1982) Androgen levels in the liquid of the canine vas deferens and peripheral plasma. J Endocr 93:109–115

Boulanger P, de Saulniers M, Bleau G, Roberts KD and Chapdelaine A (1983) Sex steroid concentrations in plasma from the canine deferential vein. J Endocr 96:223–228

Bouthillier M, Chapdelaine A, Bleau G and Roberts KD (1981) Steroid sulfotransferase in hamster epididymis. Steroids 38:523–535

Bouthillier M, Bleau G, Chapdelaine A and Roberts K (1984) Distribution of steroid sulfotransferase in the male hamster reproductive tract. Biol Reprod 31:936–941

Bouthillier M, Bleau G, Chapdelaine A and Roberts KD (1985a) The assay and partial characterization of 3β-hydroxysteroid sulfotransferase of the hamster epididymis. Can J Biochem Cell Biol 63:71–76

Bouthillier M, Bleau G, Chapdelaine A and Roberts KD (1985b) The purification of 3β-hydroxysteroid sulfotransferase of the hamster epididymis. J Steroid Biochem 22:733–738

Briggs MH (1973) Steroid hormones and the fertilizing capacity of spermatozoa. Steroids 22:547–551

Brooks DE (1976) Control of glycolytic enzymes by androgens in the rat epididymis. J Endocr 71:355–365

Brooks DE (1979a) Biochemical Environment of Maturing Spermatozoa. In: The Spermatozoon. Eds DW Fawcett and JM Bedford, Urban and Schwarzenberg, Baltimore. pp 23–34

Brooks DE (1979b) Influence of androgens on on the weights of the male accessory reproductive organs and on the activities of mitochondrial enzymes in the epididymis of the rat. J Endocr 82:293–303

Brown DV and Amann RP (1984) Inhibition of testosterone metabolism in cultured rat epididymal principal cells by dihydrotestoerone and progesterone. Biol Reprod 30:67–73

Brown DV, Amann RP and Wagley LM (1983) Influence of rete testis fluid on the metabolism of testosterone by cultured principal cells isolated from the proximal or distal caput of the rat epididymis. Biol Reprod 28:1257–1268

Calandra RS and Blaquier JA (1973) Receptors for androgens in rat epididymis. Interactions with 5α-androstane-3α,17β-diol. J Steroid Biochem 4:525–531

Calandra RS, Blaquier JA, del Castillo EJ and Rivarola MA (1975) Androgen dependence of the androgen receptors in rat epididymis. Biochem Biophys Res Commun 67:97–102

Calandra RS, Purvis K, Attramadal A and Hansson V (1977) Androgen receptors in the rat epididymis do not disppear after castration. J Steroid Biochem 8:1205–1206

Carreau S, Drosdowsky MA and Courot M (1984a) Androgen-binding proteins in sheep epididymis: characterization of a cytoplasmic androgen receptor in the ram epididymis. J Endocr 103:273–279

Carreau S, Drosdowsky MA and Courot M (1984b) Androgen-binding proteins in sheep epididymis: age-related effects on androgen-binding protein, cytosolic androgen receptor and testosterone concentrations. Correlations with histological studies. J Endocr 103:281–286

Cheng CY and Boettcher B (1979) The effect of steroids in the in vitro migration of washed human spermatozoa in modified Tyrodes solution or in fasting human blood serum. Fert Steril 32:566–570

Cheng CY and Boettcher B (1981) Effects of steroids on the forward migration of human spermatozoa. Contraception 24:183–194

Cheng CY, Boettcher B, Tinneberg HR and Buxton J (1980) Activation of spermatozoa forward migration in vitro by hydrocortisone. Int J Androl 3:654–670

Cheng CY, Boettcher B, Rose RJ, Kay DJ and Tinneberg HR (1981a) The binding of sex steroids to human spermatozoa. An autoradiographic study. Int J Androl 4:1–17

Cheng CY, Rose RJ and Boettcher B (1981b) The binding of oestradiol-17β to human spermatozoa – an electron microscope autoradiographic study. Int J Androl 4:304–313

Chubb C and Desjardins C (1982) Vasculature of the mouse, rat, and rabbit testis-epididymis. Am J Anat 165:357–372

Cooke GM and Robaire B (1985) Modulation of epididymal Δ4-steroid 5α-reductase activity in vitro by the phospholipid environment. J Biol Chem 260:7789–7495

Cooper TG (1980) The general importance of proteins and other factors in the transfer of steroids into the rat epididymis. Int J Androl 3:333–348

Cooper TG and Waites GMH (1979a) Factors affecting the entry of testosterone into the lumen of the cauda epididymidis of the anaesthetized rat. J Reprod Fert 56:165–174

Cooper TG and Waites GMH (1980) Role of proteins in the transport of steroids into the epididymis. In: Testicular Development, Structure, and Function. Eds A Steinberger and E Steinberger, Raven Press, New York. pp 299–304

Cunningham GR, Tindall DJ and Means AR (1979) Differences in steroid specificity for rat androgen binding protein and the cytoplasmic receptor. Steroids 33:261–276

Dacheux JL, Paquignon M and Combarnous Y (1983) Head-to-head agglutination of ram and boar epididymal spermatozoa and evidence for an epididymal antagglutinin. J Reprod Fert 67:181–189

Danzo BJ and Eller BC (1978) Androgen metabolism by and binding to mature rabbit epididymal tissue: studies on cytosol. J Steroid Biochem 9:209–217

Danzo BJ and Eller BC (1979a) Androgen metabolism by and binding to mature rabbit epididymal tissue: the effects of castration. J Steroid Biochem 10:87–94

Danzo BJ and Eller BC (1979b) The presence of a cytoplasmic estrogen receptor in sexually mature rabbit epididymides: comparison with the estrogen receptor in immature rabbit epididymal cytosol. Endocrinology 105:1128–1134

Danzo BJ and Eller BC (1980) Androgen metabolism by mature rabbit epididymal tissue: the effects of castration and androgen replacement. J Steroid Biochem 13:661–667

Danzo BJ and Eller BC (1985) The ontogeny of biologically active androgen-binding protein in rat plasma, testis, and epididymis. Endocrinology 217:1380–1388

Danzo BJ, Eller BC and Orgebin-Crist MC (1973) Characterization of a cytoplasmic receptor for 5α-dihydrotestosterone in the caput epididymidis of intact rabbits. Endocrinology 92:310–317

Danzo BJ, Eller BC, Orgebin-Crist MC and Cooper TG (1975a) Effects of efferent duct ligation on 5α-dihydrotestosterone binding and sperm maturation. In: Control of Male Fertility. Eds JJ Sciarra, C Markland and JJ Speidel, Harper and Row, Hagerstown. pp 86–96

Danzo BJ, Eller BC, Judy LA, Trautman JR and Orgebin-Crist MC (1975b) Estradiol binding in cytosol from epididymides of immature rabbits. Mol Cell Endocrinol 2:91–105

Danzo BJ, Eller BC and Orgebin-Crist MC (1976) Approaches to fertility regulation through an understanding of the molecular control of epididymal function. In: Sperm Action. Progress in Reproductive Biology Volume 1. Eds PO Hermite, M L'Hermite and J Schirren, Karger, Basel. pp 74–88

Danzo BJ, Cooper TG and Orgebin-Crist MC (1977a) Androgen binding protein (ABP) in fluids collected from the rete testis and cauda epididymidis of sexually mature and immature rabbits and observations on morphological changes in the epididymis following ligation of the ductuli efferentes. Biol Reprod 17:64–77

Danzo BJ, Wolfe MS and Curry JB (1977b) The presence of an estradiol binding component in cytosol from immature rat epididymides. Mol Cell Endocrinol 6:271–279

Danzo BJ, Sutton W and Eller BC (1978) Analysis of (^{3}H)estradiol binding to nuclei prepared from epididymides of sexually immature intact rabbits. Mol Cell Endocrinol 9:291–301

Danzo BJ, Taylor CA and Schmidt WN (1980) Binding of the photoaffinity ligand 17β-hydroxy-4,6-androstadien-3-one to rat androgen-binding protein: comparison with the binding of 17β-hydroxy-5α-androstan-3-one. Endrocrinology 107:1169–1175

Danzo BJ, St Raymond PA and Davies J (1981) Hormonally responsive areas of the reproductive system of the male guinea pig. III. Presence of cytoplasmic estrogen receptors. Biol Reprod 25:1159–1168

Danzo BJ, Dunn JC and Davies J (1982) The presence of androgen-binding protein in the guinea-pig testis, epididymal fluid. Mol Cell Endocrinol 28:513–528

Danzo BJ, Eller BC and Hendry WJ (1983) Identification of cytoplasmic estrogen receptors in the accessory sex organs of the rabbit and their comparison with the cytoplasmic estrogen receptor in the epididymis. Mol Cell Endocr 33:197–209

de Larminat M, Monsalve A, Charreau CH, Calandra RS and Blaquier JA (1978) Hormonal regulation of 5α-reductase activity in rat epididymis. J Endocr 79:157–165

Dhabuwala CB and Pierrepoint CG (1977) Venous drainage and function of the canine prostate gland. J Endocrinol 75:105–108.

Dhabuwala CB, Roberts EE and Pierrepoint CG (1978) The radioautographic demonstration of the dynamic transfer of radio-opaque material from the deferential vein to the prostate in the dog. Invest Urol 15:346–347

Djøseland O (1976) Androgen metabolism by rat epididymis. 3. Effect of castration and anti-androgens. Steroids 27:47–64

Djøseland O (1976) Androgen metabolism by rat epididymis. 4. The formation of conjugates. Steroids 27:617–636

Djøseland O, Hansson V and Haugen HN (1973) Androgen metabolism by rat epididymis. 1. Metabolic conversion of ^{3}H-testosterone in vivo. Steroids 21:733–783

Djøseland O, Hansson V and Haugen HN (1974) Androgen metabolism by rat epididymis. 2. Metabolic conversion of ^{3}H-testosterone in vitro. Steroids 23:397–410

Djøseland O, Hastings CD and Hansson V (1976) Androgen metabolism by rat epididymis. 5. Metabolic conversion and nuclear binding after injection of 5α-androstane-3α,17β-diol, in vivo. Steroids 28:585–596

Djøseland O, Høglo S, Nyberg K, Hastings CD and Attramadal, A (1978) Uptake and metabolism of androgens by bovine spermatozoa. Steroids 31:307–317

Djøseland O, Bruchovsky N, Rennie PS, Otal N and Høglo S (1983) 5α-reductase activity in stroma and epithelium of rat prostate and epididymis. A contribution to elucidation of the mechanism for development of hyperplastic growth of prostatic tissue. Acta Endocrinol 103:273–281

Dupuy GM, Boulanger P, Roberts KD, Bleau G and Chapdelaine A (1979) Metabolism of sex steroids in the human and canine vas deferens. Endocrinology 104:1553–1558

Einer-Jensen N (1974) Local recirculation of injected (^{3}H)testosterone from the testis to the epididymal fat pad and the corpus epididymidis in the rat. J Reprod Fert 27:145–148

El Eltreby MF, Beier HM, Elger W, Mahrous AT and Topert M (1983) Immunocytochemical localization of uteroglobin in the genital tract of male rabbits. Cell Tissue Res 229:61–74

Ericsson RJ, Cornette JC and Buthala DA (1967) Binding of sex steroids to rabbit sperm. Acta Endocr 56:424–432

Ewing L and Brown B (1975) Formation and secretion of 5α-androstan-17β-ol-3-one, 5α-androstan-3α,17β-diol and 5α-androstan-3β,17β-diol by the perfused rabbit testis epididymis.Endocrinology 96:479–485

Ewing L, Brown B, Irby DR and Jardine I (1975) Testosterone and 5α-related reduced androgen secretion by rabbit testes-epididymis perfused in vitro. Endocrinology 96:610–617

Feigelson M, Noske IG, Goswami AK and Kay E (1977) Reproductive tract proteins and their hormonal control. Ann New York Acad Sci 286:273–286

Foldesy RG and Leathem JH (1981) Pubertal changes in androgen composition of rat rete testis fluid and cauda epididymal fluids. J Steroid Biochem 14:109–110

Frankel AI and Chapman JC (1984) Nuclear androgen binding sites in the male rat. III. Late spermatids and spermatozoa in the testis with an introduction to epididymal spermatozoa. J Steroid Biochem 20:1301–1311

Frankel AI and Eik-Nes KB (1970a) Testosterone and dehydroepiandrosterone in the epididymis of the rabbit. J Reprod Fert 23:441–445

Free MJ and Jaffe RA (1978) Target organs for testosterone transferred from vein to artery in the pampiniform plexus: the epididymis. Biol Reprod 18:639–642

French FS and Ritzén EM (1973a) A high-affinity androgen-binding protein (ABP) in rat testis: evidence for its secretion into efferent duct fluid and absorption by epididymis. Endocrinology 93:88–95

Galena HJ, Pillai AK and Terner C (1974) Progesterone and androgen receptors in non-flagellate germ cells of the rat testis. J Endocr 63:223–237

Gallagher TF (1928) Distribution of testicular comb growth stimulating principles in tissues. Am J Physiol 87:447–449

Ganjam VK and Amann RP (1973) Testosterone and dihydrotestosterone concentrations in the fluid milieu of spermatozoa in the reproductive tract of the bull. Acta Endocr 74:186–200

Ganjam VK and Amann RP (1976) Steroids in fluids and sperm entering and leaving the bovine epididymis, epididymal tissue, and accessory sex glands. Endocrinology 99:1618–1630

Gloyna RE and Wilson JD (1969) A comparative study of the conversion of tstosterone to 17β-hydroxy-5α-androstan-3-one (dihydrotestosterone) by prostate and epididymis. J Clin Endocr 29:970–

Gwatkin RBL and Williams DT (1970) Inhibition of sperm capacitation in vitro by contraceptive steroids. Nature (Lond) 227:182–183

Haltmeyer GC and Eik-Nes KB (1974) Transport of 5α-dihydrotestosterone by testicular lymph in the dog. J Reprod Fert 36:41–47

Hamilton DW (1971) Steroid function in the mammalian epididymis. J Reprod Fert Suppl 13:89–97

Hamilton DW, Jones AL and Fawcett, DW (1969) Cholesterol biosynthesis in the mouse epididymis and ductus deferens: a biochemical and morphological study. Biol Reprod 1:167–184

Hamilton DW and Fawcett DW (1970) In vitro synthesis of cholesterol and testosterone from acetate by rat epididymis and vas deferens. Proc Soc exp Biol Med 133:693–695

Hammerstedt RH and Amann RP (1976a) Effects of physiological levels of exogenous steroids on metabolism of testicular, cauda epididymal and ejaculated bovine sperm. Biol Reprod 15:678–685

Hammerstedt RH and Amann RP (1976b) Interconversion of steroids by intact bovine sperm and epididymal tissue. Biol Reprod 15:686–694

Hansson V, Djøseland O, Reusch E, Attramadal A and Tørgersen O (1973b) Intracellular receptors for 5α-dihydrotestosterone in the epididymis of adult rats. Comparison with the androgen receptors in the ventral prostate and the androgen binding protein (ABP) in the testicular and epididymal fluid. Steroids 22:19–33

Hansson V, Trygstad O, French FS, McLean WS, Smith AA, Tindall DJ, Weddington SC, Petrusz P, Nayfeh SN and Ritzén EM (1974) Androgen transport and receptor mechanisms in testis and epididymis. Nature (Lond) 250:387–391

Hansson V, Ritzén EM and French FS (1975) Androgen transport and receptor mechanisms in testis and epididymis. In: Handbook of Physiology, Section VII Endocrinology, Volume 5 Male Reproductive System. Eds DW Hamilton and RO Greep, Amer Physiol Soc, Washington DC. pp 173–202

Harris ME and Bartke A (1975) Maintenance of rete testis fluid testosterone and dihydrotestosterone levels by pregnenolone and other C_{21} steroids in hypophysectomized rats. Endocrinology 96:1396–1402

Hastings CD and Djøseland O (1977) Androgen metabolism by rat epididymis. 6. Metabolic conversion of 5α-androstane-3α,17β-diol in vitro. Steroids 30:5341–539

Hastings CD and Hansson V (1979) Physico-chemical characterization of the NADPH dependent soluble 3α-hydroxysteroid oxidoreductase in the rat epididymis. Int J Androl 2:263–274

Hastings CD, Brekke I, Purvis K, Attramadal A and Hansson V (1980) Cofactor dependency of soluble 3α-hydroxysteroid oxidoreductase in rat testis, prostate, and epididymis. Endocrinology 107:1762–1766

Hees H, Leiser R, Kohler T and Wrobel KH (1984) Vascular morphology of the bovine spermatic cord and testis. I. Light- and scanning electron microscopic studies on the testicular artery and pampiniform plexus. Cell Tiss Res 237:31–38

Hendry WJ, Eller BC, Orgebin-Crist MC and Danzo BJ (1985) Hormonal effects on the estrogen receptor system in the epididymis and accessory sex organs of sexually immature rabbits. J Steroid Biochem 23:39–49

Hernández-Pérez O, Ballesterios LM and Rosado A (1979) Binding of 17-β-estradiol to the outer surface and nucleus of human spermatozoa. Archs Androl 3:23–29

Hinton BT and Keefer DA (1985) Binding of (^{3}H)aldosterone to a single population of cells within the rat epididymis. J Steroid Biochem 23:231–233

Huang HFS and Nieschlag E (1984) Alteration of free sulphydryl content of rat sperm heads by suppression of intratesticular testosterone. J Reprod Fert 70:31–38

Hyne RV and Boettcher B (1977) The selective binding of steroids by human spermatozoa. Contraception 15:163–174

Hyne RV and Boettcher B (1978) Binding of steroids to human spermatozoa and its possible role in contraception. Fert Steril 30:322–328

Hyne RV, Murdoch RN and Boettcher B (1978) The metabolism and motility of human spermatozoa in the presence of steroid hormones and synthetic progestagens. J Reprod Fert 53:315–322

Inano H, Machino A and Tamaoki BI (1969) In vitro metabolism of steroid hormones by cell-free homogenates of epididymides of adult rats. Endocrinology 84:997–1003

Jean-Faucher C, Berger M, De Turckheim M, Veyssiere G and Jean C (1985) Testosterone and dihydrotestosterone levels in epididymis, vas deferens, seminal vesicle and preputial gland of mice after hCG injection. J Steroid Biochem 23:201–205

Jégou B, Dacheux JL, Garnier DH, Terqui M, Colas G and Courot M (1979) Biochemical and physiological studies of androgen-binding protein in the reproductive tract of the ram. J Reprod Fert 57:311–318

Jones R, von Glos KI and Brown CR (1980a) The synthesis of a sperm-coating protein in the initial segment of the rat epididymis is stimulated by factors in testicular fluid. IRCS Med Sci 8:56

Jones R, Brown CR, von Glos KI and Parker MG (1980b) Hormonal regulation of protein synthesis in the rat epididymis: characterization of androgen-dependent and testicular fluid-dependent proteins. Biochem J 188:667–676

Kesserü E, Camacho-Ortega P, Laudahn G and Schöpflin G (1975) In vitro action of progesterone on sperm migration in human cervical mucus. Fert Steril 26:59–61

Kinoshita Y (1981) Studies on the human epididymis: partial characterization of 3α- and 3β-hydroxysteroid dehydrogenase, regional distribution of 5α-reductase and inhibitory effect of $^4Δ$-3-oxysteroids on 5α-reductase. Endocr Jap 28:499–513

Kirchoff J, Soffie M and Rousseau GG (1979) Differences in the steroid-binding site specificities of rat prostate androgen receptor and epididymal androgen-binding protein (ABP). J Steroid Biochem 10:487–497

Klinefelter GR and Amann RP (1980) Metabolism of testosterone by principal cells and basal cells isolated from the rat epididymal epithelium. Biol Reprod 22:1149–1154

Klinefelter GR, Amann RP and Hammerstedt RH (1982) Culture of principal cells from the rat caput epididymidis. Biol Reprod 26:885–901

Kumari GL, Allag IS, Das RP and Datta JK (1980) Regional differenes in steroidogenesis and hormone levels in the epididymis and vas deferens of adult rats. Int J Androl 3:267–281

Larreau F, Musto N, Gunsalas G and Bardin CW (1981) The microheterogeneity of rat androgen-binding protein from the testis, rete testis fluid, and epididymis, as demonstrated by immunoelectrophoresis and photoaffinity labelling. Endocrinology 109:1212–1220

Lau IF and Saksena SK (1979) Steroids in the rete testis fluid of fertile male rabbits. Archs Androl 2:49–52

Lau IF and Saksena SK (1981) Effects of Colprone on in vitro release of androgens from the reproductive organs of the male rat. Int J Androl 4:291–296

Lawless JJ (1931) Effects of castration leaving the epididymis intact in the rat. Proc Soc exp Biol and Med 29:232

Legault Y, vandenHeuvel WJA, Arison BH, Bleau G, Chapdelaine A and Roberts KD (1978) 5α-cholesta-7,24 -dien-3β-ol as a major sterol of the male hamster reproductive tract. Steroids 32:649–658

Legault Y, Bouthillier M, Bleau G, Chapdelaine A and Roberts KD (1979a) The sterol and sterol sulfate content of the male hamster reproductive tract. Biol Reprod 20:1213–1219

Leinonen P, Hammonis GL and Vihko R (1980) Testosterone and some of its precursors and metabolites in the human epididymis. J Clin Endocr Metab 51:423–428

Lewis MH and Moffat DB (1975) The venous drainage of the accessory reproductive organs of the rat with special reference to prostatic metabolism. J Reprod Fert 42:497–502

Lipschultz LZ, Tsai YH, Sanborn BM and Steinberger E (1977) Androgen-binding activity in the human testis and epididymis. Fert Steril 28:947–951

Lincoln GA (1975) An effect of the epididymis on the growth of antlers of castrated red deer. J Reprod Fert 42:159–161

Lindahl PE (1978a) Effects of some steroid hormones on head-to-head association in bovine spermatozoa. Exp Cell Res 111:73–81

Lobl TJ (1981) Androgen transport proteins: physical properties, hormonal regulation and possible mechanims of TeBG and ABP action. Archs Androl 7:133–151

Lobl TJ, Campbell JA, Tindall DJ, Cunningham GR and Means AR (1980) A model for the mechanism of androgen binding, transport, and translocation to the nucleus. In: Testicular Development, Structure, and Function. Eds A Steinberger and A Steinberger, Raven Press, New York. pp 323–330

Lubicz-Nawrocki CM (1973a) Anomalous effects of progesterone on the maturation and survival of spermatozoa in the epididymis of the golden hamster. J Endocr 58:199–205

Lubicz-Nawrocki CM (1973b) The effect of low doses of human chorionic gonadotrophin on the fertilizing ability of hamster epididymal spermatozoa. J Endocr. 58:185–192

Lubicz-Nawrocki CM (1974) The inhibition of fertilizing ability of epididymal spermatozoa by the administration of oestradiol benzoate to testosterone-maintained hypophysectomised or castrated hamsters. J Endocr 61:133–138

Lubicz-Nawrocki CM and Glover TD (1973) The effects of hypophysectomy and the paradoxical effect of human chorionic gonadotrophin on the viability of hamster spermatozoa. J Endocr 58:177–184

Lubicz-Nawrocki CM, Hooker RH and Chang MC (1973) Fertilizing ability of hamster spermatozoa following subcutaneous implantation or insertion into seminal vesicles of silastic tubules containing estradiol benzoate. Fert Steril 24:872–876

McGadey J, Baillie AH and Ferguson MM (1966) Histochemical utilization of hydroxysteroids by the hamster epididymis. Histochemie 7:211–217

Mercado E and Reyes A (1974) Interaction of hormonal steroids with human spermatozoa. J Steroid Biochem 5:380 (abst)

Milner AJ and Hamilton DW (1971) Ultrastrucutrual criteria for assessing the functional integrity of endocrine cells in vitro. Acta Endocri Suppl 153:62–80

Mock EJ, Kamel F, Wright WW and Frankel AI (1975) Plasma testosterone levels in vasectomized rats. J Reprod Fert 44:575–578

Moniem KA (1972) Histochemical localization of 3β-hydroxysteroid dehydrogenase in the rat epididymis. J Reprod Fert 28:461–462

Monsalve A and Blaquier JA (1977) Partial characterization of epididymal 5α-reductase in the rat. Steroids 30:41–51

Morfin RF, Leav I, Orr JC, Picart D and Öfner P (1980) C_{19} steroid metaboism by canine prostate, epididymis and perianal glands. Applications of the twin-ion technique of gas chromatography/mass spectrometry to establish 7α-hydroxylation. Eur J Biochem 109:119–127

Mulder E, van Beurden-Lamers WMO, de Boer W, Brinkman AO and van der Molen HJ (1974) Testicular estradiol receptors in the rat. In: Hormone Binding and Target Cell Activation in the Testis. Eds ML Dufau and AR Means, Current Topics in Endocrinology volume 1, Plenum Press, New York. pp 343–355

Murphy JB, Emmott RC, Hicks LL and Walsh PC (1980) Estrogen receptors in the human prostate, seminal vesicle, epididymis, testis, and genital skin: a marker for estrogen-responsive tissues. J Clin Endocr Metab 50:938–948

Noske IG and Fiegelson M (1976) Immunological evidence of uteroglobin (blastokinin) in the male reoproductive tract and in non-reproductive ductal tissues and their secretions. Biol Reprod 15:704–713

Noske IG and Gooding M (1978) Evidence of a uteroglobin-like protein in epithelial cells of reproductive and non-reproductive tissues of the rabbit. J Reprod Fert 54:193–196

O'Connor JJ (1950) Cutting a horse "Proud". In: Dollar's Veterinary Surgery, General, Operational and Regional, Part II, 4th Edition. Ed JJ O"Connor, Bailliere, Tindall and Cox, London. p 376

Öfner PO, Vena RL and Morfin RF (1974a) Acetylation and hydroxylation of 5α-androstane-3β,17β-diol by prostate and epididymis. Steroids 24:261–279

Öfner O, Vena RL and Morfin RF (1974b) In vivo metaboloism of C_{19}-steroids in prostatic tissue. In: Normal and Abnormal Growth of the Prostate. Ed M Golan, CC Thomas, Springfield Illinois. pp 111–124

Öfner P, Leav I and Cavazos LF (1974c) C_{19}-steroid metabolism in male accessory sex glands. Correlation of changes in fine structure and radiometabolite patterns in the prostate of the androgen-deprived dog. In: Male Accessory Organs. Structure anf function in mammals. Ed DE Brandes, Acad Press, New York. pp 267–305

Öfner P, Vena RL, Leav I and Hamilton DW (1979) Metabolism of C_{19}-radiosteroids by explants of canine prostate and epididymis with disposition as hydroxylated products: a possible mechanism for androgen inactivation. J Steroid Biochem 11:1367–1379

Pearce PT, Lipkevic OR and Funder JM (1986) High-affinity (Type-1) aldosterone-binding sites in rat epididymis. Endocrinology 118:2072–2075

Pedron N, Hicks JJ, Lara-Ricalde R and Rosado (1973) Effect of some contraceptive steroids upon human spermatozoa. Contraception 7:213–226

Peyre A and Laporte P (1966) Action de la testosterone et de l'oestradiol sur les sialo-proteins de la queue de l'épididyme de rat impuibere castre. CR Soc de Biol 160:2178–2180

Pierrepoint CG (1974) Does hormone transfer along the vasa deferentia contribute to the control of prostatic function and could it be a factor in the etiology of prostatic hyperplasia? In: Normal and Abnormal Growth of the Prostate. Ed M Golan, CC Thomas, Springfield. pp 517–529

Pierrepoint CG and Davies P (1973) The effect of vasectomy on the activity of prostate RNA Polymerase. J Reprod Fert 35:149–152

Pierrepoint CG and Davies P (1984) A study of the androgenic function of the epididymis. J Steroid Biochem 20:1105–1111

Pierrepoint CG and John BM (1974a) A direct control of the prostate and seminal vesicles *via* the vas deferens. J Steroid Biochem 5:383

Pierrepoint CG, Davies P and Wilson DW (1974) The role of the epididymis and ductus deferens in the direct and unilateral control of the prostate and seminal vesicles of the rat. J Reprod Fert 41:413–423

Pierrepoint CG, Davies P, Millington D and John B (1975a) Evidence that the deferential vein acts as a local transport system for androgen in the rat. J Reprod Fert 43:293–303

Pierrepoint CG, Davies P, Lewin MH and Moffat DR (1975b) Examination of the hypothesis that a direct control system exists for the prostate and seminal vesicles. J Reprod Fert 44:395–409

Podesta EJ, Calandra RS, Rivarola MA and Blaquier JA (1975) The effect of castration and testosterone replacement in specific proteins and androgen levels of the rat epididymis. Endocrinology 97:399–405

Prakash A and Moore HDM (1982) Localization of Δ^5-3β- and 17β-hydroxysteroid dehydrogenase activity in the efferent ducts, epididymis and vas deferens of the rabbit, hamster and marmoset monkey. J Reprod Fert 66:95–100

Pujol A and Bayard F (1978) 5α-reductase and 3β-hydroxysteroid oxidoreductase enzyme activities in epididymis and their control by androgen and rete testis fluid. Steroids 31:485–493

Pujol A and Bayard F (1979) Androgen receptors in the rat epididymis and their hormonal control. J Reprod Fert 56:217–222

Pujol A, Bayard F, Louvet JP and Boulard C (1976) Testosterone and dihydrotestosterone concentrations in plasma, epididymal tissues, and seminal fluid of adult rats. Endocrinology 98:111–113

Purvis K and Hansson V (1978) Androgens and androgen-binding protein in the rat epididymis. J Reprod Fert 52:59–63

Purvis K, Calandra R, Sander S and Hansson V (1978) Androgen binding proteins and androgen levels in the human testis and epididymis. Int J Androl 1:531–548

Rajalakshmi M, Leask JTS and Waites GMH (1978) In vitro metabolism of androgens by mammalian cauda epididymal spermatozoa. Steroids 31:747–760

Rastogi RK, Saksena PK and Iela L (1976) Histochemical localization of some hydroxysteroid dehydrogenases in the mouse epididymis. Horm Res 7:43–48

Ridley AJ and Blasco L (1981) Testosterone and gossypol effects on human sperm motility. Fert Steril 36:638–642

Ritzén EM (1984) cited in Pierrepoint and Davies, 1984

Ritzén EM and French FS (1974) Demonstration of an androgen binding protein (ABP) in rabbit testis: secretion in efferent duct fluid and passage into epididymis. J Steroid Biochem 5:151–154

Ritzén EM, Nayfeh SN, French FS and Dobbins MC (1971) Demonstration of androgen-binding components in rat epididymis cytosol and comparison with binding components in prostate and other tissues. Endocrinology 89:143–151

Ritzén EM, Hagenäs L, Hansson V, Weddington SC, French FS and Nayfeh SN (1975) Androgen binding and transport in testis and epididymis. Vitam and Horm 33:283–295

Robaire B (1979) Effects of unilateral orchidectomy on rat epididymal Δ^4-5α-reductase and 3α-hydroxysteroid dehydrogenase. Can J Physiol Pharmacol 57:998–1003

Robaire B and Buchanan LJ (1985) The use of epididymal cell and organ culture in toxicology. Concepts in Toxicology 3:114–129.

Robaire B and Zirkin BR (1981) Hypophysectomy and simultaneous testosterone replacement: effects on male reproductive tract and epididymal Δ^4-5α-reductase and 3α-hydroxysteroid dehydrogenase. Endocrinolgy 109:1225–1233

Robaire B, Ewing LL, Zirkin BR and Irby DC (1977) Steroid Δ^4-5α-reductase and 3α-hydroxysteroid dehydrogenase in the rat epididymis. Endocrinology 101:1379–1390

Robaire B, Scheer H and Hachey C (1981) Regulation of epididymal metabolizing enzymes. In: Bioregulators of Reproduction. Eds G Jagielle and HJ Vogel, Acad Press, London. pp 487–498

Rommerts FFG, Grootegoed JA and Van Der Molen HJ (1976) Physiological role for androgen binding protein-steroid complex in testis? Steroids 28:43–49

Rousseau GG, Quivy JI, Kirchoff J, Bui XH and Devis R (1980) Nonsteroidal compounds which bind epididymal androgen-binding protein but not the androgen receptor. Nature (Lond) 284:458–459

Rousseau GG, Quivy JJ and Crabbé P (1981) Interaction of A-nor, A,19-dinor, and A-homo-5α-androstane derivatives with the androgen binding receptor and epididymal androgen binding protein. Steroids 37:383–392

Rousseau GG, Quivy JI, Rolin Jacquemyns CF, Street DAN, de Reviers MT, Viguer-Martinez MC, Delpech S and Colas G (1985) Non-steroidal inhibitors of androgen transport and metabolism. In: Male Fertility and its Regulation. Eds TJ Lobl and ESE Hafez, MTP Press Ltd, Lancaster. pp 331–342

Saksena SK, Lau IF and Chang M (1976) The inhibition of the conversion of testosterone into 5α-dihydrotestosteron in the reproductive organs of the male rat. Steroids 27:751–757

Sanborn BM, Elkington JSH, Steinberger A, Steinberger E and Meistrich L (1975) Androgen binding in the testis: in vitro production of androgen binding protein (ABP) by Sertoli cell cultures and measurement of nuclear bound androgen by a nuclear exchange assay. In: Hormonal regulation of Spermatogenesis. Eds FS French, V Hansson and EM Ritzén, Plenum Press, New York USA. pp 293–309

Scheer H and Robaire B (1982) Characterization of epididymal Δ^4-5α-reductase. Ann NY Acad Sci 383:496–498

Schleicher G, Drews U, Stumpf WE and Sar M (1984) Differential distribution of dihydrotestosterone and estradiol binding sites in the epididymis of the mouse. An autoradiographic study. Histochemistry 81:139–147

Schmidt WN, Taylor CA and Danzo BJ (1981) The use of a photoaffinity ligand to compare androgen binding protein (ABP) present in rat Sertoli cell culture media with ABP present in epididymal cytosol. Endocrinology 108:786–794

Skinner JD and Rowson LEA (1968a) Some effects of unilateral cryptorchidism and vasectomy on sexual development of the pubescent ram and bull. J Endocr 42:311–321

Skinner JD and Rowson LEA (1968b) Effects of testosterone injected unilaterally down the vas deferens and the accessory glands on the ram. J Endocr 42:355–356

Sommerville IF (1971) Some data on steroid transformation in human epididymal and prostatic tissue. In: Basic Actions of Sex Steroids in Target Organs. Eds PO Hubinot, F LeRoy and P Galand, Karger, Basel. pp 265–266

Sowell JG and Eik-Nes KB (1972) Formation in vivo of 5α-dihydrotestosterone by the canine epididymis. Proc Soc exp Biol Med 141:827–829

Stumpf WE and Sar M (1976) Autoradiographic localization of estrogen, androgen, progestin, and glucocorticosteroid in "target tissues" and "non target tissues". In: Receptors and Mechanisms of Action of Steroid Hormones, Part I. Ed JR Pasqualini, Marcel Dekker Inc, New York USA. pp 41–84

Sulková J and Starká L (1973) The metabolism of androgens in normal human testis and epididymis in vitro. Endocrinologia Experimentalis 7:113–118

Taylor CA, Smith HE and Danzo BJ (1980) Characterization of androgen binding protein in rat epididymal cytosol using a photoaffinity label. J Biol Chem 255:7769–7773

Terner C (1977) Progesterone and progestins in the male reproductive system. Proc Nat Acad Sci 286:313–320

Tezón JG and Blaquier J (1983) Androgens control androgen-binding sites in rat epididymis. Endocrinooogy 113:1025–1030

Tezón JG, Vazquez MH and Blaquier JA (1982) Androgen-controlled subcellular ditribution of its receptor in the rat epididymis: 5α-dihydrotestosterone-induced translocation is blocked by anti-androgens. Endocrinology 111:2039–2045

Tezón J, Vazquez M, Pineiro L, De Larminat MA and Blaquier J (1983) The effect of androgens in the human epididymis in organ culture. In: Therapy in Andrology. Pharmacological Surgical and Pathological Aspects. Eds Menchini-Fabris, Pasini and Martini. Elsevier Press, New York. pp 53–60

Tindall DJ, French FS and Nayfeh SN (1972) Androgen uptake and binding in rat epididymal nuclei in vivo. Biochem Biophys Res Commun 49:1391–1397

Tindall DJ, Hansson V, Sar M, Stumpf WE, French FS and Nayfeh SN (1974) Further studies on the accumulation and binding of androgen in rat epididymis. Endocrinology 95:1119–1128

Tindall DJ, Cunningham GR and Means AR (1978) Structural requirements for steroid binding to androgen binding proteins. Int J Androl Suppl 2:434–448

Tindall DJ, French FS and Nayfeh SN (1981) Estradiol-17β inhibition of androgen uptake, metabolism and binding in epididymis of adult male rats in vivo: a comparison with cyproterone acetate. Steroids 37:257–268

Tsuji M, Takeyama M, Takatsuka D, Yabumoto H, Terada N and Matsumoto K (1984) 5α- and 5β-reductases for 4-ene-3-ketosteroids and 17β-ol-dehydrognease in epididymis and testis of golden hamster during sexual development. J Steroid Biochem 21:179–183

Tuohimaa P and Niemi M (1974) Cell renewal and mitotic activity of testosterone in male acccessory organs. In: Male Accessory Organs. Structure and Function in Mammals. Eds DE Brandes, Acad Press, New York. pp 329–343

Tuovinen PI and Pohjola R (1949) The importance of the epididymis in castration. Castration experiments performed on rats by means of orchidoepididymectomy and orchiectomy, using the prostate as the criterion of response. Acta Endocrinol 3:1–16

Turner TT, Cochran RC and Howards SS (1981) Transfer of steroids across the hamster blood testis and blood epididymal barriers. Biol Reprod 25:342–348

Turner TT, Jones CE, Howards SS, Ewing LL, Zegeye B and Gunsalas GL (1984) On the androgen microenvironment of maturing spermatozoa. Endocrinology 115:1925–1932

Vigersky RA, Loriaux DL, Howards SS, Hodgen GB, Lipsett MP and Chramback A (1976) Androgen binding proteins of testis, epididymis and plasma in man and monkey. J Clin Invest 58:1061–1068

Voglmayr JK (1971) The effect of testosterone and related steroids on the oxidative and glycolytic activity and on lipid synthesis of testicular and ejaculated bull spermatozoa. Acta Endocr 68:793–804

Voglmayr JK (1975) Metabolic changes in spermatozoa during epididymal transit. In: Handbook of Physiology, Sect VII Endocrinology, Vol 5 Male Reproductive System. Eds DW Hamilton and RO Greep, Amer Physiol Soc, Washington DC. pp 437–451

Voglmayr JK and Amann RP (1973a) Glucose metabolism and lipid synthesis of cauda epididymal and ejaculated bull spermatozoa in the presence of selected androgens. Acta Endocr 73:196–208

Voglmayr JK and Amann RP (1973b) The distribution of free myo-inositol in fluid, spermatozoa, and tissues of the bull genital tract and observations on its uptake by the rabbit epididymis. Biol Reprod 8:504–573

Voglmayr JK, Murdoch RN and White I (1970) Metabolism of ram testicular spermatozoa in the presence of selected steroids. Acta Endocr 65:565–576

Voglmayr JK, Musto NA, Saksena SK, Brown-Woodman PDC, Marley PB and White IG (1977) Characteristics of semen collected from the cauda epididymidis of conscious rams. J Reprod Fert 49:245–251

Vreeburg JTM (1975) Distribution of testosterone and 5α-dihydrotestosterone in rat epididymis and their concentrations in efferent duct fluid. J Endocr 67:203–210

Vreeburg JTM and Aafjes JH (1974) Dihydrotestosterone (5α-androstan-17β-ol-3-one) in the epididymis of rats. In: Current Problems in Fertility. Eds A Ingelman-Sundberg and NO Lunell, Plenum Press, New York. pp 203–206

Vreeburg JTM and Scholte HR (1973) Localization of 5α-steroid reductase in the epididymis of the rat. Acta Endocr Suppl 177:67

Vreeburg JTM, Bielske and Ooms M (1976) Maturation and survival of spermatozoa in the epididymis of pregnenolone treated hypophysectomised rats. Endocrinology 99:824–830

Weddington SC, McLean WS, Nayfeh SN, French FS, Hansson V and Ritzén EM (1974) Androgen binding protein (ABP) in rabbit testis and epididymis. Changes with age in comparison with the androgen binding protein (TeBG) in serum and separation from the intracellular receptor in epididymis. Steroids 24:123–134

Wenstrom JC and Hamilton DW (1980) Dolichol concentration and biosynthesis in rat testis and epididymis. Biol Reprod 23:1054–1059

Wester RC and Foote RH (1972) Ouabain effect on bovine spermatozoan motility and testosterone binding. Proc Soc exp Biol Med 141:26–30

Wilson EM and French FS (1976) Binding properties of androgen receptors. Evidence for identical receptors in rat testis, epididymis and prostate. J Biol Chem 251:5620–5629

Younes MA and Pierrepoint CG (1981a) Estrogen steroid-receptor binding in the canine epididymis. Andrologia 13:562–572

Younes MA and Pierrepoint CG (1981b) Androgen steroid receptor binding in the canine epididymis. The Prostate 2:133–142

Younes MA, Evans BAJ, Chaisiri N, Valotaire Y and Pierrepoint CG (1979) Steroid receptors in the canine epididymis. J Reprod Fert 56:45–52

C. Resorption and Secretion of Ions by the Epididymis

1. Luminal Contents

Techniques of micropuncture and electron microprobe analysis have located ions within different compartments of the epididymis. There are increasing intraluminal sodium and potassium concentrations distally in tissue slices from the mouse epididymis (Hamilton et al., 1976), but these measurements would include the contributions from the increasing numbers of sperm cells. In other species during maturation in the epididymis the concentration of sodium ions within spermatozoa decline in parallel with removal of Na^+ from the surrounding fluid and intracellular K^+ in sperm is always maintained above that of epididymal fluid (Table 12). Ion concentrations in epididymal fluid differ markedly from blood or testicular fluid and regional differences are present (see Table 13); most noticeable are a loss of Na^+ and an increase in K^+ with high phosphate concentrations in the rat (Hinton et al., 1980a) and Mg^{2+} in the rabbit (Jones, 1978). Despite increased concentrations, a net resorption of potassium along the length of the duct is observed when the resorption of fluid is taken into account (Turner, 1984). pH decreases sharply between the testis and the caput epididymidis of the rat (Levine and Kelly, 1978).

2. Transporting Activities

The transporting activity of the testicular excurrent duct system has been demonstrated in the efferent ducts (Burgos, 1974; Hohlbrugger, 1980), initial segments (Hohlbrugger and Pfaller, 1983), caput, corpus and cauda (Wong et al. 1978, 1979). The transport of Na^+ from the lumen is an active process requiring metabolic energy (Wong and Yeung, 1976, 1977a) and is coupled to secretion of K^+ (Wong and Yeung, 1978; Wong et al., 1980). In-vitro perfusion has shown that this K^+ arises from an extratubular source (Wong et al., 1980) and as expected, ouabain and amiloride, inhibitors of transport ATPase and passive Na^+ transport respectively, block these ion exchanges (Wong and Yeung, 1976, 1977a, 1978; Wong et al., 1980). Bicarbonate is removed from the lumen coupled to chloride secretion (Au and Wong, 1980). Absorption of water from the lumen is associated with absorption of sodium or chloride and linked to the production of iso- or hypo-tonic resorbates in the cauda and caput, respectively (Wong et al., 1978, 1979). Rates of transport of water, Na^+ and K^+ in the caput are all less than those occurring in the cauda when measured by luminal perfusion (Wong et al., 1978, 1979), although these estimates may not reflect the rates obtaining in-situ, because the normal concentrations of Na^+ in the caudal lumen are much lower than in the caput (see Table 13) and so the driving force for resorption would be lower than that provided in the experiments cited.

The enlarged intercellular spaces observed between epididymal epithelial cells in-vitro (YC Wong et al., 1978) and in-vivo (Cooper and Yeung, 1980; Yeung and Cooper, 1982) when fluid resorption from the lumen is occurring implicate the

ATPases detected on the basolateral membranes of rat epididymal epithelial cells (Brandes, 1974) in Na^+ transport. Other Mg^{2+}-dependent ATPases found in light particulate fractions of the bovine epididymis (Garbers et al., 1970), and originating from the microvilli (Reed and Takahashi, 1973), may be responsible for maintaining ionic gradients agross the apical membrane. Active electrogenic pumps in the epithelium maintain K^+-dependent membrane potentials across the apical membrane of epithelial cells from the rat epididymis (Cheung et al., 1978). The values of these potentials (–21.1mV in the caput and –30.6mV in the cauda) are less than those of excitable cells (Williams, 1970) and thus reflect a "leaky" epithelium towards ions. Initial attempts failed to detect transepithelial potential differences (Wong et al., 1980), but they have been measured with microelectrodes (Wu et al., 1985).

The retention of injected Li^+ within epididymal tissue (Svejnohova et al., 1980) and its higher concentration in epididymal fluid than testicular fluid (Voglmayr and Gandhi, 1978) suggest that this ion is also transported by the Na^+ transport ATPase. Zinc is present in porcine epididymal fluid (Bowen et al., 1973) and can be transported across the rat epididymal epithelium (Gunn and Gould, 1970), possibly on the divalent metal ion transport ATPase. The carbonic anhydrase complement of apical and clear cells in the rat epididymis (Cohen et al., 1976) may be responsible for the acidification of the caput (Levine and Kelly, 1978) and cauda (Jones, 1978).

3. Control of Transporting Activities

a. Androgens

Castration and androgen replacement studies reveal that ion concentrations in epididymal fluid are androgen-dependent in the rabbit (Jones and Glover, 1973b), although degenerating spermatozoa also contribute to luminal potassium (Jones, 1974). The effect of androgens on ion transport has been confirmed in-vitro (Wong and Yeung, 1977b) and anti-androgens cause a detectable decrease in transport by the perfused epididymis in-vivo (Wong et al., 1978, 1979) although cyproterone acetate has no discernible effect on the ion concentrations in epididymal fluid from rats or rabbits (Back et al., 1977; Jones, 1977). Changes in intraluminal ion concentrations following administration of oestrogens to rats (Jenkins et al., 1978, 1983b) are thought to reflect their anti-androgenic activity.

The membrane potential across epididymal epithelial cells falls after castration and this reflects an increased permeability towards Na^+ and decreased permeability towards K^+ (Cheung et al., 1978).

b. Other Factors

Adrenal steroids control the extent of water resorption by the perfused rat epididymis (Au et al., 1978) and treatment with an aldosterone antagonist, spironolactone, can increase intraluminal sodium in the rat caput (Jenkins et al., 1983a) and, as a consequence, decreases in the spermatocrit (packed cell volume) are observed (Wong and Lee, 1982; Turner and Cesarini, 1983). The presence in the epididymis of active metabolites of vitamin D (Kidroni et al., 1983) and receptors for them (Walters et al., 1982, 1983) are suggestive of a regulatory role for the vitamin in Ca^{2+} and phosphate transport. Prolonged exposure to abdominal temperature also compromises ion transport in the rat epididymis (Wong et al., 1982; Rasweiler and Bedford, 1982), although the gross composition of epididymal fluid in the rabbit cauda is unaltered by this treatment (Jones and Glover, 1975).

4. Role of Monovalent and Divalent Cations

a. On Spermatozoa

i. Motility

The normal ion transporting activities of the epididymis should act to reduce the motility of sperm by removing Na^+, which is required for sperm motility, and providing K^+, which inhibits motility (see Section II.A.4.d.i). In-vitro lithium inhibits the motility of rat spermatozoa (Lee et al., 1981) but its action on human spermatozoa is equivocal (MacLeod et al., 1949; Levin et al., 1981b). A role for Li^+, which in other tissues inhibits myoinositol phosphate phosphatase and can control the gating of Ca^{2+} into cells (Berridge, 1984), has not been examined in the epididymis.

If the concentration of Ca^{2+} within the epididymal lumen is directly related to sperm motility in-situ and inversely related to the concentration required for subsequent initiation of motility in-vitro, as suggested by Morton et al. (1978), then the epididymis may also regulate the motility of spermatozoa after release from the tract by controlling intraluminal Ca^{2+}. In mice the stimulating effect of Ca^{2+} on motility was considered to be mediated by an action on the plasma membrane (Heffner and Storey, 1981). In the bull, Ca^{2+} can be surface-active and induce agglutination of spermatozoa (Lindahl, 1973, 1974) and can dictate their membrane potential (McGradey and Nelson, 1974), but it is not inhibitory to bovine sperm motility (Carr and Acott, 1984). Calcium uptake into the cells is a two-step process, with binding to the membrane preceding entry into the cell (Singh et al., 1978) and intracellular calcium may control interactions with cyclic nucleotides (see Section II.D.3.b) how pH may depress the motility of spermatozoa in the epididymis (Carr et al., 1985).

ii. Capacitation

By maintaining low concentrations of Ca^{2+} in the lumen, the epididymis acts to prevent the occurrence of Ca^{2+}-dependent events, *e.g.* capacitation (see Section II.B.7) and the acrosome reaction (see Section II.D.3.a) within the epididymal canal. A similar effect would be achieved through the maintenance of high concentrations of ions which prevent the induction of the acrosome reaction in-vitro, *e.g.* K^+ (Rogers et al., 1981) and Mg^{2+} (Rogers and Yanagimachi, 1976). The acidic nature of the epididymal fluid (Bishop and Matthews, 1952; Jones, 1978; Carr et al., 1985) may permit the supply of protons for maintaining the acrosomal pH low.

iii. FertilisingAbility

Despite subtle changes in epithelial transport of ions and fluid demonstrated in the perfused cauda of rats treated with antifertility agents α-chlorohydrin (Wong and Yeung, 1977; Wong et al., 1977) and 6-chloro-6-deoxyglucose (Wong et al., 1980) these are not reflected in gross changes in the composition of epididymal fluid surrounding the spermatozoa in this species (Back et al., 1975).

iv. Indirect Effects

The water resorption occurring as a result of active Na^+ pumping has other consequences for the stored spermatozoa (see Fig 3). By increasing the packed cell volume (Table 14; see Wong and Lee, 1982; Turner and Cesarini, 1983) motility is inhibited by cell crowding (Cascieri et al., 1976); by increasing protein concentration, viscosity is increased, and motility is reduced in media of high viscosity in-vitro (Rikmenspoel et al., 1973; Chulavatnatol, 1982; Rikmenspoel, 1984; see Section III.B.12.b.ii); by removing water from their immediate environment, sperm cells experience dehydration (Lindahl and Kihlstrom, 1952; Lavon et al., 1966; Benedict et al., 1967) and so their metabolism may be inhibited in-situ (Salisbury, 1956).

b. On the Epithelium

Altering the concentrations of sodium and potassium ions in the epididymal lumen has little effect on the secretion of proteins by the perfused epididymis (Wong et al., 1980b,d; Yeung and Cooper, 1982) and changing the calcium concentration in the lumen does not alter the secretion of monovalent ions (Wong et al., 1980a, 1982a). Preventing the normal flow of testicular fluid through the epididymis, by ligating the testicular efferent ducts, has no effect on secretion or resorption of ions by the cauda epididymidis in-vivo (Wong et al., 1978, 1979).

5. Interactions of Ions with Maturing Spermatozoa

a. Potassium

Porcine caput sperm cells accumulate more ^{42}K than caudal cells after injection of the isotope (Crabo, 1967) and sperm from the proximal corpus accumulate more K^+ and Na^+ than mature cells when incubated in their normal bathing fluid in-vitro. The uptake of these ions is reduced when protein-free biological salt solutions are used and proteins from the cauda reduce the uptake of both ions into cells from the corpus (Zimmermann et al., 1979).

b. Calcium

Mature sperm are not freely permeable to Ca^{2+} and possess a outwardly-directed transport ATPase (Babcock et al., 1976) and so only accumulate the ion slowly in rams (van Eerten et al., 1979), bulls (Babcock et al., 1976; Bradley et al., 1979) and rabbits (Storey, 1975). Ca^{2+} uptake into epididymal spermatozoa in exchange for Na^+ can be reduced by extracellular proteins in seminal plasma (Babcock et al., 1979; Rufo et al., 1982, 1984) which differ from a previously described calcium-binding protein (Lubac et al., 1976). This protein may act by promoting the Ca^{2+}-ATPase activity of the membrane (Babcock et al., 1979). Electron-microprobe analysis (Babcock et al., 1978) has confirmed studies with ionophores (Storey and Keyhani, 1973, 1974; Storey, 1975) which indicate that Ca^{2+} is accumulated by sperm in the mid-piece mitochondria, where fluxes can be controlled by cholinergic agents in-vitro (Stewart and Forrester, 1979). Calcium uptake into immature bovine sperm cells exceeds that into mature cells, and may be altered by FMP (Hoskins et al., 1983; see Section II.A.6.c.ii).

6. ATPase Activity in Spermatozoa

a. Mature Spermatozoa

Most Na^+-dependent ATPase activity has been localised in the sperm mid-piece in rams (Voglmayr et al., 1969) and boars (Uesugi and Yamazo, 1966) and could be of mitochondrial origin. Na^+/K^+-ATPase is present in hamster sperm heads (Mrsny and Meizel, 1983) but little activity is found in homogenates of ovine sperm heads (Voglmayr et al., 1969) or membranes from the boar sperm head (Ashraf et al., 1982). The latter finding would implicate the flagellar ATPase in controlling ion concentrations in the sperm head.

Calcium-dependent or calcium-inhibitable Mg^{2+}-ATPases have been localised in sperm heads from the rabbit (Gordon, 1973; Gordon and Dandekar, 1977), guinea pigs (Gordon and Barrnett, 1967; Gordon et al., 1978; Yanagimachi and Usui, 1974) and expressed biochemically in the hamster (Working and Meizel, 1982). Activity is also present in the flagellum of rams and boars (Voglmayr et al., 1969; Ashraf et al., 1982) but it is not detectable histochemically in the rabbit (Gordon, 1977).

b. Immature Spermatozoa

The total ATPase activity of ovine testicular sperm homogenates is less than that of mature epididymal cells and there is a change in pH optimum between mature and immature cells (Voglmayr et al., 1969). Surface ATPase activity of intact rat sperm is greater in sperm from the caput than the cauda and there are also differences in sensitivity towards inhibitors (Chulavatnatol and Yindepit, 1976) and temperature (Chulavatnatol et al., 1978). Ca^{2+}-ATPase is observed in sperm heads from the rabbit caput and cauda (Gordon, 1977; Gordon and Dandekar, 1977).

7. Polycations

There is more spermine in rat caudal tissue that the caput but the reverse is true for spermidine (Brooks, 1976); acetylhomospermidine is present in the hamster epididymis in greater concentration in the caput than cauda (Matsuzaki et al., 1982). Both spermine and spermidine are found in rat epididymal spermatozoa (MacIndoe and Turkington, 1973) and there is more in spermatozoa from the caput than those from the cauda in the bull (Bamberg et al., 1975).

While rat sperm do not contain the enzymes for synthesising spermidine (Majumder and Turkington, 1972) the epididymis does (Majumder et al., 1974). Numerous reviews have summarised the actions of these compounds in other tissue (Tabor and Tabor, 1964, 1972, 1976, 1984; Goyns, 1982) of which the polycationic nature of the molecules with affinity for polyanions is most prevalent. Some actions of these compounds, therefore, are similar to those of divalent metal ions but acheived at lower concentration. The binding to phospholipids (Goyns, 1982) and nucleic acids may explain non-specific effects on protein synthesis, and this could underlie the beneficial action of spermine and spermidine in tissue culture (Ham, 1964).

Specific roles for polycations in reproduction are not established (Williams-Ashman and Lockwood, 1971) but the above effects could have importance for the maintenance of the epididymal epithelial cells, as would their antimicrobial activity (Clarke and Shankel, 1975).

Spermine has a dramatic effect on the motility of spermatozoa from the vas deferens, promoting a high frequency, low amplitude non-progressive flagellation (termed "hyperactivated" by Tabor and Rosenthal, 1956), but it is unrelated to the motility of capacitated sperm cells (see Section II.E). Whereas spermine phosphate does not improve the motility of ejaculated human sperm cells (Harrison, 1931), it does so in rabbits (White, 1954). Its action in stimulating aerobic fructolysis and inhibiting pyruvate oxidation by rat epididymal spermatozoa (Pulkinnen et al., 1975) is lost in cell-free systems (Pulkinnen et al., 1978), suggesting an action on the cell membrane: spermine is known to alter the permeability properties of other cell types (Lockwood and East, 1974) and spermine and spermidine bind both to caput and cauda spermatozoa (Paonessa et al., 1984). Spermine activates both the adenylate cyclase (Shar et al., 1975; Casillas et al., 1980) and phosphodiesterase (Shah and Sheth, 1978) of human spermatozoa and spermine is associated with, and is an activator of, bovine sperm protein kinase (Atmar et al., 1981). These actions could affect motility by known mechanisms (see Section II.A.c).

In-vitro spermine promotes capacitation (Fleming and Armtrong, 1975), sperm-egg binding (Saling et al., 1978) and fertilisation (Stanger and Quinn, 1982), whereas spermidine and putrescine inhibit fertilisation (Stanger and Quinn, 1982). It is by no means clear if these agents function in-vivo or whether the epididymis would provide any polyations found so high in the female tract.

In-vitro polyamines activate acrosin while they inhibit the autocatalytic activation of proacrosin (Parrish and Polakoski, 1977) and thus could act to prevent premature release of acrosomal enzymes from spermatozoa in the epididymis.

8. Summary

The resorption of ions entering the lumen from the testis drives the removal of water which concentrates the sperm cells, while the secretion of other ions is expected to depress the potential for motility and acrosome reactions in-situ. By providing or depriving the sperm cells of particular ions, the epididymal epithelium can exert control over the activity of spermatozoa or their interaction with secreted compounds.

9. References

Ashraf M, Peterson RN and Russell LD (1982) Activity and localization of cation-dependent ATPase on the plasma membranes of boar spematozoa. Biochem Biophys Res Commun 107:1273–1278

Atmar VJ, Kuehn GD and Casillas ER (1981) A polyamine-dependent protein kinase from bovine epididymal spermatozoa. J Biol Chem 256:8275–8278

Au CL and Wong PYD (1980) Bicarbonate transport in the perfused rat cauda epididymidis. J Physiol 301:29p

Au CL, Ngai HK, Yeung CH and Wong PYD (1978) Effect of adrenalectomy and hormone replacement on sodium and water transport in the perfused rat cauda epididymis. J Endocr 77:265–266

Babcock DF, First NL and Lardy HA (1976) Action of ionophore A23187 at the cellular level. Separation of effects at the plasma and mitochondrial membranes. J Biol Chem 251:3881–3886

Babcock DF, Stamerjohn DS and Hutchison T (1978) Calcium redistribution in individual sperm cells correlated with ionophore activity on motility. J exp Zool 204:391–400

Babcock DF, Singh JP and Lardy HA (1979) Alteration of membrane permeability to calcium ions during maturation of bovine spermatozoa. Develop Biol 69:85–93

Back DJ, Glover TD, Shenton JC and Boyd GP (1975) The effect of α-chlorohydrin on the composition of rat and rabbit epididymal plasma. A possible explanation of species difference. J Reprod Fert 45:117–128

Back DJ, Glover TD, Shenton JC and Boyd CP (1977) Some effects of cyproterone acetate on the reproductive physiology of the male rat. J Reprod Fert 49:237–243

Bamberg E, Weiser M and Desser H (1975) Polyamines in bovine epididymal spermatozoa. J Reprod Fert 45:363–365

Benedict RC, Schumaker VN and Davis RE (1967) The buoyant density of bovine and rabbit spermatozoa. J Reprod Fert 13:237–249

Berridge MJ (1984) Inositol triphosphate and diacyl glycerol as second messengers. Biochem J 220:345–360

Bishop DW and Matthews HP (1952) The significance of intravas pH in relation to sperm motility. Science 115:209–211

Bowen RE, Graham EF and Crabo B (1973) A technique for collecting epididymal spermatzoa and fluids from the boar virtually free from accessory sex gland secretions. J Reprod Fert 33:319–322

Bradley MP, van Eerten MTW and Forrester IT (1979) The energy-dependent uptake of Ca^{2+} by mammalian spermatozoa. Proc U Otago Med Sch 57:5–7

Brandes D (1974) Fine structure and cytochemistry of male sex accessory organs. In: Male Accessory Sex Organs. Structure and function in mammals. Ed D Brandes, Acad Press, New York. pp 17–113

Brooks DE (1976) Control of glycolytic enzymes by androgens in the rat epididymis. J Endocr 71:355–365

Burgos MH (1974) Biochemical and functional properties related to sperm metabolism and fertility. In: Male Accessory Organs. Structure and Function in Mammals. Eds D Brandes. Acad Press, New York. pp 151–160

Carr DW and Acott TS (1984) Inhibition of bovine spermatozoa by cauda epididymal fluid: 1. Studies of a sperm motility quiescence factor. Biol Reprod 30:913–925

Carr DW, Usselman MC and Acott TS (1985) Effects of pH, lactate, and viscoelastic drag on sperm motility – a species comparison. Biol Reprod 33:588–595

Cascieri M, Amann RP and Hammerstedt RH (1976) Adenine nucleotide changes at initiation of bull sperm motility. J Biol Chem 251:787–793
Casillas ER, Elder CM and Hoskins DD (1980) Adenylate cyclase activity of bovine spermatozoa during maturation in the epididymis and the activation of sperm particulate adenylate cyclase by GTP and polyamines. J Reprod Fert 59:297–302
Cheung YM, Hwang JC and Wong PYD (1978) Membrane potentials of epithelial cells in the epididymis. Ionic and castration effects. Comp Biochem Physiol 59:403–407
Chulavatnatol M and Yindepit S (1976) Changes in surface ATPase of rat spermatozoa in transit from the caput to the cauda epididymidis. J Reprod Fert 48:91–97
Chulavatnatol M, Eksittikul T and Toowicharanont P (1978) Control of epididymal sperm motility: an approach to male fertility regulation. Int J Androl Suppl 2:629–638
Clarke CH and Shankel DM (1975) Antimutagenesis in microbial systems. Bacteriol Rev 39:33–53
Cohen JP, Hoffer AP and Rosen S (1976) Carbonic anhydrase localization in the epididymis and testis of the rat: histochemical and biochemical analysis. Biol Reprod 14:339–346
Cooper TG and Yeung CH (1980) Epithelial structure of the rat cauda epididymidis after luminal perfusion. Int J Androl 3:361–374
Crabo B (1967) Chemophysiological aspects of epididymal function. Nord Med 78:1009–1040
Fleming AD and Armstrong DT (1985) Effects of polyamines upon capacitation and fertilization in the guinea pig. J exp Zool 233:93–100
Garbers DL, Wakayabashi T and Reed, PW (1970) Enzyme profile of the cytoplasmic droplet rom bovine epididymal spermatozoa. Biol Reprod 3:327–337
Gordon M (1973) Localization of phosphatase activity on the membranes of the mammalian sperm head. J exp Zool 185:111–120
Gordon M (1977) Cytochemical analysis of the membranes of the mammalian sperm head. In: Male Reproductive System. Fine Structure Analysis by Scanning and Transmission Electron Microscopy. Eds RD Yates and M Gordon, Abacus Press, Tunbridge Wells. pp 15–33
Gordon M and Barrnett RJ (1967) Fine structural cytochemical localizations of phosphatase activities of rat and guinea pig. Exp Cell Res 48:395–412
Gordon M and Dandekar PV (1977) Fine-structural localization of phosphatase activity on the plasma membrane of the rabbit sperm head. J Reprod Fert 49:155–156
Gordon M, Dandekar PV and Eager PR (1978) The identification of phosphatases on the membranes of guinea pig sperm. Anat Rec 191:123–134
Goyns MH (1982) The role of polyamines in animal cell physiology. J Theoret Biol 95:577–589
Gunn SA and Gould TC (1975) Vasculature of the testes and adnexia. In: Handbook of Physiology, Section VII Endocrinology, Volume 5 Male Reproductive System, Eds DW Hamilton and RO Greep, American Physiological Society, Washington DC. pp 117–142
Ham RG (1964) Putrescine and related amines as growth factors for a mammalian cell line. Biochem Biophys Res Commun 14:34–38
Hamilton DW, Olson GE and Beeuwkes R (1976) Epididymal physiology and sperm maturation. In: Progeess in Reproductive Biology Vol 1. Ed PO Hubinot, S Karger, Basel. pp 62–73
Harrison GA (1931) Spermine in human tissues. Biochem J 25:1885–1892
Heffner LJ and Storey BT (1981) The role of calcium in maintaining motility in mouse spermatozoa. J exp Zool 218:427–434
Hinton BT and Setchell BP (1980) Concentrations of glycerophosphocholine, phosphocholine and free inorganic phosphate in the luminal fluid of the rat testis and epididymis. J Reprod Fert 58:401–406
Hohlbrugger G (1980) A micropuncture study of transepithelial water reabsorption in rat ductuli efferentes. Fert Steril 34:50–54
Hohlbrugger G and Pfaller K (1983) Post-vasectomy impairment of transepithelial water reabsorption in the initial segment of the epididymis. Archs Androl 11:265–270
Hoskins DD, Acott TS, Critzchlow L and Vijayaraghavan S (1983) Studies on the roles of cyclic AMP and calcium in the development of bovine sperm motility. J Submicroscop Cytol 15:21–27
Jenkins AD, Lechene CP and Howards SS (1978) The effect of estrogen on the concentration of seven elements in the intraluminal fluids of the seminiferous tubules, rete testis, and epididymides. Fert Steril 30:732–733

Jenkins AD, Lechene CP and Howards SS (1983a) Effect of spironolactone on the elemental composition of the intraluminal fluids of the seminiferous tubules, rete testis and epididymis of rat. J Urol 129:851–854

Jenkins AD, Lechene CP and Howards SS (1983b) The effect of estrogen administration in vivo on the elemental composition of intraluminal fluids of the seminiferous tubules, rete testis and epididymis of the rat. J Androl 4:272–275

Jones R (1974) Absorption and secretion in the cauda epididymidis of the rabbit and the effects of degenerating spermatozoa on epididymal plasma after castration. J Endocr 63:157–165

Jones R (1977) Effects of testosterone, testosterone metabolites and anti-androgens on the function of the male accessory glands in the rabbit and rat. J Endocr 74:75–78

Jones R (1978) Comparative biochemistry of mammalian epididymal plasma. Comp Biochem Biophys 61B:365–370

Jones R and Glover TD (1973) The effects of castration on the composition of rabbit epididymal plasma. J Reprod Fert 34:405–411

Jones R and Glover TD (1975) Interrelationships between spermatozoa, the epididymis and epididymal plasma. In: The Biology of the Male Gamete. Eds JG Duckett and PM Racey, Suppl No 1 to the Biol J Linnaen Soc 7:367–384

Kidroni G, Har-Nir R, Menczel J, Frutkoff IW, Palti Z and Ron M (1983) Vitamin D_3 metabolites in rat epididymis: high 24,25-dihydroxyvitamin D_3 levels in the cauda region. Biochem Biophys Acta 113:982–989

Lavon U, Volcani R, Amir D and Danon D (1966) The specific gravity of bull spermatozoa from different parts of the reproductive tract. J Reprod Fert 12:597–599

Lee WM, Tsang AYF and Wong PYD (1981) Effects of divalent and lanthanide ions on motility initiation in rat caudal epididymal spermatozoa. Brit J Pharmac 73:633–638

Levin RM, Amsterdam JD, Winokur A and Wein AJ (1981) Effects of psychotropic drugs on human sperm motility. Fert Steril 36:503–506

Levine N and Kelly H (1978) Measurement of pH in the rat epididymis in vivo. J Reprod Fert 52:333–335

Lindahl PE (1973) Activators of the ATP-dependent surface reaction in the apical cell cell membrane of the bull-sperm head, causing head-to-head association. Exp Cell Res 81:413–431

Lindahl PE (1974) Nucleotide specificity of the ATP-dependent surface reaction of the acrosomal region of the bull sperm head. Exp Cell Res 87:47–54

Lindahl PE and Kihlstrom JE (1952) Alterations in specific gravity during the ripening of bull spermatozoa. J Dairy Sci 35:393–402

Lockwood DH and East LE (1974) Studies on the insulin-like actions of polyamines on lipid and glucose metabolism in adipose tissue cells. J Biol Chem 249:7717–7722

MacIndoe JH and Turkington RW (1973) Hormonal regulation of spermidine formation during spermatogenesis in the rat. Endocrinology 92:595–605

Majumder GC, MacIndoe JH and Turkington RW (1974) Hormonal Control of ornithine decarboxylase and S-adenosyl-L-methionine decarboxylase during development of the rat epididymis. Life Sci 15:45–55

Lukac J, Pribanic M and Karen E (1976) Calcium-binding protein in bull seminal vesicle secretion and seminal plasma. J Reprod Fert 48:77–81

Matsuzaki S, Haman K, Imai K and Matsuhura K (1982) Occurrence in high concentrations of N-acetylspermidine and sym-homospermidine in the hamster epididymis. Biochem Biophys Res Commun 107:307–313

McGradey AV and Nelson L (1974) Effect of calcium on bull spermatozoan membrane potential. Exp Cell Res 87:398–400

McLeod J, Swan RC and Aitken GA (1949) Lithium: its effect on human spermatozoa, rat testicular tissue and upon rats in vivo. Am J Physiol 157:177–183

Morton BE, Sagadraca R and Fraser C (1978) Sperm motility within the mammalian epididymis: species variation and correlation with free calcium levels in epididymal plasma. Fert Steril 29:695–698

Mrsny RJ and Meizel S (1983) Initial evidence for the modification of hamster sperm Na^+, K^+-ATPase activity by cyclic nucleotide-mediated processes. Biochem Biophys Res Commun 112:132–138

Paonessa G, Metafora S, Tajana G, Abrescia P, de Santis A, Gentile V and Porta R (1984) Transglutaminase-mediated modifications of the rat sperm surface in vitro. Science 226:852–825
Parrish RF and Polakoski KL (1977) Effect of polyamines on the activity of acrosin and the activation of proacrosin. Biol Reprod 17:417–422
Pulkkinen P, Sinervirta R and Janne J (1975) Modification of the metabolism of rat epididymal spermatozoa by spermine. Biochem Biophys Res Commun 67:714–722
Pulkkinen P, Piik K, Koso P and Janne J (1978) Effect of polyamines and their oxidized derivatives on the metabolism of rat epididymal spermatozoa. Acta Endocr 87:845–854
Rasweiler JJ and Bedford JM (1982) Biology of the scrotum.III. Effects of abdominal temperature upon the epithelial cells of the rat cauda epididymidis. Biol Reprod 26:691–705
Reed PW and Takahashi Y (1973) A high activity divalent cation ATPase from the distal, caudal epididymis of bulls. Biol Reprod 9:46–56
Rikmenspoel R (1984) Movements and active moments of bull sperm flagella as a function of temperature and viscosity. J exp Biol 108:205–230
Rikmenspoel R, Jacklett AC, Orris SE and Lindemann CB (1973) Control of bull sperm motility. Effects of viscosity, KCN and thiourea. J Mechanochem Cell Motility 2:7–24
Rogers BJ and Yanagimachi R (1976) Competitive effect of magnesium on the calcium-dependent acrosome reaction in guinea pig sperm. Biol Reprod 15:614–619
Rogers BJ, Ueno M and Yanagimachi R (1981) Fertilization by guinea pig spermatozoa requires potassium ions. Biol Reprod 25: 639–648
Rufo GA, Singh JP, Babcock DF and Lardy HA (1982) Purification and characterization of a calcium transport inhibitor protein from bovine seminal plasma. J Biol Chem 257: 4627–4632
Rufo GA, Schoff PK and Lardy HA (1984) Regulation of calcium content in bovine spermatozoa. J Biol Chem 259:2547–2552
Saling PM, Storey BT and Wolf DP (1978) Calcium-dependent binding of mouse epididymal spermatozoa tzo the zona pellucida. Develop Biol 65:515–525
Salisbury GW (1956) The function of the epididymis of the bull. I. A theory for the activation of spermatozoan motility. Tijdschr. Diergeneesk 87:616–623
Shah GV and Sheth AR (1978) Inhibition of phosphodiesterase activity of human spermatozoa by spermine. Experientia 34:980–981
Shah GV, Sheth AR, Mugatwala PP and Rao SS (1975) Effect of spermine on adenyl cyclase activity of spermatozoa. Experientia 31:631–632
Singh JP, Babcock DF and Lardy HA (1978) Increased calcium -ion influx is a component is capacitation of spermatozoa. Biochem J 172:549–556
Stanger JD and Quinn P (1982) Effect of polyamines on fertilization of mouse ova in vitro. J exp Zool 220:377–380
Stewart TA and Forrester IT (1979) Acetylcholine-induced calcium movements in hypotonically washed ram spermatozoa. Biol Reprod 21:109–115
Storey BT (1975) Energy metabolism of spermatozoa.IV. Effect of calcium on respiration of mature epididymal sperm of the rabbit. Biol Reprod 13:1–9
Storey BT and Keyhani E (1973) Interaction of calcium ion with the mitochondrion of rabbit spermatozoa. FEBS Letts 37:33–36
Storey BT and Keyhani E (1974b) Energy metabolism of spermatozoa: III. Energy-linked uptake of calcium ion by the mitochondria of rabbit epididymal spermatozoa. Fert Steril 25:976–984
Svejnohova D, Janousek I and Suva J (1980) Study of lithium levels in the serum and genital organs of male rats. Activitis Nervosa Superior 22:178–180
Tabor CW and Rosenthal SM (1956) Pharmacology of spermine and spermidine. Some effects on animals and bacteria. J Pharm exp Therap 116:139–155
Tabor H and Tabor CW (1964) Spermidine, spermine, and related amines. Pharm Rev 16:245–300
Tabor H and Tabor CW (1972) Biosynthesis and metabolism of 1,4-diaminobutane, spermidine, spermine and related amines. Adv Enz 36:213–268
Tabor CW and Tabor H (1976) 1,4-diamobutaine (putrescine), spermidine and spermine. Ann Rev Biochem 45:285–306

Tabor CW and Tabor H (1984) Polyamines. Ann Rev Biochem 53:749–790
Turner TT (1984) Resorption versus secretion in the rat epididymis. J Reprod Fert 72:509–514
Turner TT and Cesarini DM (1983) The ability of the rat epididymis to concentrate spermatozoa. Responsiveness to aldosterone. J Androl 4:197–202
Uesugi S and Yamazoe S (1966) Presence of sodium-potassium-stimulated ATPase in boar epididymal spermatozoa. Nature (Lond) 209:403
van Eerten MTW, Bradley MP and Forrester IT (1979) Ca^{2+} transport across the plasma and mitochondrial membranes of mammalian spermatozoa 11th Int Congr Biochem, Toronto (abst)
Voglmayr JK and Gandhi JS (1978) Survival of ram testicular spermatozoa in vitro: effects of glucose, glucose metabolites, rete testis fluid-proteins, selected androgens and phospholipids. Theriogenology 9:463–478
Voglmayr JK, White IG and Quinn PJ (1969) A comparison of adenosine triphosphatase activity in testicular and ejaculated spermatozoa of the ram. Biol Reprod 1:121–129
Walters MR, Hunziker W and Norman AW (1982) 1,25-dihydroxyvitamin D_3 receptors: exchange assay and presence in reproductive tissues. In: Vitamin D, Chemical, Biochemical and linical Endocrinolgy of Calcium Metabolism, Walter de Gruyter and Co, Berlin. pp 91–93
Walters MR, Cuneo DL and Jamieson AP (1983) Possible significance of new target tissues for 1,25-dihydroxyvitamin D_3. J Steroid Biochem 19:913–920
White IG (1954) The effect of some seminal constituents and related substance on diluted mammalian spermatozoa. Aust J Biol Sci 7:379–390
Williams JA (1970) Origin of transmembrane potentials in non-excitable cells. J Theoret Biol 28:287–296
Williams-Ashman HG and Lockwood DH (1971) Role of polyamines in reproductive physiology and sex hormone action. Ann NY Acad Sci 171:882–894
Working PK and Meizel S (1982) Preliminary charactrization of a Mg^{2+}-ATPase in hamster sperm head membranes. Biochem Biophys Res Commun 104:1060–1065
Wong PYD and Lee WM (1982) Effects of spironolactone (aldosterone antagonist) on electrolyte and water content of the cauda epididymidis and fertility of male rats. Biol Reprod 27:771–777
Wong PYD and Yeung CH (1976) Inhibition by amiloride of sodium-dependent fluid resorption in the rat isolated caudal epididymis. Brit J Pharm 58:529–531
Wong PYD and Yeung CH (1977a) Fluid reabsorption in the isolated duct of the rat cauda epididymidis. J Reprod Fert 49:77–81
Wong PYD and Yeung CH (1977b) Hormonal regulation of fluid resorption in isolated rat cauda epididymidis. Endocrinology 101:1391–1397
Wong PYD and Yeung CH (1978) Absorptive and secretory functions of the perfused rat cauda epididymidis. J Physiol 275:13–26
Wong PYD, Yeung CH and Ngai HK (1977) Effect of α-chlorohydrin on transport processes in perfused rat cauda epididymidis. Contraception 16:637–644
Wong PYD, Au CL and Ngai HK (1978) Electrolyte and water transport in rat epididymis. Its possible role in sperm maturation. Int J Androl Suppl 5:608–628
Wong PYD, Au CL and Ngai HK (1979) Some characteristics of salt and water transport in the rat epididymis. In: The Spermatozoon. Eds DW Fawcett and JM Bedford, Urban and Schwarzenberg, Baltimore. pp 57–63
Wong PYD, Au CL and Ngai HK (1980a) Effects of 6-chloro-6-deoxyglucose on electrolyte and water transport in the epididymis and fertility of male rats. Int J Androl 3:82–86
Wong PYD, Au CL and Ngai HK (1980b) The isolated duct of the rat cauda epididymidis as a model for isosmotic transport studies. Jap J Physiol 30:1–15
Wong PYD, Tsang AYF, Lee WM and Li CM (1980c) Secretion of the rat cauda epididymidis. Archs Androl 5:327–336
Wong PYD, Tsang AYF and Lee WM (1982a) Secretion of macromolecules by the rat epididymis. Int J Androl Suppl 5:34–47
Wong PYD, Au C and Bedford JM (1982b) Biology of the scrotum. II. Suppression by abdominal temperature of transepithelial ion and water transport in the cauda epididymidis. Biol Reprod 26:683–689
Wong YC, Wong PYD and Yeung CH (1978) Ultrastructural correlation of water resorption in isolated cauda epididymides. Experientia 34:485–487

Wu YW, Xie YK and Yuan D (submitted) Electrophysiological and ultrastructural studies of the blood-epididymis barrier of the rat. Int J Androl

Yanagimachi R and Usui N (1974) Calcium dependence of the acrosome reaction and activation of guinea pig spermatozoa. Exp Cell Res 89:161–174

Yeung CH and Cooper TG (1982) Ultrastructure of the perfused rat epididymis: effect of luminal sodium ion concentration. Cell Tiss Res 226:407–425

Zimmerman KJ, Crabo BG, Moore R, Weisberg S, Diebel FC and Graham EF (1979) Movements of sodium and potassium into epididymal boar spermatozoa. Biol Reprod 21:173–179

D. Epididymal Secretion of Glycerophosphocholine (GPC)

1. Concentration in Epididymal Tissue

Concentrations of GPC in the caput epididymidis are similar to those in the cauda epididymidis in rats (Riar et al., 1973; Brooks et al., 1974; Brown-Woodman et al., 1976) and rabbits (Holtz and Foote, 1978), others estimate slightly more in the cauda (Dawson and Rowlands, 1959; Dacheux et al., 1970; Setty et al., 1979). Epithelial cells from the caput contain more GPC than those from the corpus or cauda (Hoffmann and Killian, 1981) and the principal cells contain more than basal cells (Killian and Chapman, 1980; Hoffmann and Killian, 1981). Within the epididymal lumen there is more GPC in fluid than sperm cells (Brooks, et al., 1974). The concentration in epididymal fluid is high in the caput and is maintained distally (Table 15) whereas concentrations in spermatozoa increase as they mature (Quinn and White, 1967).

2. Origin of Epididymal GPC

In man GPC arises proximal to the vas deferens because its concentration in seminal plasma declines after vasectomy (Frenkel et al., 1974; Arrata et al., 1978; Sade et al., 1978; Naik et al., 1979) and its highest concentration in split ejaculates is found in the sperm-rich fraction (Brown-Woodman et al., 1980).

a. Testicular Fluid

The concentration of GPC entering the epididymis is lower than that leaving it (Hinton and Setchell, 1980) and the concentration observed cannot be explained by the removal of water (Setchell and Hinton, 1981). Other indices that the exocrine testicular contribution of GPC is minimal is the maintenance of epididymal GPC after eliminating spermatozoa by vitamin E deficiency (Dawson and Rowlands, 1959) or preventing the inflow of testicular fluid by ligating the testicular efferent ducts (Dawson and Rowlands, 1959; Scott et al., 1963a; Dacheux et al., 1970; Brown-Woodman et al., 1976; Dacheux and Laporte, 1977).

This may indicate that concentrations found distally arise from the flow of fluid through the lumen from the caput where it is synthesised. The concentration of GPC in cauda epididymal fluid from the bull (Gustaffson, 1966) and rabbit (Jones, 1977) does not decrease after this operation, implicating the cauda in GPC synthesis in these species. The normal flow of fluid and spermatozoa through the epididymis has been suggested to promote the synthesis of GPC (Dachuex and Laporte, 1971; Scott et al., 1963b) and there is evidence from in-vitro studies to support an interaction between spermatozoa and epithelial cells in GPC synthesis (see Section III.D.2.b).

b. Epididymal Synthesis

Radioactive phosphate injected into rabbits is incorporated into GPC in the epididymis of rabbits (Scott et al., 1963b) and rats (Dawson and Rowlands, 1959) and into seminal GPC in rams (Dawson, 1958). In the rabbit testicular GPC is not labelled (Scott et al., 1963b), confirming the insignificant contribution from the testis; the incorporation of injected phosphate into ovine seminal GPC occurs even if the caput is isolated from the testis (Dawson, 1958).

Phosphate is also incorporated into epididymal lecithin in-vitro (phosphatidylcholine: Wallace et al., 1966) and labelled choline injected into the caput epididymides is incorporated into lecithin, which can be converted into GPC in-vitro (Bjerve and Reitan, 1978). The sequential labelling of precursor and products of lipid metabolism is not necessarily reflected in aqueous sytems (Verger et al., 1973) and arguments against lecithin being the precursor of GPC (Scott et al., 1963b; Wallace et al., 1966) are not strong (see Brooks, 1979) and now plasma lecithin is the accepted as the precursor of epididymal GPC (Hammerstedt and Rowan, 1982). Recently another pathway of GPC synthesis, *via* choline transfer from phosphatidylcholine to glycero 3-phosphate, has been postulated, but not yet proved to exist in the epididymis (Infante and Huszagh, 1985).

Phosphatidylcholine is present in epididymal fluid (Evans and Setchell, 1979) and its concentration increases towards the cauda (Quinn and White, 1967). In-vitro GPC and lecithin synthesis from phosphate is greater in the rat caput than cauda (Dawson and Rowlands, 1959; Scott et al., 1963a), even after ligating the efferent ducts, but after labelling of epididymal lecithin, its conversion to GPC is greater in the cauda than caput and corpus (Bjerve and Reitan, 1978).

i. By Epithelial Cells

The concentration of GPC in principal cells isolated from the epithelium from the rat caput is higher than that in basal or other somatic cells (Killian and Chapman, 1980). The suggestion that GPC is synthesised by the epithelium has received support from in-vitro studies where cells from the caput epididymidis converted lecithin into GPC and the conversion was accentuated by the presence of spermatozoa (Wang et al., 1981).

ii. By Spermatozoa

Ejaculated spermatozoa from rams and bulls metabolise exogenous lecithin slowly and there is no liberation of GPC (Dawson et al., 1957; Jones and Mann, 1976); similarly, testicular and mature epididymal spermatozoa from rams also hydrolyse lecithin slowly (Scott and Dawson, 1968). Epididymal spermatozoa from the rat, however, bind lecithin and convert it to GPC, and caput sperm are more active than cauda sperm in this regard (Wang et al., 1981). Lysophospholipase and phospholipase activities are also more active in spermatozoa from the caput than cauda in the goat (Atreja and Anand, 1985).

3. Androgen Dependence

The tissue content of GPC decreases after castration of rats (Dawson and Rowlands, 1959; Brooks et al., 1974), rabbits and monkeys (Setty et al., 1979) and is promoted by hCG administration in rats (Dacheux and Laporte, 1977). The GPC concentration in luminal fluid also decreases after castration of rabbits (Jones and Glover, 1973, 1975; Jones, 1977) but the anti-androgen cyproterone acetate reduces neither the tissue content of GPC (Brooks et al., 1974) nor its concentration in epididymal fluid of rats (Back et al., 1977; Jones, 1977) or rabbits (Jones, 1977). The activity of phospholipase A, which is responsible for the first hydrolytic step in the conversion of lecithin to GPC is an androgen-dependent enzyme in the rat epididymis (Bjerve and Reitan, 1978; Beck, 1980).

4. Role of GPC in Epididymal Function

a. Non-Specific

GPC may play no specific role in the epididymis; its presence in the lumen merely reflecting the inevitable partition into epididymal fluid of water-soluble products of lecithin hydrolysis, however, it would contribute to the high osmotic pressure of epididymal fluid by virtue of its high concentration. The fatty acids released from the hydrolysis may be important as a fuel for sperm (see Bjerve and Reitan, 1978; Section III.E).

Whether, as in other tissues, phosphorylcholine derived from GPC regulates protein synthesis (Mookerjea et al., 1969) or modulates the activity of galactosyltransferase (Mookerjea and Ying, 1974) remains unstudied but worthy of investigation in view of the importance of these processes in epididymal function (see Section III.G).

b. Sperm Metabolism

GPC is not metabolised by sperm (Dawson et al., 1957; Storey and Keyhani, 1974) so cannot provide a source of energy for sperm in the epididymis. The diesterase activity in the female genital tract, however, is able to liberate glycerol and its phosphate ester, both of which can be metabolised by spermatozoa, so in this respect GPC may act as a reserve of substrate. Just how much epididymal GPC accompanies sperm in their ascent of the female tract is not known.

c. Sperm Motility

A decreased incorporation of phosphate into seminal GPC occurred in one ram displaying poor sperm motility (Dawson, 1958), however, there is little evidence yet to support a role for GPC in sperm motility. The ability of GPC (73mM) to depress the induction of epididymal sperm motility on dilution (Turner et al., 1978) was largely due to contaminating cadmium ions (Turner et al. 1979) and a subsequent demonstration of depression by 40mM GPC of the maintenance of sperm motility in parallel with inhibition of pyruvate oxidation (Turner and Giles, 1981) was irreversible and thus unlikely to mimic the depression of motility in the epididymis before ejaculation. There is no effect of GPC on the motility of sperm from hamsters (0.4–3.6mM: Morton and Chang, 1973) or bulls (40 or 80 mM: Carr and Acott, 1984a) but 1mM GPC was sufficient to inhibit the surface ATPase of rat spermatozoa (Chulavatnatol and Yindepit, 1976).

d. Capacitation and Fertilisation

GPC (1.8–3.6mM) does not prevent capacitation or fertilisation of rat eggs in-vitro (Davis, 1976).

5. Summary

Glycerophoshocholine, synthesised in the epithelium, is encountered by spermatozoa at high concentration along the length of the duct but its function is not known. Direct actions on spermatozoa in-vitro have not always been dissociated from the effects of contaminating cadmium in commercial preparations and the possibility remains that it serves no function, being a secondary hydrolysis product of lecithin metabolism that accumulates in epididymal fluid owing to its water soluble properties.

6. References

Arrata WSM, Bunt T and Corder S (1978) The role of phosphate esters in male fertility. Fert Steril 30:329–333

Atreja SK and Anand SR (1985) Phospholipase and lysophospholipase activities of goat spermatozoa in transit from the caput to the cauda epididymidis. J Reprod Fert 74:687–691

Back DJ, Glover TD, Shenton JC and Boyd CP (1977) Some effects of cyproterone acetate on the reproductive physiology of the male rat. J Reprod Fert 49:237–243

Beck B (1980) Tissue specificity of the epididymal androgen dependent phospholipase A. Int J Androl 3:349–362

Bjerve KS and Reitan LJ (1978) The presence of an androgen controlled phospholipase A in rat epididymis. Its possible function in the synthesis of sn-3-glycerophosphate and the supply of fatty acids to spermatozoa. Int J Androl Suppl 2:574–591

Brooks DE (1979) Biochemical Environment of Maturing Spermatozoa. In: The Spermatozoon. Eds DW Fawcett and JM Bedford, Urban and Schwarzenberg, Baltimore. pp 23–34

Brooks DE, Hamilton DW and Mallek AH (1974) Carnitine and glycerylphosphorylcholine in the reproductive tract of the rat. J Reprod Fert 36:141–146

Brown-Woodman PDC, Sales D and White IG (1976) The glycerylphosphorylcholine content of the rat epididymis after injecting α-chlorohydrin or ligating the vasa efferentia. Acta Europea Fertilitatis 7:155–162

Brown-Woodman PDC, Marley PB, Morris S, Rodger JC and White IG (1980) Origin of glycerylphosphorylcholine, inositol, N-acetyl-aminosugar, and prostaglandins in human seminal plasma and their effects on sperm metabolism. Archs Androl 4:149–155

Carr DW and Acott TS (1984) Inhibition of bovine spermatozoa by cauda epididymal fluid: I. Studies of a sperm motility quiescence factor. Biol Reprod 30:913–925

Chulavatnatol M and Yindepit S (1976) Changes in surface ATPase of rat spermatozoa in transit from the caput to the cauda epididymidis. J Reprod Fert 48:91–97

Dacheux JL and Laporte P (1977) Influence du fluide testiculaire, des spermatozoides et des androgens sur l'activite physiologique de l'épididyme de rat. CR Acad Sci D285:175–178

Dacheux JL, Laporte P and Peyre A (1970) Mise en evidence des correlations testiculo-épididymaires d'aprés les variations de la glycerylphosphorylcholine. I. Etude lors d'une cryptorchide unilaterale. CR Soc de Biol 164:2074–2079

Davis BK (1976) Inhibitory effect of synthetic phospholipid vesicles containing cholesterol on the fertilizing ability of rabbit spermatozoa. Proc Soc exp Biol Med 152 257–261

Dawson RMC (1958) The labelling of ram semen in vivo with radioactive phosphate and (carboxy-^{14}C)stearic acid. Biochem J 68:512–519

Dawson RMC and Rowlands IW (1959) Glycerylphosphorylcholine in the male reproductive organs of rats and guinea pigs. Quart J exp Biol 44:26–34

Dawson RMC, Mann T and White IG (1957) Glycerylphosphorylcholine and phosphorylcholine in semen, and their relation to choline. Biochem J 65:627–634

Evans RW and Setchell BP (1979) Lipid changes in boar spermatzoa during epididymal maturation with some observations on the flow and composition of boar rete testis fluid. J Reprod Fert 57:189–196

Frenkel G, Peterson RN, Davis JE and Freund M (1974) GPC and carnitine in normal human semen and in post vasectomy semen: differences in concentration. Fert Steril 25:84–87

Gustafsson B (1966) Luminal contents of the bovine epididymis under conditions of reduced spermatogenesis, luminal blockage and certain sperm abnormalities. Acta Vet Scand Suppl 17.

Hammerstedt RH and Rowan WA (1982) Phosphatidylcholine of blood lipoproteins is the precursor of glycerophosphorylcholine found in seminal plasma. Biochem Biophys Acta 710:370–376

Hinton BT and Setchell BP (1980) Concentrations of glycerophosphocholine, phosphocholine and free inorganic phosphate in the luminal fluid of the rat testis and epididymis. J Reprod Fert 58:401–406

Hoffmann DS and Killian GJ (1981) Isolation of epithelial cells from the corpus epididymidis and analysis for glycerylphosphorylcholine, sialic acid and protein. J exp Zool 217:93–102

Holtz W and Foote RH (1978) Composition of rabbit semen and the origin of several constituents. Biol Reprod 18:286–292

Infante JP and Huszagh VA (1985) Synthesis of highly unsaturated phosphatidyl-cholines in the development of sperm motility: a role for epididymal glycerol-3-phosphorylcholine. Mol Cell Biochem 69:3–6

Jones R (1977) Effects of testosterone, testosterone metabolites and anti-androgenson the function of the male accessory glands in the rabbit and rat. J Endocr 74:75–78

Jones R and Glover TD (1973) The effects of castration on the composition of rabbit epididymal plasma. J Reprod Fert 34:405–411

Jones R and Glover TD (1975) Interrelationships between spermatozoa, the epididymis and epididymal plasma. In: The Biology of the Male Gamete. Eds JG Duckett and PM Racey, Suppl No 1 to the Biol J Linnaen Soc 7:367–384

Jones R and Mann T (1976) Lipid peroxides in spermatozoa; formation, role of plasmalogen, and physiological significance. Proc Roy Soc (Lond) B 193:317–333

Killian GJ and Chapman DA (1980) Glycerylphosphorylcholine, sialic acid and protein in epithelial cells isolated from the rat caput epididymidis by elutriation. Biol Reprod 22:846–850

Mookerjea S and Yung JWM (1974) A study of the effect of lysolecithin and phospholipase A on membrane-bound galactosyltransferase. Can J Biochem 52:1053–1066

Mookerjea S, Park CE and Chow A (1969) Stimulation of glycoprotein synthesis by phosphorylcholine in vivo and in liver slices. Biochem Biophys Res Commun 36:851–857

Morton B and Chang TSK (1973) The effect of fluid from the cauda epididymidis, serum components and caffeine upon the survival of diluted epididymal hamster spermatozoa. J Reprod Fert 35:255–263

Naik VK, Pardonani DS, Joshi UM and Sheth AR (1979) Seminal plasma concentration of glycerylphosphorylcholine before and after vasectomy and vas reanastomosis. Fert Steril 32:685–686

Quinn PJ and White IG (1967) Phospholipid and cholesterol content of epididymal and ejaculated spermatozoa and seminal plasmna in relation to cold shock. Aust J Biol Sci 20:1205–1215

Riar SS, Setty BS and Kar AB (1973) Studies on the physiology and biochemistry of mammalian epididymis. Biochemical composition of epididymis. A comparative study. Fert Steril 24:355–363

Sade M, Saverymuttu I, Dinsdale O and Gow JG (1978) The management of azoospermia. Brit J Urol 50:595–597

Scott TW and Dawson RMC (1968) Metabolism of phospholipids by spermatozoa and seminal plasma. Biochem J 108:457–463

Scott TW, Dawson RMC and Rowlands IW (1963a) Phospholipid relationships in rat epididymal tissue and spermatozoa. Biochem J 87:507–512

Scott TW, Wales RG, Wallace KC and White IG (1963b) Composition of ram epididymal and testicular fluid and the biosynthesis of glycerylphosphorylcholine by the rabbit epididymis. J Reprod Fert 6:49–59

Setchell BP and Hinton BT (1981) The effects on the spermatozoa of changes in the composition of luminal fluid as it passes along the epididymis. In: Epididymis and Fertility: Biology and Pathology. Eds C Bollack and A Clavert, S Karger, London. Progr Reprod Biol 8:58–66

Setty BS, Riar SS and Kar AB (1979) Androgen control of epididymal function in rhesus monkey and rabbit. Fert Steril 28:674–681

Storey BT and Keyhani E (1974) Energy metabolism of spermatozoa. II. Comparison of pyruvate and fatty acid oxidation by mitochondria of rabbit epididymal spermatozoa. Fert Steril 25:857–864

Turner TT and Giles RD (1981) The effects of carnitine, glycerylphosphorylcholine, caffeine, and egg yolk on the motility of rat epididymal spermatozoa. Gamete Res 4:283–295

Turner TT, D'Addario D and Howards SS (1978) Further observations on the initiation of sperm motility. Biol Reprod 19:1095–1101

Turner TT, D'Addario D and Howards SS (1979) Correction: Further observations on the initiation of sperm motility. Biol Reprod 21:535–536

Verger R, Mieras MCE and Haas GHDC (1973) Action of phospholipase A at interfaces. J Biol Chem 248:4023–4034

Wallace JC, Wales RG and White IG (1966) The respiration of the rabbit epididymis and its synthesis of glycerophosphorylcholine. Aust J Biol Sci 19:849–856

Wang CY, Killian G and Chapman DA (1981) Association of (^{14}C)phosphatidylcholine with rat epididymal sperm and its conversion to (^{14}C)glycerylphosphorylcholine by sperm and principal cells. Biol Reprod 25:969–976

E. Epididymal Secretion of Carnitine

1. Concentration in Epididymal Tissue

Concentrations of carnitine in epididymal tissue increase towards the distal regions in bulls, boars, monkeys (Casillas, 1972), rats (Casillas, 1972; Marquis and Fritz, 1965a,c; Brooks et al., 1974), rabbits (Casillas and Chaiyungpan, 1979) and hamsters (Casillas et al., 1984) but human epididymal tissue contains more carnitine in the caput than cauda (Bøhmer and Johansen, 1978). Estimates that carnitine concentrations in epithelial cells of the corpus were lower than in the cauda (Yeung et al., 1980) have been confirmed by direct measurement in cells isolated from the epididymis, and principal cells contain more carnitine than basal cells (Dorsey et al., 1980).

Carnitine accumulates in proximal regions of the rat epididymis after ligation of the efferent ducts (Brooks, 1979b), suggesting that the distal flow of fluid contributes to the normal regional distribution, and is thus mainly intraluminal. High concentrations within the cauda epididymal lumen were first demonstrated by Pearson and Tubbs (1967) and Marquis and Fritz (1965a) and micropuncture studies has revealed increasing concentrations in fluid from the caput to cauda (Hinton et al., 1979; Golan et al., 1982). Lower concentrations are calculated to be found in human epididymal fluid (Soufir et al., 1981) and concentrations of acetylcarnitine in the lumen of the corpus and cauda exceed those in the caput (Golan et al., 1982; Table 16).

Taking into consideration the resorption of water it can be deduced that secretion of carnitine occurs proximally, but that increases in the distal epididymis can be accounted for by water resorption alone (Hinton et al. 1979; Setchell and Hinton, 1981; Turner, 1984).

In the human an epididymal source of seminal carnitine is also implied because concentrations of carnitine in human seminal plasma fall after vasectomy (Frenkel et al., 1974; Sade et al., 1978; Wetterauer and Heite, 1978; Lewin et al., 1979, 1981; Soufir et al., 1981; Fahimi et al., 1981; Menchini-Fabris et al., 1984); concentrations are lower than normal in cases of obstructive azoospermia (Wetterauer and Heite, 1980; Lewin et al., 1976, 1981); and the highest concentration of carnitine in split ejaculates is in the sperm-rich fraction (Wetterauer and Heite, 1978; Soufir et al., 1981; Golan et al., 1983). The smaller fall in carnitine after vasectomy and the correlation with seminal fructose found by Lewin et al. (1976) most likely reflects their use of gas chromatography, which estimates total carnitine esters as well as free carnitine. Bioautographic estimates of carnitine esters confirms they are contributed by the accessory organs (Golan et al., 1983).

2. Origin of Epididymal Carnitine

Carnitine is not synthesised in the epididymis from trimethyllysine (Brooks et al., 1973) or deoxycarnitine (Casillas, 1973; Casillas and Erickson, 1975a).

a. Testicular Fluid

The testis does not substantially contribute to epididymal carnitine, since the carnitine concentration in rete testis fluid is lower than that in the tail of the epididymis and the extent of water resorption is insufficient to account for the high concentration found proximally (Setchell and Hinton, 1981). Furthermore, recent evidence indicates that carnitine in the testicular exocrine secretion is removed from the epididymal lumen in the proximal caput (Hinton and Hernandez, 1985). There is no relationship between epididymal ABP (also derived from testicular fluid; see Section III.G) and epididymal carnitine (Bøhmer et al., 1977) and there is no decrease in epididymal carnitine content after ligation of the efferent ducts (Brooks, 1979b).

b. The Bloodstream

Labelled carnitine is taken up by the epididymis from the bloodstream; initially more is taken up in proximal regions but with time the activity migrates distally (Brooks et al., 1973). Higher uptake into proximal tissue was confirmed with the micropuncture technique by Hinton et al., (1979); this work also proved that the activity was in the lumen. A luminal location of tracer in the caput is confirmed by radioautography (Bøhmer et al., 1979) and there is uptake into the perfused corpus and cauda lumina with higher rates into the corpus (Yeung et al., 1980).

In all these instances transport was against substantial concentration gradients and of various models (Brooks, 1979a), there is evidence for both apical and basal pumps within the epithelium (Yeung et al., 1980). The uptake of L-carnitine into isolated epithelial cells (James et al., 1981) is not stereoselective and uptake into tubules is not inhibited by D-carnitine (Johansen and Bøhmer, 1979a), but transport across the epithelium is stereoselective (Yeung et al., 1980) and inhibitable by the D-isomer (Cooper et al., 1986) so this suggests a stereoselective apical carrier (Brooks, 1983d).

i. Regional Differences in Uptake

Transport of carnitine to the rat corpus lumen exceeds that to the cauda in-vivo (Yeung et al., 1980), but the regional differences in uptake observed are unlikely to reflect the higher blood flow in the proximal epididymis (see Section III.A.3) because the caput and corpus tubules take up more carnitine than caudal tubules in-vitro (Johansen and Bøhmer, 1979a) and cells isolated from the corpus epithelium take up more carnitine than those from the cauda (Dorsey et al., 1980).

3. Androgen Dependence

Concentrations of carnitine in epididymal fluid decrease after castration in rats (Marquis and Fritz, 1965b,c; Brooks et al., 1974) and rabbits (Jones and Glover, 1973), although the antiandrogen cyproterone acetate has no effect in intact rats (Brooks et al., 1974; Jones, 1977). In rats plasma carnitine is reduced after castration or oestrogen-treatment (Borum, 1978, 1980) but an inverse relationship between circulating testosterone and carnitine was found in the bull (Carter et al., 1980a). Removal of the testis does not decrease the uptake of injected carnitine into heart muscle as it does in the epididymis, so a specific action of androgens on carnitine transport into the epididymis was inferred (Bøhmer and Hansson, 1975).

Carnitine uptake into the epididymis (Bøhmer and Hansson, 1975), or cells taken from it (James et al., 1981), is reduced after castration, and the reduction in carnitine transport into the epididymis of oestrogen-treated males (Bøhmer and Hansson, 1975) can be ameliorated by androgen therapy (Bøhmer, 1978; see Bøhmer et al., 1980). In cultures of proximal rabbit epididymis, DHT increases the concentration of carnitine within spermatozoa contained in its lumen (Casillas and Chaiyungpan, 1982). An ipsilateral decline in carnitine uptake after unilateral castration hinted at a role for testicular luminal fluid in carnitine uptake (Bøhmer and Hansson, 1975) but there is no correlation with carnitine uptake and another testicular fluid component ABP (Bøhmer et al., 1977).

4. Interaction of Carnitine with Maturing Spermatozoa

The concentrations of carnitine in spermatozoa increases as they mature in bulls (Casillas, 1973), rabbits (Casillas and Chaiyungpan, 1979) and hamsters (Casillas et al., 1984; see Table 16). The uptake of labelled carnitine into caput sperm in-vitro is greater than that into cauda calls (Casillas, 1973); uptake into cauda cells is slow (Day-Francesconi and Casillas, 1982) and less than that found for ejaculated spermatozoa (Johansen and Bøhmer, 1978) and it is inhibited by unlabelled carnitine. In hypotonically demembranated mature spermatozoa carnitine is taken up into the mitochondria where exchange with acetylcarnitine occurs (Calvin and Tubbs, 1976). Within the epididymal lumen immature proximal corpus spermatozoa from the rabbit acquire carnitine from the medium in the presence of androgens (Casillas and Chaiyungpan, 1982).

5. Roles of Carnitine in Epididymal Function

a. Transport of Fatty Acids

The role of acylcarnitine esters in the transport of fatty acids or carnitine across the epididymal epithelium is only speculative at present (Brooks, 1979c) and a role for transport into spermatozoa is not proven. There is little evidence that carnitine promotes fatty acid transport into spermatozoa *via* formation of acylcarnitine, as has been postulated for the erythrocyte from the presence of palmitoylcarnitine transferase in the red cell membrane (Wittles and Hochstein, 1967). The entry of palmitate into bovine spermatozoa would appear to be difficult, since it is only oxidised after sonication of spermatozoa (Casillas, 1972) and palmitoylcarnitine itself is only metabolised by bovine spermatozoa if they are rendered permeable with filipin (Hutson et al., 1977a,b) or by rabbit spermatozoa if they are demembranated (Storey and Keyhani, 1974). On the other hand, rat sperm can metabolise palmitate in the absence of exogenous carnitine (Geer et al., 1975).

The greater uptake of acetate than carnitine into bovine spermatozoa, and the sole location of acetylcarnitine transferase in the mitochondria of this species (Day-Francesconi and Casillas, 1982) rule out this enzyme as a carrier for short-chain fatty acids across the sperm cell membrane.

b. Fatty Acid Oxidation

The accepted role of carnitine in other tissues, that of aiding the transport of long-chain fatty acids as acylcarnitines into mitochondria, is established for spermatozoa, but the need for acetylcarnitine to transport acetate into the mitochondria has been questioned (see Brooks, 1979d).

i. Short Chain Fatty Acids

Carnitine acetyltransferase is present in epididymal spermatozoa from rats (Marquis and Fritz, 1965b; Brooks et al., 1974; Brooks, 1978, 1979a) and there is no change in its activity in sperm from the testis, caput or cauda (Vernon et al., 1971). Because acetylcarnitine is measurable is mature spermatozoa in bulls (Casillas, 1973) and rats (Brooks et al., 1974) and the acetylation state of carnitine is altered during in-vitro metabolism of sperm from monkeys (Casillas and Erickson, 1975b) and bulls (Casillas, 1973; Casillas and Erickson, 1975b; Milkowski et al., 1976; van Dop et al., 1977; Hutson et al., 1977a; see Carter et al., 1980), acetylcarnitine may act to buffer intracellular concentrations of acetylCoA (Casillas and Erickson, 1975b; Calvin and Tubbs, 1976; Brooks et al., 1974) or act as a store of cellular energy in the form of acetate (Milkowski et al., 1976; Hutson et al., 1977a).

In support of this, the acetylcarnitine:carnitine ratio of motile human spermatozoa is higher that that of poorly motile sperm (Golan et al., 1984) and exogenous acetylcarnitine is metabolised by spermatozoa from rabbits (Storey and Keyhani, 1974), bulls (Hutson et al., 1977b; Storey, 1980; Day-Francesconi and Casillas, 1982).

ii. Long Chain Fatty Acids

Carnitine palmitolytransferase is present in rat epididymal spermatozoa (Brooks, 1978, 1979a), but a mitochondrial location has not been confirmed: rabbit spermatozoal mitochondria do not oxidise palmitoylCoA in the presence of carnitine (Storey and Keyhani, 1974). There is no measurable palmitoylcarnitine in mature rat sperm (Brooks et al., 1974) or caput or cauda sperm from the bull (Casillas, 1973; Van Dop et al., 1977a), but large concentrations of these molecules would not be anticipated in view of the lytic nature of these compounds in-vitro towards mitochondria (Bremer, 1962), erythrocytes (Cho and Proulx, 1969, 1971) and rat spermatozoa (Cooper, 1986).

iii. Source of Fatty Acids for Spermatozoa

It is not known if the fatty acids present in epididymal fluid from rams (Scott et al., 1961) and rats (Brooks et al., 1974) arise from epididymal synthesis (Brooks, 1979b), by transport from the bloodstream, or from breakdown of phospholipids also present in epididymal fluid (Quinn and White, 1967; Voglmayr et al., 1977; Evans and Setchell, 1979b). In the ram the total content of phospholipid (Voglmayr et al., 1977) and phospholipid-associated fatty acid (Evans and Setchell, 1979b) in epididymal fluid exceeds that of rete testis fluid, so they could be available for utilisation by spermatozoa. Sperm from all parts of the male tract can metabolise phospholipids in this species (Scott and Dawson, 1968) and the bull (Atreja and Anand, 1985).

iv. Endogenous Reserves

Although the premise that ejaculated spermatozoa can utilise phospholipids in substrate-free media (Lardy and Phillips, 1941a,b; Hartree and Mann, 1959, 1961) has been challenged (Bomsteil and Sterberl, 1957; Scott and Dawson, 1968; Darin-Bennett et al., 1973; Selivonchick et al., 1980), it still fuels speculation that epididymal spermatozoa utilise their own lipid reserves on transport through the epididymis.

Support for this view rests on the following observations: (1) the total phospholipid content of sperm is reduced as they pass through the epididymis of bulls (Lavon et al., 1970; Poulos et al., 1973), rams (Dawson and Scott, 1964; Quinn and White, 1967; Scott et al., 1967; Poulos et al., 1975; Dacheux, 1977; Evans and Setchell, 1979b), boars (Grogan et al., 1966; Johnson et al., 1972; Evans and Setchell, 1979a), rats (Dawson and Scott, 1964; Dacheux, 1977) and rabbits (Teichmann et al., 1974); (2) the acylester content of both neutral and phospholipids of sperm decreases on maturation. In the ram, loss of fatty acids (mainly unsaturated) from neutral lipids is greater than the loss of saturated acids from phospholipids, whereas in the boar there is a greater loss of fatty acids from phospholipids. In the rat an increase in triglyceride lipase activity on maturation of spermatozoa (Terner et al., 1975) is accompanied by a loss of total fatty acids (Adams and Johnson, 1977); (3) there is a decrease in the relative chain length of the remaining fatty acids, *e.g.* palmitate is lost from testicular spermatozoa from boars (Evans and Setchell, 1979a; Johnson et al., 1972) and there is a relative increase in myristate in sperm from the ram (Evans and Setchell, 1979b), both of which are consistent with β-oxidation having occurred.

However, the loss of free fatty acids from spermatozoa does not necessarily imply that they are oxidised: for example, both testicular and ejaculated spermatozoa can incorporate palmitate and oleate into neutral and phospholipids in rams (Mills and Scott, 1969) and bulls (Neill and Masters, 1972) and the arachidonic acid lost from ovine and bovine spermatozoa (Poulos et al., 1973, 1975) may be converted into the prostaglandins found in epididymal fluid (Voglmayr, 1973; Poulos et al., 1975). While prostaglandin synthetase is present in rat sperm (Johnson and Ellis, 1977) there is little PGE in human epididymal fluid (Bendvol et al., 1985). Prostaglandins are antilipolytic in-vivo (Bergstrom et al., 1968) but in vitro they have no effect on the oxygen consumption of spermatozoa (Voglmayr, 1970; Pento et al., 1970).

If endogenous lipids are utilised by spermatozoa during their stay in the epididymis, they are unlikely to be structural lipids; suitable candidates would be the membranous vesicles within the cytoplasmic droplet observed in rams (Dott and Dingle, 1968), bulls (Garbers et al., 1970; Nolan et al., 1983) and boars (Bloom and Nicander, 1961; Kaplan et al., 1984), whose conformation changes as they migrate through the epididymis (Bloom and Nicander, 1961; Nicander and Bane, 1962) and which lie directly over the mitochonodria in the mid-piece. Nevertheless, calculations from utilisation rates of lipids in-vitro, suggest that there are insufficient fatty acids in sperm cells to last them their entire sojourn in the epididymis (Brooks, 1979a).

c. Sperm Respiration

Exogenous carnitine does not promote the endogenous respiration of immature bovine spermatozoa (Casillas, 1973) or mature cells from rats (Johansson and Bøhmer, 1978) or bulls (Casillas, 1973; Johansson and Bøhmer, 1978), and it inhibits the endogenous respiration of ejaculated spermatozoa (Johansson and Bøhmer, 1978). Likewise, when spermatozoa are incubated with exogenous palmitate as substrate carnitine does not affect respiration of intact bovine epididymal sperm (Casillas, 1972, 1973) and it inhibits oxidation by ejaculated sperm (Hamilton and Olson, 1976).

Taken together with the fact that carnitine does promote the oxidation of palmitate by sonicated bovine epididymal sperm preparations (Casillas, 1972), these data suggest that access of substrate is a problem (Casillas, 1973; Day-Francesconi, 1982) although this situation is reversed on ejaculation (Hamilton and Olson, 1976; Johansson and Bøhmer, 1978).

d. Other Metabolic Effects

In-vitro 5mM carnitine shifts the metabolism of ovine cauda spermatozoa from glucose utlisation and lactate production to acetate production: as ATP production is maintained, the metabolic efficiency of these cells is increased, but there is no similar action on caput or corpus sperm (Inskeep and Hammerstedt, 1982). Carnitine promotes the formation of 1,2-diglycerides by ejaculated sperm (Hamilton and Olson, 1976) and these can be incorporated into phosphatidylinositol by spermatozoa (Neill and Masters, 1971, 1972). Carnitine can increase the activity of side-chain cleavage of cholesterol in the adrenal gland (Abdel-Kadei and Wolf, 1965) but the relationship between this and pregnenolone in epididymal fluid (see Table 11) has not been investigated.

e. Sperm Motility

i. Initiation of Sperm Motility

Carnitine can promote the motility of immature caput spermatozoa (see Section II.A.6.a). At higher concentrations, or after longer incubation in substrate-free media, carnitine reduces sperm motility. Whilst the stimulation may be due to increased substrate availability at low concentrations, at high concentrations carnitine is known to inhibit the metabolism of acetyl- but not palmitoyl-carnitine (Hutson et al., 1977b).

ii. Maintaining Sperm Motility

The motility of ejaculated human spermatozoa can be maintained by carnitine and acetylcarnitine (Tanphaichitr, 1977) and a close relationship exists between motility and the cellular carnitine content (Johansen and Bøhmer, 1979b) and its acetylation state (Milkowski et al., 1976; Golan et al., 1984).

iii. Sperm Quiescence

Carnitine does not prevent the burst of motility of mature spermatozoa that occurs on their removal from the tract of rats (63mM: Turner et al., 1978; 50mM Chulavatnatol, 1982) or bulls (10mM: Carr and Acott, 1984) so is unlikely to be responsible for the quiescence of spermatozoa in the epididymal tubule. Although 63mM carnitine with 40mM GPC did prevent the oxidation of pyruvate by rat spermatozoa (Turner et al., 1980) carnitine alone did not (Turner and Giles, 1981). At 60mM carnitine has no effect on the activity of the surface ATPase of rat sperm (Chulavatnatol and Yindepit, 1976).

6. Summary

As sperm move through the epididymis they experience increasing concentrations of carnitine transported from the bloodstream. Increases in the motility of immature sperm can be brought about in-vitro by concentrations of carnitine found in the lumen in-vitro. The role of acetyl carnitine as a store of energy during motility is established but whether carnitine acts as a carrier for fatty acids into sperm remains unproven.

7. References

Abdel-Kader MM and Wolf G (1965) The distribution of carnitine and its possible function in corticosteroid synthesis. In: Recent Research on Carnitine. Its relation to Lipid Metabolism. Ed. G Wolf, MIT Press, Cambridge, Mass. pp 147–156

Adams CS and Johnson AD (1977) The lipid content of epididymal spermatozoa of Rattus Norvegicus. Comp Biochem Physiol 58B:409–411

Atreja SK and Anand SR (1985) Phospholipase and lysophospholipase activities of goat spermatozoa in transit from the caput to the cauda epididymidis. J Reprod Fert 74:687–691

Bendvold E, Svanborg K, Bygdman M and Norén S (1985) On the origin of prostaglandins in human seminal fluid. Int J Androl 8:37–43

Bergstrom S, Carlson LA and Weeks JR (1968) The prostaglandins: A family of biologically active lipids. Pharmacol Rev 20:1–48

Bloom G and Nicander L (1961) On the ultrastructure and development of the protoplasmic droplet of spermatozoa. Z Zellforsch 55:833–844

Bøhmer T (1978) Accumulation of carnitine in rat epididymis after injection of (^{3}H)butyrobetaine in vivo: quantitative aspects and the effects of androgens and anti-androgens. Mol Cell Endocr 11: 213–223

Bøhmer T and Johansen L (1978) Inhibition of sperm maturation through intervention of the carnitine system. Int J Androl Suppl 2:565–573

Bøhmer T and Hansson V (1975) Androgen-dependent accumulation of carnitine by rat epididymis after injection of (^{3}H)butyrobetaine in vivo. Mol Cell Endocr 3:103–115

Bøhmer T, Weddington SC and Hansson V (1977) Effect of testosterone propionate on levels of carnitine and testicular androgen binding protein (ABP) in rat epididymis. Endocrinology 100:835–838

Bøhmer T, Hoel P, Purvis K and Hansson V (1978) Carnitine levels in human accessory organs. Archs Androl 1:53–59

Bøhmer T, Sar M, Stumpf WE and Hansson V (1979) Autoradiographic study of the (^{3}H)carnitine distribution in rat epididymis. Int J Androl 2:62–73

Bøhmer T, Johansen L and Kjekshus E (1980) Carnitine in the epididymis and in the spermatozoa: physiological aspects and clinical applications. In: Animal Models in human reproduction. Eds M Serio and L Martini, Raven Press, New York. pp 191–207

Bomstein RA and Steberl EA (1957) The utilization of phospholipides by bovine spermatozoa. Exp Cell Res 12:254–264

Borum PR (1978) Variation in tissue carnitine concentrations with age and sex in the rat. Biochem J 176:677–681

Borum PR (1980) Regulation of the carnitine concentration in plasma. In: Carnitine. Biosynthesis, Metabolism and Function. Eds RA Frenkel and JD McGarry, Acad Press London. pp 115–126

Bremer J (1962) Carnitine in intermediary metabolism. The metabolism of fatty acid esters of carnitine by mitochondria. J Biol Chem 237:3682–3632

Brooks DE (1978) Activity and androgenic control of enzymes associated with the tricarboxylic acid cycle, lipid oxidation and mitochondrial shuttles in the epididymis and epididymal spermatozoa of the rat. Biochem J 174:741–752

Brooks DE (1979a) Biochemical Environment of Maturing Spermatozoa. In: The Spermatozoon. Eds DW Fawcett and JM Bedford, Urban and Schwarzenberg, Baltimore. pp 23–34

Brooks DE (1979a) Biochemical Environment of Maturing Spermatozoa. In: The Spermatozoon. Eds DW Fawcett and JM Bedford, Urban and Schwarzenberg, Baltimore. pp 23–24

Brooks DE (1979b) Influence of testicular secretion on tissue weight and on metabolic and enzyme activities in the epididymis of the rat. J Endocr 82:305–313

Brooks DE (1979c) Carnitine, acetylcarnitine and the activity of carnitine acetytransferases in seminal plasma and spermatozoa of men, rams and rats. J Reprod Fert 56:667–673

Brooks DE (1983) Epididymal functions and their hormonal regulation. Aust J Biol Sci 36:205–221

Brooks DE, Hamilton DW and Mallek AH (1973) The uptake of L-(methyl-^{3}H)carnitine by the rat epididymis. Biochem Biophys Res Commun 52:1354–1360

Brooks DE, Hamilton DW and Mallek AH (1974) Carnitine and glycerylphosphorylcholine in the reproductive tract of the rat. J Reprod Fert 36:141–146

Calvin J and Tubbs PK (1976) A carnitine:acetylcarnitine exchange system in spermatozoa. J Reprod Fert 48:417–420

Carr DW and Acott TS (1984) Inhibition of bovine spermatozoa by cauda epididymal fluid: 1. Studies of a sperm motility quiescence factor. Biol Reprod 30:913–925

Carter AL, Hutson SM, Stratman FW and Haning RV (1980a) Relationship of carnitine and acylcarnitine in ejaculated sperm to blood plasma testosterone of dairy bulls. Biol Reprod 23:820–825

Carter AL, Stratman FW, Hutson SM and Lardy HA (1980b) The role of carnitine and its esters in sperm metabolism. In: Carnitine Biosynthesis, Metabolism and Functions. Eds RA Frenkel and JD McGarry, Acad Press, New York. pp 251–264

Casillas ER (1972) The distribution of carnitine in male reproductive tissues and its effect on palmitate oxidation by spermatozoal particles. Biochem Biophys Acta 280:545–551

Casillas ER (1973) Accumulation of carnitine by bovine spermatozoa during maturation in the epididymis. J Biol Chem 248:8227–8232

Casillas ER and Chaipayungpan S (1979) The distribution of carnitine and acetylcarnitine in the rabbit epididymis and the carnitine content of rabbit spermatozoa during maturation. J Reprod Fert 56:439–444

Casillas ER and Chaiyungpan S (1982) Carnitine content of rabbit epididymal spermatozoa in organ culture. J Reprod Fert 65:247–251

Casillas ER and Erickson BJ (1975a) Studies on carnitine synthesis in the rat epididymis. J Reprod Fert 44:287–291

Casillas ER and Erickson BJ (1975b) The role of carnitine in spermatozoan metabolism: substrate-induced elevations in the acetylation state of carnitine and coenzyme A in bovine and monkey spermatozoa. Biol Reprod 12:275–283

Casillas ER, Villalobos R and Gonzales R (1984) Distribution of carnitine and acylcarnitine in the hamster epididymis and in epididymal spermatozoa during maturation. J Reprod Fert 72:197–201

Cho KS and Proulx P (1969) Lysis of erythrocytes by long-chain acyl esters of carnitine. Biochem Biophys Acta 193:30–35

Cho KS and Proulx P (1971) Studies on the mechanism of hemolysis by acyl carnitine, lysolecithins and acylcholines. Biochem Biophys Acta 225:214–223

Chulavatnatol M (1982) Motility initiation of quiescent spermatozoa from rat caudal epididymis: effects of pH, viscosity, osmolality and inhibitors. Int J Androl 5:425–436

Chulavatnatol M and Yindepit S (1976) Changes in surface ATPase of rat spermatozoa in transit from the caput to the cauda epididymidis. J Reprod Fert 48:91–97

Cooper TG (1986) Osmotic swelling of maturing rat spermatozoa and lysis of caput spermatozoa by acylcarnitine and acylcholines. Gamete Res 14:47–56

Cooper TG, Gudermann TW and Yeung CH (1986) Characteristics of the transport of carnitine into the cauda epididymis of the rat as ascertained by luminal perfusion in-vitro. Int J Androl (in press)

Dacheux JL (1977) Reinvestigation of the variation in total phospholipid content of the spermatozoa of the rat and the ram during epididymal transit. IRCS Med Sci 5:18

Darin-Bennett A, Poulos A and White IG (1973) A re-examination of the role of phospholipids as energy substrates during incubation of ram spermatozoa. J Reprod Fert 34:543–546

Dawson RMC and Scott TW (1964) Phospholipid composition of epididymal spermatozoa prepared by density gradient centrifugation. Nature (Lond) 202:292–293

Day-Francesconi and Casillas ER (1982) The intracellular localization and properties of carnitine acetyltransferase from ram spermatozoa. Arch Biochem Biophys 215:206–214

Dorsey WF, Chapman, DA and Killian GJ (1980) Carnitine in rat epididymal sperm and epithelial cells isolated from the caput and corpus epididymidis by centrifugal elutriation. Biol Reprod 22 Suppl 1, 4A (Abstr)

Dott HM and Dingle JT (1968) Distribution of lysosomal enzymes in the spermatozoa and cytoplasmic droplets of bull and ram. Exp Cell Res 52:523–540

Evans RW and Setchell BP (1979a) Lipid changes in boar spermatozoa during epididymal maturation with some observations on the flow and composition of boar rete testis fluid. J Reprod Fert 57:189–196

Evans RW and Setchell BP (1979b) Lipid changes during epididymal maturation in ram spermatozoa collected at different times of the year. J Reprod Fert 57:197–203

Fahimi F, Bieber L and Lewin LM (1981) The sources of carnitine in human semen. J Androl 2:339–342

Frenkel G, Peterson RN, Davis JE and Freund M (1974) GPC and carnitine in normal semen and post vasctomy semen: differences in concentration. Fert Steril 25:84–87

Garbers DL, Wakayabashi T and Reed, PW (1970) Enzyme profile of the cytoplasmic droplet from bovine epididymal spermatozoa. Biol Reprod 3:327–337

Geer BW, Kelly KR, Pohlman TH and Yemm SJ (1975) A comparison of rat and Drosophila spermatozoan metabolisms. Comp Biochem Physiol 50B: 41–50

Golan R, Setchell BP, Burrow PV and Lewin LM (1982) A comparative study of carnitine and acylcarnitine concentration in semen and male reproductive tract fluids. Comp Biochem Physiol 72B:457–460

Golan R, Soffer Y, Katz S, Weissenberg R, Wasserzug O and Lewin LM (1983) Carnitine and short-chain acylcarnitines in the lumen of the human male reproductive tract. Int J Androl 6:349–357

Golan R, Weissenberg R and Lewin LM (1984) Carnitine and acetylcarnitine in motile and immotile human spermatozoa. Int J Androl 7:484–494

Grogan DE, Mayer DT and Sikes JD (1966) Quantitative differences in phospholipids of ejaculated spermatozoa. J Reprod Fert 12:431–436

Hamilton DW and Olson GE (1976) Effect of carnitine on oxygen uptake and utilization of (U-^{14}C) palmitate by ejaculated bull spermatozoa. J Reprod Fert 46:195–202

Hartree EF and Mann T (1959) Plasmalogen in ram semen, and its role in sperm metabolism. Biochem J 71:423–434

Hartree EF and Mann T (1961) Phospholipids in ram semen: metabolism of plasmalogens and fatty acids. Biochem J 80:464–476

Hinton BT and Hernandez H (1985) Selective luminal absorption of L-carnitine from the proximal regions of the rat epididymis. Possible relationship to development of sperm motility. J Androl 6:300–305

Hinton BT, Snoswell AM and Setchell BP (1979) The concentration of carnitine in the luminal fluid of the testis and epididymis of the rat and some other mammals. J Reprod Fert 56:105–111

Hutson SM, van Dop C and Lardy HA (1977a) Mitochondrial metabolism of pyruvate in bovine spermatozoa. J Biol Chem 252:1309–1315

Hutson SM, van Dop C and Lardy HA (1977b) Metabolism of pyruvate and carnitine esters in bovine epididymal sperm mitochondria. Arch Biochem Biophys 181:345–352

Inskeep PB and Hammerstedt RH (1982) Changes in metabolism of ram sperm associated with epididymal transit or induced by endogenous carnitine. Biol Reprod 27:735–743

James MJ, Brooks DE and Snoswell AM (1981) Kinetics of carnitine uptake by rat epididymal cells. Androgen-dependence and lack of stereospecificity. FEBS Lett 126:53–56

Johansen L and Bøhmer T (1978) Carnitine-binding related supressed oxygen uptake by spermatozoa. Archs Androl 1:321–324

Johansen L and Bøhmer T (1979a) Uptake of ^{3}H-L-carnitine by isolated rat epididymal tubules. Archs Androl 2:117–121

Johansen L and Bøhmer T (1979b) Motility related to the presence of carnitine/acetylcarnitine in human spermatozoa. Int J Androl 2:202–210

Johnson JM and Ellis LC (1977) The histochemical localization of prostaglandin synthetase activity in reproductive tract of the male rat. J Reprod Fert 51:17–22

Johnson LA, Pursel VG and Gerrits RJ (1972) Total phospholipid and phospholipid fatty acids of ejaculated and epididymal semen and seminal vesicle fluid of boars. J Anim Sci 35:398–403

Jones R (1977) Effects of testosterone, testosterone metabolites and anti-androgens on the function of the male accessory glands in the rabbit and rat. J Endocr 74:75–78

Jones R and Glover TD (1973) The effects of castration on the composition of rabbit epididymal plasma. J Reprod Fert 34:405–411

Kaplan M, Russell LD, Peterson RN and Martan J (1984) Boar sperm cytoplasmic droplets: their ultrastructure, their numbers in the epididymis and at ejaculation and their removal during isolation of sperm plasma membranes. Tissue and Cell 16:455–468

Kohengkul S, Tanphaichitr V, Muangmun V and Tanphaichitr N (1977) Levels of L-carnitine and L-*O*-acetylcarnitine in normal and infertile human semen: a lower level of L-*O*-acetylcarnitine in infertile semen. Fert Steril 28:1333–1336

Lardy HA and Phillips PH (1941a) The interrelation of oxidative and glycolytic processes as sources of energy for bull spermatozoa. Am J Physiol 133:602–609

Lardy HA and Phillips RH (1941b) Phospholipids as a source of energy for motility of bull spermatozoa. Am J Physiol 134:542–548

Lavon U, Volcani R and Danon D (1970) The lipid content of bovine spermatozoa during maturation and ageing. J Reprod Fert 23:215–222

Lewin LM, Beer R and Lunenfeld B (1976) Epididymis and seminal vesicle as source of carnitine in human seminal fluid: the clinical significance of carnitine concentration in human seminal fluid. Fert Steril 27:9–13

Lewin LM, Shalev AP, Weissenberg R and Soffer Y (1981) Carnitine and acylcarnitine in semen from azoospermic patients. Fert Steril 36:214–218

Lewin LM, Holzmann G, Fahimi F, Choi YU and Bilber LL (1979) Carnitine acyl esters of human semen. Int J Androl 2:542–548

Marquis NR and Fritz IB (1965a) The distribution of carnitine, acetylcarnitine, and carnitine acetyl transferase in rat tissues. J Biol Chem 240:2193–2196
Marquis NR and Fritz IB (1965b) Effects of testosterone on the distribution of carnitine, acetylcarnitine, and carnitine acetyl transferase in tissues of the reproductive system in the male rat. J Biol Chem 240:2197–2200
Marquis NR and Fritz IB (1965c) Effects of testosterone on the distribution of carnitine, acetylcarnitine, and carnitine acetyltransferase in tissues of the reproductive system of male rats. In: Recent Research on Carnitine. Its relation to lipid metabolism. Ed G Wolf, MIT Press, Cambridge. pp 31–34
Menchini-Fabris GF, Canale D, Izzo PL, Olivieri L, Barteloni M (1984) Free L-carnitine in human semen. Its variability in different androgen pathologies. Fert Steril 42:263–267
Milkowski AL, Babcock DF and Lardy HA (1976) Activation of bovine epididymal sperm respiration by caffeine. Its transient nature and relationship to the utilization of acetylcarnitine. Arch Biochem Biophys 176:250–256
Mills SC and Scott TW (1969) Metabolism of fatty acids by testicular and ejaculated ram spermatozoa. J Reprod Fert 18:367–369
Neill AR and Masters CJ (1971) Incorporation of (U-^{14}C)palmitic acid into the phospholipids of bovine semen. J Reprod Fert 24:295–297
Neill AR and Masters CJ (1972) Metabolism of fatty acid by bovine spermatozoa. Biochem J 127:375–385
Nicander L and Bane A (1962) Fine structure of boar spermatozoa. Z Zellforsch 57:390–405
Noland TD, Olson GE and Garbers DL (1983) Purification and partial characterization of plasma membranes from bovine spermatozoa. Biol Reprod 29:987–998
Pearson DJ and Tubbs PK (1967) Carnitine and derivatives in rat tissues. Biochem J 105:953–963
Pento JT, Cenedella RJ and Inskeep EK (1970) Effects of prostaglandins E_1 and $F_{1\alpha}$ upon carbohydrate metabolism of ejaculated and epididymal ram spermatozoa in vitro. J Anim Sci 30:469–471
Poulos A, Voglmayr JK and White IG (1973) Phospholipid changes in spermatozoa during passage through the genital tract of the bull. Biochem Biophys Acta 306:194–202
Poulos A, Brown-Woodman PDC, White IG and Cox RI (1975) Changes in phospholipids of ram spermatozoa during migration through the epididymis and possible origin of prostaglandin $F_{2\alpha}$ in testicular and epididymal fluid. Biochem Biophys Acta 388:12–18
Quinn PJ and White IG (1967) Phospholipid and cholesterol content of epididymal and ejaculated spermatozoa and seminal plasma in relation to cold shock. Aust J Biol Sci 20:1205–1215
Sade M, Saverymuttu I, Dinsdale O and Gow JG (1978) The management of azoospermia. Brit J Urol 50:595–597
Scott TW and Dawson RMC (1968) Metabolism of phospholipids by spermatozoa and seminal plasma. Biochem J 108:457–463
Scott TW, White IG and Annison EF (1961) Fatty acids in semen. Biochem J 78:740–742
Scott TW, Voglmayr JK and Setchell BP (1967) Lipid composition and metabolism in testicular and ejaculated ram spermatozoa. Biochem J 102:456–461
Selivonchick DP, Schmid PC, Natarajan V and Schmid HHO (1980) Structure and metabolism of phospholipids in bovine epididymal spermatozoa. Biochem Biophys Acta 618:242–254
Setchell BP and Hinton BT (1981) The effects on the spermatozoa of changes in the composition of luminal fluid as it passes along the epididymis. In: The Epididymis. Biology and Pathology. Eds CA Clavert and C Bollack, S Karger, Munich. Progr Reprod Biol 8:58–66
Soufir JC, Marson J and Jouannet P (1981) Free L-carnitine in human seminal plasma. Int J Androl 4:388–397
Storey BT (1980) Strategy of oxidative metabolism in bull spermatozoa. J exp Zool 212:61–67
Storey BT and Keyhani E (1974) Energy metabolism of spermatozoa. II. Comparison of pyruvate and fatty acid oxidation by mitochondria of rabbit epididymal spermatozoa. Fert Steril 25:857–864
Tanphaichitr N (1977) In vitro stimulation of human sperm motility by acetyl carnitine. Int J Fert 22:85–91

Teichmann RJ, Cummins JM and Takei GH (1974a) The characterization of a malachite green stainable, glutaraldehyde extractable phospholipid in rabbit spermatozoa. Biol Reprod 10:565–577

Terner C, MacLaughlin J and Smith BR (1975) Changes in lipase and phosphatase activities of rat spermatozoa in transit from the caput to the cauda epididymidis. J Reprod Fert 45:1–8

Turner TT (1984) Resorption versus secretion in the rat epididymis. J Reprod Fert 72:509–514

Turner TT, D'Addario D and Howards SS (1978) Further observations on the initiation of sperm motility. Biol Reprod 19:1095–1101

Turner TT and Giles RD (1981) The effects of carnitine, glycerylphosphorylcholine, caffeine, and egg yolk on the motility of rat epididymal spermatozoa. Gamete Res 4:283–295

van Dop C, Hutson SM and Lardy HA (1977a) Pyruvate metabolism in bovine epididymal spermatozoa. J Biol Chem 252:1303–1308

Vernon RG, Go VWL and Fritz IB (1971) Studies on the spermatogenesis in rats. II. Evidence that carnitine acetyltransferase is a marker enzyme for the investigation of germ cell differentiation. Can J Biochem 49:761–769

Voglmayr JK (1973) Prostaglandin $F_{2\alpha}$ concentration in genital tract secretions of dairy bulls. Prostaglandins 4:673–678

Voglmayr JK (1974) Alpha-chlorohydrin-induced changes in the distribution of free myo-inositol and prostaglandin $F_{2\alpha}$, and synthesis of phosphatidylinositol in the rat epididymis. Biol Reprod 11:593–600

Voglmayr JK, Musto NA, Saksena SK, Brown-Woodman PDC, Marley PB and White IG (1977) Characteristics of semen collected from the cauda epididymidis of conscious ram. J Reprod Fert 49:245–251

Wetterauer U and Heite HJ (1978) Carnitine in seminal plasma as parameter for the epididymal funktion. Andrologia 10:203–210

Wetterauer U and Heite HJ (1980) Carnitine in seminal plasma: its significance in diagnostic andrology. Archs Androl 4:137–143

Wittels B and Hochstein P (1967) The identification of carnitine palmityltransferase in erythrocyte membranes. J Biol Chem 242:126–130

Yeung CH, Cooper TG and Waites GMH (1980) Carnitine transport into the perfused epididymidis of the rat: regional differences, stereospecificity, stimulation by choline and effects of other luminal compounds. Biol Reprod 23:294–304

F. Epididymal Secretion of *myo*-Inositol

1. Concentration in Epididymal Tissue

Concentrations of inositol decrease along the length of the epididymis of rabbits and bulls (Voglmayr and Amann, 1973) but increase in the rat (Voglmayr, 1974). *Scyllo*-inositol and glycerophosphoinositol are also found in epididymal tissue (Seamark et al., 1968) and concentrations of phosphatidylinositol are greater in the head than the tail of the rat epididymis (Scott et al., 1963). Although present in the epididymal lumen in all species, concentrations of free inositol in farm animals (2mM boar: Hartree, 1957; 1–3.7mM ram: Voglmayr et al., 1977; Hinton et al., 1980; 2.5mM bull: Voglmayr and Amann, 1973) are considerably lower than those of rodents (30mM rat; 90 mM hamster: Hinton et al., 1980) and man (Setchell and Hinton, 1981; Hinton and Setchell, 1981). In man the concentration of *myo*-inositol in human seminal plasma does not vary in individual fractions of a split ejaculate (Brown-Woodman et al., 1980), so it is thought to arise from all accessory glands, not just the prostate (Lewin and Beer, 1973). Changes along the length of the organ are given in Table 17.

From estimates of water resorption in the epididymis of rats, derived from packed cell volumes, it is known that inositol is secreted proximally and absorbed or removed distally in the epididymis (Setchell and Hinton, 1981; Pholpramool et al., 1982).

2. Origin of Epididymal Inositol

a. Testicular Fluid

Inositol enters the lumen of the epididymis directly with spermatozoa and its concentration in rete testis fluid (RTF) from farm animals (5.8mM boar: Evans and Setchell, 1979a; 5.6–7.3 mM ram: Setchell et al., 1968; Hinton et al., 1980; 3.9mM bull: Voglmayr and Amann, 1973) exceeds that in the tail of the epididymis (see Table 17), so loss by removal or metabolism must account for the differences in concentrations in these fluids in these species. In rodents, however, the testis is not the sole source of epididymal inositol as concentration in RTF (2.0–5mM rat; 3mM hamster : Hinton et al., 1980) are much lower than those in epididymal fluid and ligating the testicular efferent ducts to prevent the inflow of testicular fliud does not reduce the concentrations found in the lumen (Pholpramool et al., 1982).

b. The Bloodstream

Labelled inositol injected into the bloodstream of rats is taken up into epididymal tissue (Lewin and Sulimovici, 1975; Lewin et al., 1978) and appears in luminal samples taken by micropuncture (Hinton and Howards, 1982) or microperfusion in-vivo (Cooper, 1982; Sujarit et al., 1985) and in-vitro (Cooper et al., 1985). Transport is against considerable concentration gradients in-vivo and raising the concentration of inositol but not glucose in plasma or peritubular fluid reduces transport to the lumen (Hinton and Howards, 1982c; Cooper et al., 1985), suggesting that saturable carriers are present, presumably in the basal aspect of the epithelium.

i. Regional Differences in Uptake

Entry of labelled inositol into caput tissue and lumen exceeds that into the cauda (Hinton and Howards, 1982) but in-vitro the extent of entry into the corpus epididymidis is similar to that into the cauda, although, in contrast to the cauda, a steady-state is achieved with respect to the bathing medium (Cooper et al., 1985).

c. Epididymal Biosynthesis

There is disagreement in the literature as to the extent and site of inositol synthesis in the epididymis, while the testis is known to synthesise luminal inositol from circulating glucose (Middleton and Setchell, 1968).

i. By Epididymal Tissue

The high level of biosynthetic enzymes found in the rat epididymis by Robinson and Fritz (1979) was not confirmed by Maeda and Eisenberg (1980) or Hasegawa and Eisenberg (1981). Inositol is not synthesised in homogenates of rat epididymis whether glucose (Eisenberg and Bolden, 1964) or glucose 6-phosphate (WCL Ford, personal communication) is used as substrate. Peritubular glucose was not converted into luminally secreted inositol in short term perfusions of the rat cauda or corpus in-vitro (Cooper et al., 1985) and the ovine epididymis has a low potential for insositol biosynthesis in-vitro (Loewus et al., 1983).

ii. By Spermatozoa

Roberts et al. (1976) showed that epididymal spermatozoa from the rat, and cytoplasmic droplets isolated from them, converted glucose to inositol, but Eisenberg and Bolden (1964), Eisenberg (1967) and Robinson and Fritz (1979) could show no inositol synthesis by sperm in-vitro. Epididymal spermatozoa are more active than epididymal tissue in the ram (Loewus et al., 1983).

Correlations between the numbers of spermatozoa in RTF and its inositol content led Setchell et al. (1971) to postulate that testicular spermatozoa were capable of converting glucose to inositol, and this has been demonstrated in-vitro (Voglmayr and White, 1971). Label from precursor appeared over the cytoplasmic droplet and the reduced synthetic activity of spermatozoa from heated testes was associated with disrupted membranous vesicles within the cytoplasmic droplet (Voglmayr et al., 1971). The failure of ejaculated spermatozoa to synthesise inositol from glucose under conditions in which testicular spermatozoa do so (Voglmayr and White, 1971), may reflect the loss of the cytoplasmic droplet on ejaculation, since epididymal sperm from this species are active at inositol synthesis (Loewus et al., 1983).

In the rat, Roberts et al. (1976) claimed that droplets from immature spermatozoa were more active at inositol synthesis than those from mature cells, but no account was taken of possible loss of activity during the long-term incubations required to obtain the mature droplets.

3. Androgen Dependence

While it has long been known that the inositol content of other male reproductive tissues is androgen-dependent (Melampy and Mason, 1957; Nixon, 1970) and that cyproterone acetate reduces inositol transport into the seminal vesicle (Lewin et al., 1979), it has only recently been shown that the concentration of inositol in epididymal fluid of the rat decreases after castration and is maintained by androgens (Pholpramool et al., 1982).

4. Metabolism of Inositol

a. By Epididymal Tissue

Slices of rabbit epididymis and rat cauda homogenates as well as epithelium from the rat vas deferens can convert inositol into phosphatidylinositol (Voglmayr and Amann, 1973; Voglmayr, 1974). The presence of glycerophosphoinositol in the epididymis (Seamark et al., 1968) suggests that some hydrolysis of phosphatidylinositol occurs in the tissue.

b. By Spermatozoa

Endogenous reserves of inositol decrease on incubation of testicular spermatozoa from rams (Voglmayr and White, 1971) and bulls (Voglmayr and Amann, 1973) but the metabolites were not identified in these studies. Ejaculated bovine spermatozoa can incorporate fatty acids into phosphatidylinositol by *via* synthesis of 1,2-diglycerides (Neill and Masters, 1971, 1972). Phosphatidylinositol, absent from ejaculated porcine spermatozoa, is extractable from them after incubation in the uterus (Snider and Clegg, 1975), but in this case it was not clear if the phospholipid was synthesised by the cells or taken up from uterine tissue.

Since phosphatidylinositol is present in the cytoplasmic droplet of mature bovine spermatozoa (Clegg and Foote, 1973; Clegg et al., 1975), the decreased content of phosphatidylinositol in ovine spermatozoa on maturation in the epididymis (Scott et al., 1967) may reflect loss of droplets during handling of the cells or increased metabolism of phospholipid. Hydrolysis to inositol monophosphate occurs in spermatozoa from all regions of the tract (Scott and Dawson, 1968) and polyphosphoinositides have been found in increasing amounts in the membranes of maturing porcine spermatozoa (Nikolopoulou et al., 1985).

Inositol is the acceptor of galactose from UDPgalactose in the presence of testicular galactosyltransferase and epididymal a-lactalbumin (Hamilton, 1981b; Jones and Brown, 1982; Qasba et al., 1983; see Section II.C.6.b). mRNA coding for a-lactalbumin is present in the epididymis (Qasba et al., 1983) but there are no reports yet of galactinol in epididymal fluid.

5. Interaction of Inositol with Maturing Spermatozoa

How inositol interacts with spermatozoa is uncertain although erythrocytes are impermeable to the cyclitol (Widdas, 1954; Lefevre and Davies, 1951). The five equatorial hydroxyl groups give an apparent molal volume similar to that of hydrated electrolytes (Suggett, 1975) and its protective action against damage done by dehydration to bacteria and viruses has been interpreted in terms of its imitating bound cell water (Webb, 1963).

In contrast to other phospholipids, phosphatidylinositol does not bind to testicular spermatozoa in-vitro (Evans and Setchell, 1978). Specific phospholipid transfer proteins (Akeroyd and Wirtz, 1982), however, may be required.

6. Role of Inositol in Epididymal Function

The roles of inositol are speculative since most studies have produced negative results.

a. Sperm Motility

The reduced incorporation of fatty acids into phosphatidylinositol by sperm displaying poor motility may be related to a role in flagellar function (Neill and Masters, 1972), but in-vitro a concentration of 27.8mM does not maintain the motility of ejaculated spermatozoa (White, 1954) and 30mM does not prevent the initiation of motility of quiescent mature epididymal spermatozoa (Chulavatnatol, 1982). Phosphatidyl inositol is said to improve the motility of ejaculated bovine spermatozoa (Voglmayr, 1973).

b. Sperm Metabolism

The metabolism of glucose by testicular spermatozoa is inhibited after long term incubation in 1mg/ml phosphatidylinositol (Voglmayr and Gandhi, 1978), whereas 250μg/ml promotes glucose uptake and conversion into lipid by epididymal sperm (Voglmayr, 1973). However, it is not known whether these differences reflect the different concentrations used or the different maturational states of the sperm used.

c. Sperm-Egg Binding

By accepting galactose from UDP-galactose (see Section III.F.4.b) inositol may modulate the extent of sperm surface galactosylation (see Section II.C.6.b) by competition with surface carbohydrates. Inositol hexaphosphate does not block sperm-egg binding in the pig (Peterson et al., 1984).

d. Capacitation

Inositol can H-bond to proteins (Bruce et al., 1958) and by lowering the ionic strength of epididymal fluid it could prevent the premature loss of proteins bound to the cell surface electrostatically (see Section III.G.10.a.i). Whether phosphatidyinositol acts as an anchor for membrane-bound proteins on spermatozoa (see Low and Kincade, 1985) has not been examined. At 5.6mM it does not alter the release of hyaluronidase from the guinea pig acrosome (Lewin et al., 1978).

e. Sperm Quiescence

By mimicking the action of bound cell water (Webb, 1963), inositol may achieve preservation of the sperm cell surface during the dehydration that is known to occur in the epididymis (see Section III.C.4.a.iv). The concentration of inositol itself increases as a consequence of water resorption in the proximal epididymis (Setchell and Hinton, 1981).

f. Other Possibilities

At the high concentrations found in epididymal fluid inositol promotes release of newly secreted protein in the pancreas (Slaby and Bryan, 1976) but this aspect of protein synthesis in the epididymis has not been examined. Inositol is an essential growth factor for cultured cells (Eagle et al., 1957) and it could act to maintain the viablity of epithelial and sperm cells in the epididymis.

If phosphatidylinositol is part of a gating mechanism for Ca^{2+} entry into cells (see Hawthorne, 1982) it could directly influence the calcium levels within the sperm cells or epithelium. By maintaining the membrane potentials, it could maintain consistent secretion by epithelial cells (Berridge, 1984). Polyphosphoinositides may also be important in the control of intracellular calcium because they bind divalent metal ions (see Hendrickson and Ballou, 1964; Hendrickson and Fuington, 1965; Dawson, 1965; Hendrickson, 1969) and the dephosphorylation of triphosphoinositide to diphosphoinositide can release up to 70% of bound calcium (Hendrickson and Reinertson, 1971).

Another aspect of cellular physiology of current interest is the role of inositol triphosphate as a second messenger (see Berridge and Irvine, 1984; Michel, 1986). The control of Ca^{2+} transport may be affected by lithium which inhibits myoinositol phosphate phosphatase in other tissues (Berridge, 1984) and is present in the epididymis (see Section III.C).

The interactions of inositol and its metabolites with mitotic spindles (Chargraff et al., 1948; Murphy et al., 1951) and tubulin (see Kirazov and Lagnado, 1977) may be relevant with respect to the maintenance of both the epithelial cell cytoskeleton, and its role in cell secretion and resorption, and to the spermatozoon, which has cytoskeletal elements within the acrosome and tubulin in the tail (see Section II.A.4.a.i).

7. Summary

Inositol is both synthesised by the epididymal epithelium and transported from the bloodstream into the lumen. Concentrations in the lumen increase distally but a role in sperm maturation remains elusive because in-vitro experiments have proved largely negative. While undoubtedly contributing to the osmotic pressure of epididymal fluid it remains speculation whether the low ionic strength environment it prevents the loss of proteins held electrostatically to the sperm surface.

8. References

Akeroyd R and Wirtz KWA (1982) Properties of Phospholipid transfer proteins. In: Membranes and Transport, Vol 1. Ed AN Martonosi, Plenum Press, London. pp 93–98

Berridge MJ (1984) Inositol triphosphate and diacyl glycerol as second messengers. Biochem J 220:345–360

Berridge MJ and Irvine RF (1984) Inositol triphosphate, a novel second messenger in cellular signal transduction. Nature (Lond) 312:315–327

Brown-Woodman PDC, Marley PB, Morris S, Rodger JC and White IG (1980) Origin of glycerylphosphorylcholine, inositol, N-acetylaminosugar, and prostaglandins in human seminal plasma and their effects on sperm metabolism. Archs Androl 4:149–155

Bruce ICL, Giles CH and Jain SK (1958) Studies in hydrogen-bond formation. Part VII. Reactions between carbohydrates and proteins in water and their relation to transfer of carbohydrates across red-cell membranes. J Chem Soc 1610–1613

Chargraff E, Stewart RN and Magasanik B (1948) Inhibition of mitotic poisoning by *meso*-inositol. Science 108:556–558

Chulavatnatol M (1982) Motility initiation of quiescent spermatozoa from rat caudal epididymis: effects of pH, viscosity, osmolality and inhibitors. Int J Androl 5:425–436

Clegg ED and Foote RH (1973) Phospholipid composition of bovine sperm fractions, seminal plasma and cytoplasmic droplets. J Reprod Fert 34:379–383

Clegg ED, Morré DJ and Lunstra DD (1975) Porcine sperm membranes: in vivo phospholipid changes, isolation and electron microscopy. In: The Biology of the Male Gamete. Eds JG Duckett and PA Racey, Biol J Linnaen Soc, Suppl 1, Vol 7 Acad Press. pp 321–335

Cooper TG (1982) Secretion of inositol and glucose by the perfused rat cauda epididymidis. J Reprod Fert 64:373–379

Cooper TG, Yeung CH, Lui W and Yang CZ (1985) Luminal secretion of *myo*-inositol by the rat epididymis perfused in vitro. J Reprod Fert 74:135–144

Dawson RMC (1965) "Phosphatido-peptide"-like complexes formed by the interaction of calcium triphosphoinositol with protein. Biochem J 97:134–138

Dawson RMC and Scott TW (1964) Phospholipid composition of epididymal spermatozoa prepared by density gradient centrifugation. Nature (Lond) 202:292–293

Eagle H, Oyama VI, Levy M and Freeman AE (1957) *Myo*-inositol as an essential growth factor for normal and malignant human cells in tissue culture. J Biol Chem 226:191–206

Eisenberg F (1967) D-*myo*inositol 1-phosphate as product of cyclization of glucose 6-phosphate and substrate for a specific phosphatase in rat testis. J Biol Chem 242:1375–1382

Eisenberg F and Bolden AH (1964) Reproductive tract as site of synthesis and secretion of inositol in the male rat. Nature (Lond) 202:599–600

Esbenshade KL and Clegg ED (1980) Surface proteins of ejaculated porcine sperm and sperm incubated in the uterus. Biol Reprod 23:530–537

Evans RW and Setchell BP (1978) Association of exogenous phospholipids with spermatozoa. J Reprod Fert 53:357–362

Evans RW and Setchell BP (1979a) Lipid changes in boar spermatozoa during epididymal maturation with some observations on the flow and composition of boar rete testis fluid. J Reprod Fert 57:189–196

Evans RW and Setchell BO (1979b) Lipid changes during epididymal maturation in ram spermatozoa collected at different times of the year. J Reprod Fert 57:197–203

Hamilton DW (1981) Evidence for α-lactalbumin-like activity in reproductive tract fluids of the male rat. Biol Reprod 25:385–392

Hamilton DW and Olson GE (1976) Effect of carnitine on oxygen uptake and utilization of (U-^{14}C) palmitate by ejaculated bull spermatozoa. J Reprod Fert 46:195–202

Hartree EF (1957) Inositol in seminal plasma. Biochem J 66:131–137

Hasegawa R and Eisenberg F (1981) Selective hormonal control of *myo*-inositol biosynthesis in reproductive organs and liver of the male rat. Proc Natn Acad Sci 78:4863–4866

Hawthorne JN (1982) Inositol phospholipids. In: Phospholipids. Eds JN Hawthorne and GB Ansell, Elsevier Biomedical Press, Amsterdam. pp 263–278

Hendrickson HS (1969) Physical properties and interactions of phosphoinositides. Ann NY Acad Sci US 165:608–676

Hendrickson HS and Ballou CE (1964) Ion exchange chromatography of intact brain phosphoinositides on diethylaminoethyl cellulose by gradient salt elution in a mixed solvent system. J Biol Chem 239:1369–1373

Hendrickson HS and Fullington JG (1965) Stabilities of metal complexes of phospholipids: Ca(II), Mg(II), an Ni(II) complexes of phosphatidylserine and triphosphoinositide. Biochem 4:1599–1605

Hendrickson HS and Reinertsen JL (1969) Comparison of metal-binding properties of trans-1,2-cyclohexanediol diphosphate and deacylated phosphoinositides. Biochem 8:4855–4858

Hinton BT and Howards SS (1982) Rat testis and epididymis can transport (^{3}H)3-*O*-methyl-D-glucose, (^{3}H)inositol and (^{3}H)α-aminoisobutyric acid across its epithelia in vivo. Biol Reprod 27:1181–1189

Hinton BT and Setchell BP (1981) Micropuncture and microanalytical studies of Rhesus monkey and baboon epididymis and the human ductus deferens. Am J Primatol 1:251–256

Hinton BT, White RW and Setchell BP (1980) Concentrations of *myo*-inositol in the luminal fluid of the mammalian testis and epididymis. J Reprod Fert 58:395–399

Hinton BT, Pryor JP, Hirsch AV and Setchell BP (1981) The concentration of some inorganic ions and organic compounds in luminal fluid of the human ductus deferens. Int J Androl 4:457–461

Jones R and Brown CR (1982) Association of epididymal secretory proteins showing α-lactalbumin-like activity with the plasma membrane of rat spermatozoa. Biochem J 206:161–164

Kirazov EP and Lagnado JR (1977) Interaction of *myo*-inositol with brain microtubules. FEBS Letts 81:173–178

Lefèvre PG and Davies RI (1951) Active transport into the human erythrocyte: evidence from comparative kinetics and competition among monosaccharides. J Gen Physiol 34:515–524

Lewin LM and Beer R (1973) Prostatic secretion as the source of *myo*-inositol in human seminal fluid. Fert Steril 24:666–670

Lewin LM and Sulimovici S (1975) The distribution of radioactive *myo*inositol in the reproductive tract of the male rat. J Reprod Fert 43:355–358

Lewin LM, Nevo Z, Marcus Z, Gabsu A and Nebel L (1978a) Hyaluronidase release from guinea pig spermatozoa as affected by reproductive tract secretions and metabolic inhibitors. Archs Androl 1:163–167

Lewin LM, Yannai Y and Kraicer PF (1978b) The distribution of radioactive *myo*-inositol in the reproductive tract of the male rat. The blood-testis barrier to *myo*-inositol. In: Cyclitols and Phosphoinositides. Eds WW Wells and F Eisenberg, Acad Press, New York. pp 317–322

Lewin LM, Yannai Y and Kraiger P (1979) The effect of cyproterone acetate on *myo*-inositol uptake and secretion in the reproductive tract of the male rat. Int J Androl 2:171–181

Loewus MW, Wright RW, Bondioli KR, Bedgak DL and Karl A (1983) Activity of *myo*-inositol-1-phosphate synthase in the epididymal spermatozoa of rams. J Reprod Fert 69:215–220

Low MG and Kincade PW (1985) Phosphatidylinositol is the membrane-anchoring domain of the Thy1-glycoprotein. Nature (Lond) 318:62–64

Maeda T and Eisenberg F (1980) Purification, structure, and catalytic properties of L-*myo*-inositol-1-phosphate synthase from rat testis. J Biol Chem 255:8458–8464

Melampy RM and Mason RB (1957) Androgen and the *myo*-inositol content of male accessory organs of the rat. Proc Soc exp Biol Med 96:405–408

Michell B (1986) Profusion and confusion. Nature (Lond) 319:176–177

Middleton A and Setchell BP (1972) The origin of inositol in the rete testis fluid of the ram. J Reprod Fert 30:473–475

Murphy MR, De Lam HH and Chargraff E (1951) Specific inhibition by *meso*-inositol of the colchicine effect on rat fibroblasts. Exp Cell Res 2:165–177

Neill AR and Masters CJ (1971) Incorporation of (U-^{14}C)palmitic acid into the phospholipids of bovine semen. J Reprod Fert 24:295–297

Neill AR and Masters CJ (1972) Metabolism of fatty acid by bovine spermatozoa. Biochem J 127:375–385

Nikolopoulou M, Souck DA and Vary JC (1985) Changes in the lipid content of boar sperm plasma membranes during epididymal maturation. Biochem Biophys Acta 815:486–498

Nixon DA (1970) The effect of sex hormones upon the inositol concentration of the rat seminal vesicle. J Reprod Fert 21:187–189

Peterson RN, Russell LD and Hunt WP (1984) Evidence for specific binding of uncapacitated boar spermatozoa to porcine zonae pellucidae in vitro. J exp Zool 231:137–147

Pholpramool C, White RW and Setchell BP (1982) Influence of androgens on inositol secretion and sperm transport in the epididymis of rats. J Reprod Fert 66:547–553

Qasba PK, Hewlett IK and Byers S (1983) The presence of the milk protein, α-lactalbumin, and its mRNA in the rat epididymis. Biochem Biophys Res Commun 117:306–312

Roberts ML, Scouten WH and Nyquist SE (1976) Isolation and characterization of the cytoplasmic droplet in the rat. Biol Reprod 14:421–424

Robinson R and Fritz IB (1979) *Myo*inositol biosynthesis by Sertoli cells and levels of *myo*inositol biosynthetic enzymes in the testis and epididymis. Can J Biochem 57:962–967

Scott TW and Dawson RMC (1968) Metabolism of phospholipids by spermatozoa and seminal plasma. Biochem J 108:457–463

Scott TW, Dawson RMC and Rowlands IW (1963) Phospholipid relationships in rat epididymal tissue and spermatozoa. Biochem J 87:507–512

Scott TW, Voglmayr JK and Setchell BP (1967) Lipid composition and metabolism in testicular and ejaculated ram spermatozoa. Biochem J 102:456–461

Seamark RF, Tate ME and Smeaton TC (1968) The occurrence of *scyllo*inositol and D-glycerol 1-(L-*myo*inositol 1-hydrogen phosphate) in the male reproductive tract. J Biol Chem 243:2424–2428

Setchell BP and Hinton BT (1981) The effects on the spermatozoa of changes in the composition of luminal fluid as it passes along the epididymis. In: The Epididymis: Biology and Pathology. Eds A Clavert and C Bollack, S Karger, Basel. Progr Reprod Biol 8:58–66

Setchell BP, Dawson RMC and White RW (1968) The high concentration of free *myo*-inositol in rete testis fluid from rams. J Reprod Fert 17:219–220

Setchell BP, Voglmayr JK and Hinks NT (1971) The effect of local heating on the flow and composition of rete testis fluid. J Reprod Fert 24:81–89

Slaby F and Bryan J (1976) High uptake of *myo*-inositol by rat pancratic tissue in vitro stimulates secretion. J Biol Chem 251:5078–5986

Snider DR and Clegg ED (1975) Alteration of phospholipids in porcine spermatozoa during in vivo uterus and oviduct incubation. J Anim Sci 40:269–274

Suggett A (1975) Polysaccharides. In: Water. A comprehensive treatise. Volume 4. Aqueous solutions of amphiphiles and macromolecules. Ed F Franks, Plenum Press, London. pp 519–567

Sujarit S, Chaturapanich G and Pholpramool C (1985) Evidence for blood *myo*-inositol as a source of the epididymal secretion in the perfused cauda epididmidis of the rat. Andrologia 17:321–326

Voglmayr JK (1973) Prostaglandin $F_{2\alpha}$ concentration in genital tract secretions of dairy bulls. Prostaglandins 4:673–678

Voglmayr JK (1974) Alpha-chlorohydrin-induced changes in the distribution of free *myo*-inositol and prostaglandin $F_{2\alpha}$, and synthesis of phosphatidylinositol in the rat epididymis. Biol Reprod 11:593–600

Voglmayr JK and Amann RP (1973) The distribution of free *myo*-inositol in fluid, spermatozoa, and tissues of the bull genital tract and observations on its uptake by the rabbit epididymis. Biol Reprod 8:504–573

Voglmayr JK and Gandhi JS (1978) Survival of ram testicular spermatozoa in vitro: effects of glucose, glucose metabolites, rete testis fluid proteins, selected androgens and phospholipids. Theriogenology 9:463–478

Voglmayr JK and White IG (1971) Synthesis and metabolism of *myo*inositol in testicular and ejaculated spermatozoa of the ram. J Reprod Fert 24:29–37

Voglmayr JK, Setchell BP and White IG (1971) The effects of heat on the metabolism and ultrastructure of ram testicular spermatozoa. J Reprod Fert 24:71–80

Voglmayr JK, Musto NA, Saksena SK, Brown-Woodman PDC, Marley PB and White IG (1977) Characteristics of semen collected from the cauda epididymidis of conscious ram. J Reprod Fert 49:245–251

Webb SJ (1963) Possible role for water and inositol in the structure of nucleoproteins. Nature (Lond) 198:785–786

White IG (1954) The effect of some seminal constituents and related substance on diluted mammalian spermatozoa. Aust J Biol Sci 7:379–3909

Widdas WF (1954) Facilitated transfer of hexoses across the human erythrocyte membrane. J Physiol 125:163–180

G. Epididymal Secretion and Resorption of Proteins

1. Luminal Proteins

Epididymal proteins obtained by shaking slices of tissue or from tissue minces (Gupta et al., 1974; Arora et al., 1975; Bose et al., 1975; Vreeburg, 1975; Huang and Johnson, 1976) necessarily contain proteins from blood and lymph in addition to those from the epididymal lumen. Bloodless micropuncture of the tubules provides uncontaminated samples with lower protein concentrations than blood (Table 18) and a variety of different proteins separable by gel electrophoresis (Koskimies and Kormano, 1975; Turner, 1979; Turner and Plesums, 1979; Hamilton, 1981b; Wong et al., 1981; Olson and Hinton, 1985) that are not present in serum.

2. Origin of Epididymal Proteins

a. The Bloodstream

Most major epididymal tissue proteins differ both from those in blood plasma and those entering the epididymal lumen in testicular fluid (Cameo and Blaquier, 1976; Brooks and Higgins, 1980) but the presence in uncontaminated epididymal fluid of proteins similar in molecular weight to serum albumin in mice (del Villano et al., 1977), rats (Koskimies and Kormano, 1975; Wong et al., 1981), rabbits (Jones and Dott, 1980; Moore, 1981b), boars (Sedláková et al.. 1968; Dostál and Veselský, 1972), rams (Voglmayr et al., 1980; Dacheux and Voglmayr, 1983) and bulls (Amann et al., 1973; Stánek and Dostál, 1974) raises the question of the source of luminal albumin. Whilst blood was considered the source of this protein in ram epididymal fluid by Alumot et al. (1971), del Rio (1979) considered the presence of albumin in guinea pig epididymal fluid to indicate contamination with blood.

Albumin persists in fluid obtained from the lumen of the rat epididymis from which all blood has been cleared by vascular perfusion (Brooks and Higgins, 1980), although here contamination with lymph cannot be ruled out. Short-term experiments in laboratory animals have shown that the epididymal epithelium excludes from the lumen a number of macromolecules circulating in the bloodstream, including proteins, but there is some evidence that in the long-term, proteins are able to cross the epithelium from the blood (see Section III.A.8).

b. Testicular Fluid

Testicular fluid is another source of plasma proteins entering the epididymal lumen, because Sertoli cells in culture secrete several proteins found in blood plasma (Wright et al., 1981). Proteins in testicular fluid that are not degraded by proteolytic enzymes in the epididymal lumen (see Table 21) and which do not pass between the epithelial cells will be concentrated by the continuous removal of water from the lumen (see Section III.C.2) and will disappear from the epididymis after ligating the testicular efferent ducts. These include androgen-binding protein (ABP: see Section III.B.3.c) and a variety of protein secretion products of the Sertoli cell (Wilson and Griswold, 1979; Wright et al., 1981) including transferrin (Skinner and Griswold, 1980, 1982; Kissinger et al., 1982), caeruloplasmin (Skinner and Griswold, 1983), "clusterin" (Fritz et al., 1983; Blaschko et al., 1983; Blaschuk and Fritz, 1984), retinol binding-protein (Huggenvik and Griswold, 1981) and growth factor (Brown et al., 1982a,b). Insulin is also present in the testes (Rosenweig et al., 1980) but a luminal location has not been confirmed.

The concentration of ABP decreases along the length of the rat epididymis (Aafjes and Vreeburg, 1972; Vreeburg, 1975; Pelliniemi et al., 1981; Purvis and Hansson 1982) and the concentration of ABP in epididymal fluid is less than that in rete testis fluid in rats (Turner, 1979; Turner et al., 1979) and rabbits (Danzo et al., 1976, 1977). In the rat there is evidence that ABP is taken up by the principal cells of the caput epididymidis (see Section II.G.14). Concentrations of ABP in ovine epididymal fluid are higher than those in rete testis fluid (Voglmayr et al., 1977; Jegou et al., 1979), most likely as a result of water resorption.

c. Epithelial Biosynthesis

Proteins that persist in the epididymis after ligation of the efferent ducts, namely acidic epididymal glycoprotein (AEG: Lea et al., 1978) and specific epididymal protein (SEP: Kohane et al., 1980) in the rat, SMA-4 (Vernon et al., 1982) and acrosin inhibitors (Poirier and Nicolson, 1984) in the mouse, or whose synthesis in-vitro does not decline after this operation (proteins B–E: Brooks and Higgins, 1980) are clearly of epididymal origin. The depletion of proteins B and C from cauda epididymal fluid after ligation of the efferent ducts (Brooks and Higgins, 1980) or AEG from caudal fluid after ligation of the corpus epididymidis (Lea et al., 1978) locates the site of synthesis of these proteins in more proximal regions of the duct and this is confirmed by the accumulation in the caput of proteins B and C after efferent duct ligation (Brooks and Higgins, 1980) or SEP after isolation of this region (Kohane et al., 1980).

The synthesis and secretion of proteins in the absence of the testis provides most evidence for this function of the epididymis (see Section III.G.3.a.ii).

Some human seminal proteins (*e.g.* glucosidase) originate from the epididymis because their concentration falls after vasectomy (Chapdelaine et al., 1978; Tremblay et al., 1979, 1981; Rivas et al., 1981; Guerin et al., 1981) and is lower than normal in cases of obstructive azoospermia (Guerin et al. 1981). Glucosidase has been identified in luminal fluid distal to the caput (Guerin et al., 1981). Some other proteins disappear from the human ejaculate after epididymovasostomy (Wong et al., 1982b), suggesting a synthesis distal to the site of the anastomosis.

Principal cells from the rat caput epididymidis have a higher protein content than basal cells or fibroblasts (Killian and Chapman, 1980) and this may reflect their biosynthetic activity. Principal cells from the corpus epididymidis are more active at protein synthesis than those from the caput (Hoffmann and Killian, 1981).

3. Protein Synthesis and Precursors in the Epididymis

a. Amino Acids

A reserve of free amino acids is disposed along the length of the epididymis (Kochakian, 1975, 1976), some within the epididymal lumen (Huang and Johnson, 1975; Jones, 1978). The tissue content of glutamate in the rat caput is dependent on continued flow of testicular fluid and circulating androgens and converse changes in the content of glutamine after castration and androgen replacement led Shimizuki et al. (1976) to postulate that glutamate aminotransferase activity was androgen-dependent in the epididymis.

Amino acids can arise from the circulation. The non-metabolisable α-aminoisobutyric acid (α-AIB) is taken up into the lumen of the rat caput more rapidly than into the cauda after injection. Its uptake is depressed by the simultaneous injection of large amounts of unlabelled α-AIB, which indicates the operation of carrier-mediated transport (Hinton and Howards, 1982). Gamma-glutamyltranspeptidase, one of several enzymes held responsible for amino acid transport across cell membranes (see Meister, 1973, 1974; Meister and Tate, 1972) is present in the epididymis of several species (Goldberg et al., 1960; Glenner et al., 1962; Albert et al., 1964; Matsuzumi, 1981). It is located in the stereocilia (Glenner et al., 1962; Rutenberg et al., 1969), a site more usually associated with movement of substances from the lumen to the bloodstream and between the cells (Lojda, 1975). More of the enzyme is found in the caput than cauda along with other enzymes of the postulated transport cycle (DeLapp et al., 1977).

The major amino acid free in epididymal fluid from bulls and rams, glutamate, is not metabolised by spermatozoa from rams (Setchell et al., 1967) or bulls (Sexton et al., 1971, 1972) and it does not stimulate respiration of spermatozoa. Taurine and hypotaurine have membrane-active actions (see Section II.A.4.g), and may be important for sperm after ejaculation from the male tract. All amino acids present in high concentration in the epididymal lumen would act osmotically and could serve as precursors of proteins.

i. Incorporation into Proteins In-Vivo

After injection of labelled amino acids some incorporation of label into luminal epididymal proteins was found in the guinea pig (del Rio, 1979) and many radioautographic studies have confirmed incorporation of injected amino acids into luminal proteins (see Section III.G.4). During short-term in-vivo perfusion of the rat cauda epididymidis ^{3}H-lysine was transferred to the lumen but no association with macromolecules was found (Fig. 4).

ii. Incorporation into Proteins In-Vitro

Many in-vitro studies have demonstrated the incorporation of amino acids into epididymal proteins in-vitro in the rat (Cameo and Blaquier, 1976; Kohane et al., 1980; Brooks and Higgins, 1980; Brooks, 1981, 1982, 1983a, 1984; Brooks and

Tiver, 1984), rabbits (Jones et al., 1981a), rams (Jones et al., 1982) and hamsters (González Echeverría et al., 1982). A variety of techniques, including incubation of tissues pieces (Brooks, 1981, 1982, 1983a, 1984), cultured tubules in rats (Blaquier, 1973; Blaquier and Breger, 1974; Blaquier et al., 1974) and man (Tezón and Blaquier, 1981; Tezón et al., 1981, 1982, 1983), perifused epididymal explants of rabbits (Orgebin-Crist and Ménézo, 1980) and rats (Klinefelter and Hamilton, 1984, 1985), stop-flow perfusion of the anterior pig epididymis (Dacheux, 1980), epithelial strips (Wenstrom and Hamilton, 1984) and cells isolated from the epithelium (Higgins et al., 1981) have indicated that synthetic capabilities and protein products of the epididymis vary in different regions.

Protein synthesis in-vitro is a sensitive technique and detects synthesis of certain proteins (named B and C) in the initial segment, where none is observed in the lumen (Brooks and Higgins, 1980) and a 22kDa protein in all segments of the epididymis, where secretion is detected only in the cauda (Brooks and Tiver, 1984).

Cell-free systems employing translation of mRNA have revealed that the molecular weight of nascent androgen-dependent, specific epididymal proteins synthesised in the rat caput is about 2000 greater than those of proteins secreted into the medium from intact cells (d'Agnostino et al., 1980); however, procaine and esterase did not provide confirmatory evidence for signal peptidase activity (Brooks and Tiver, 1984). The composition of epididymal secretory proteins has come from molecular cloning of cDNA coding for them (Brooks et al., 1986).

b. Sugars

Most early work centred on the presence of neuraminic (sialic) acid in the epididymis. Its presence in the cauda exceeds that in the caput (Bose et al., 1966; Fournier, 1966; Bose and Kar, 1968; Setty et al., 1971; Riar et al., 1973) and an epithelial source was indicated by its presence in the epithelium before the arrival of sperm and fluid from the testis (Peyre and Laporte, 1966b; Rajalakshmi and Prasad, 1969; Laporte, 1970), the failure of EDL to reduce the epididymal content (Rajalakshmi and Prasad, 1968, 1969) and the androgen-dependence of sialic acid in the epididymis (Fournier, 1966; Bose and Kar, 1968; Rajalakshmi and Prasad, 1968; Setty et al., 1971). However, the sialic content of principal cells from the rat caput epididymidis is no higher than that of basal or other somatic cells (Killian and Chapman, 1980).

In-vivo non-metabolisable 3-*O*-methylglucose enters the perfused cauda epididymal lumen (Cooper and Waites, 1979a) and accumulates in tissue and luminal fluid in the caput more than the cauda after injection (Hinton and Howards, 1982b). Intraluminal levels do not exceed those of blood, transport is reduced when blood glucose in raised, and the transfer of D-glucose exceeds that of the L-isomer, so specific, saturable carriers are thought to facilitate diffusion through the epithelium (Hinton and Howards, 1982b; Brooks, 1979).

Much of the sialic acid is present in sialoproteins found in the epididymal lumen (Fournier-Delpech et al., 1973a, 1977; Toowicharanont and Chulavatnatol, 1983a). The high galactosyltransferase activity of the epididymis (Fleischer et al., 1969), sialyltransferases with greater activity in the caput than cauda (Bernals et al., 1980), glycosyltransferases (Letts et al., 1974; Tadolini et al., 1977; Hamilton, 1981a; see Section II.C.6.c.1) and α-lactalbumin (Byers et al., 1984) may be responsible for the incorporation of sugars into proteins. The epididymis also possesses and synthesises dolichols (Wenstrom and Hamilton, 1980) which are involved in the synthesis of long-chain oligosaccharides prior to their transfer to proteins (see Schachter, 1978). Tunicamycin, an antibiotic that prevents dolichol-mediated transfer of completed oligosaccharide chains to glycoproteins (Tkacz and Lampen, 1975), reduces the incorporation of sugars into specific epididymal glycoproteins and promotes secretion of non-glycosylated forms (Brooks, 1981, 1983b). Vitamin A (Velásquez et al., 1975) and retinol- and retinoic acid-binding proteins in the epididymis (Porter et al., 1985; Kato et al., 1985) may also be involved in the glycosylation of proteins.

c. Phosphate

Phosphate is incorporated by tissue slices into specific proteins secreted into medium in-vitro (Mongkolsirikirieat and Chulavatnatol, 1984).

4. Protein Secretion in the Epididymis

Proteins are secreted into the lumen of the sperm-free perfused epididymis (Wong et al., 1981; Yeung and Cooper, 1982), where higher secretion rates are found proximally (Wong et al., 1978, 1979), and into the lumen of the perifused organ (Klinefelter and Hamilton,1985). Caput cells also secrete proteins into media in-vitro (Killian and Chapman, 1980; Klinefelter et al., 1982) including specific proteins as AEG (Kierszenbaum et al., 1981; White et al., 1982).

Radioautography has permitted the movement of protein precursors to the epididymal lumen to be monitored. Initial studies did not distinguish between the movement to the lumen *via* testicular fluid entering the caput (Glucksmann et al., 1955) but shorter incubation studies indicated that movement was across the epithelium in mice (Vendrely and Durliat, 1968; Kopečný, 1971; Kanka and Kopečný, 1977; Flickinger, 1983), guinea pigs (del Rio, 1979) and rats (Neutra and Lebond, 1966; Bennett, 1970; Bennett and Leblond, 1970). More detail has come from electron microcopical studies which reveal movement of tracer from rough endoplasmic reticulum (RER) to Golgi to the luminal brush border and luminal content in rats (Bennett et al., 1974) and mice (Flickinger, 1979, 1981, 1983, 1984, 1985; Dadoune, 1981; Fain-Maurel et al., 1981, 1983). Regional differences in rates of transport are observed and may reflect differing amounts of protein-synthetic machinery in cells. Morphometric analysis applied to cells from the rat caput epididymidis reveals larger volumes of the cell occupied by Golgi than RER (Rohr et al., 1976; Bartsch et al., 1978) and there is less Golgi in the caudal cells than those from the caput in rats (Moore and Bedford, 1979b) and primates (Moore and Pyor, 1981).

Because there was no good morphological evidence for the secretion of protein in the initial segment of the rat epididymis, Hoffer et al. (1979) raised the possibility that normal secretory pathways were inoperative; however, in the mouse the pathway of amino acids and sugars into secreted luminal proteins has been shown to be that found in other tissues and the Golgi is not bypassed (Flickinger, 1985; Flickinger et al., 1984). Merocrine secretion has been observed in the initial segment of the rabbit epididymis (Nicander and Malmqvist, 1977).

5. Androgen Dependence

a. DNA

The incorporation of labelled thymidine into epididymal epithelial DNA is very low (Kanka and Kopečńy, 1977) as would be expected from the low mitotic index (see Section III.A.7), but is detectable in cells (Kierszenbaum et al., 1981; Olson et al., 1983) and tubules (Tezón et al., 1981) in culture. There is little loss of DNA from the epididymis after castration (Brooks, 1977) and androgen replacement has little effect on the incorporation of precursors into DNA in the rat in-vivo (Niemi and Tuohimaa, 1971; Durliat and Vendrely, 1972). In-vitro, however, androgens stimulate the incorporation of nucleotides into human epididymal DNA and cyproterone acetate inhibits this incorporation (Tezón and Blaquier, 1981).

b. RNA

The incorporation into RNA of injected precursors is more rapid in the corpus than caput and least in the initial segment (Kanka and Kopečńy, 1977) and the epididymal RNA content falls after castration (Brooks, 1977). Androgens control the extent of the macromolecular synthesis in the epididymis of castrated mice (Kanka and Kopečńy, 1977) and rats (Charreau et al., 1970) and in cultured epididymal tubules from the rat (Blaquier et al., 1974, 1975; Blaquier and Breger, 1974, 1975). Aggregation of ribosomes (poly-ribosome) follows androgen stimulation (Blaquier et al., 1975) and androgens stimulate the cell-free translation of mRNA in the rat caput (d'Agnostino et al., 1980). However, in cells isolated from the castrated rat caput epididymidis, incorporation of uridine into RNA is as rapid as in cells from intact animals and androgen action was therefore considered to be on RNA degradation (Higgins et al., 1981). Messenger RNA coding for specific epididymal proteins declines, but does not disappear, from the rat epididymis after castration (Brooks et al., 1986), so translation may be inhibited.

c. Amino Acids

Androgens promote the incorporation of amino acids into total epididymal proteins by tubules in culture and their action is blocked by antiandrogens in rats (Blaquier, 1975) and man (Tezón and Blaquier, 1981; Tezón et al., 1985a). More information on the type of proteins synthesised has come from the use of $^{3}H/^{14}C$- or ^{35}S-labelled amino acids followed by 1-D or 2-D polyacrylamide gel electrophoresis. Castration reduces the subsequent incorporation of amino acids into proteins in rats (Kohane et al., 1980; Brooks and Higgins, 1980; Brooks, 1981a, 1983a), hamsters (González Echeverría et al. 1982), rabbits (Jones et al., 1981a) and rams (Jones et al., 1982). For most epididymal proteins androgen replacement therapy reverses the deleterious effect of androgen withdrawal, but for some proteins synthesised in the proximal part of the epididymis there is evidence that the deprivation of testicular fluid alone restricts synthetic capacity of the tissue (see Section III.G.6).

d. Sugars

The incorporation of labelled sugars into tissue and secreted proteins of the rat epididymis is reduced by castration (Brooks, 1981a, 1983b) and the reduction is greater than the reduction of amino acid incorporation into proteins. The androgen-dependent *N*-glycosylation of proteins by the rat epididymis does not reflect a decline in the tissue pool dolichol, but a reduction in transferase activity (Iusem et al., 1984).

e. Proteins

The total protein content of epididymal fluid decreases after bilateral castration in rabbits (Jones and Glover, 1973; Jones, 1974, 1977). Whereas anti-androgens do not affect protein concentration in epididymal fluid in either rats (Back et al., 1975) or rabbits (Jones, 1977) cyproterone acetate reduces protein secretion into the perfused epididymis (Wong et al., 1978, 1979). The increased albumin content of the rat epididymis (Cameo and Blaquier, 1976; Brooks and Higgins, 1980) and epididymal fluid (Jones and Dott, 1980) after castration may reflect the decreased effectiveness of the tight junctions (Moore and Bedford, 1979b; see Section III.A.8.d).

The synthesis of specific epididymal proteins is also reduced after castration of rats (Cameo and Blaquier, 1976; Garberi et al., 1979; Lea et al., 1978; Kohane et al., 1979, 1983; Brooks and Higgins, 1980; Bayard et al., 1981; Rajalakshmi, 1985), rabbits and hamsters (Moore, 1981a) and maintained by androgens in cultures of tubules from hamsters (Cuasnicú et al., 1984a) and men (Tezón et al., 1985a). One widely distributed tissue protein found only in luminal secretions of the testis and epididymis is not androgen-dependent (Brooks, 1985).

6. Control of Protein Synthesis by Luminal Fluid

The in-vitro synthesis and secretion of certain specific epididymal proteins is reduced after castration but not restored by androgens and also reduced by merely ligating the testicular efferent ducts. This has been observed in rats (Jones et al., 1980a; Brooks, 1983a), rams (Jones et al., 1982) and rabbits (Jones et al., 1981a) and is thought to reflect the dependence of the epithelial cells in that region on a continuous flow of testicular fluid (see Section III.A.7). Whether this is due to its constituent steroids (see Section III.B.2.a), polyanions (see Section III.C.7), GPC (Section III.D.4.a), *myo*-inositol (Section III.F.6.f) or growth factors

(Brown et al. 1982a,b), all of which affect some aspects of protein synthesis, secretion or cell maintenance in other tissues, is not established yet for the epididymis.

7. Site of Synthesis of Specific Proteins

Some proteins are secreted by more than one region of the epididymis, others are solely located within one restricted area (see Section III.G.9.a). To date more specific proteins have been found to be secreted by the proximal epididymal regions (Table 20).

8. Evidence for Interactions of Luminal Proteins with Spermatozoa

a. Direct Observations

Transmission electron microscopy has shown that a "Variable Affinity Glycoprotein" (VAG) adheres to the plasma membrane of rat sperm over the acrosome (Suzuki and Nagano, 1980).

b. Similarities Between Proteins in Fluid and from Sperm

There are differences in the total protein spectrum of solubilised mature and immature spermatozoa from bulls (Lavon et al., 1971; Yoo et al., 1982) and rats (Jones et al., 1983c), but as some of these proteins are of intracellular origin it is not clear whether the epididymis is directly responsible for the changes observed. A decrease in the total protein content of maturing spermatozoa has been observed in rats (Vermouth et al., 1986) and bulls (Amir and Lavon, 1976).

There are some similarities between the proteins free in epididymal fluid and those on the surface of spermatozoa from rats (Fournier-Delpech et al., 1977; Brooks and Higgins, 1980; Jones et al., 1980a; Olson and Orgebin-Crist, 1982; Jones and Brown, 1982; Brown et al., 1983; Brooks and Tiver, 1984; Klinefelter and Hamilton, 1985), rams (Voglmayr et al., 1980), mice (Devine and Carroll, 1985) and man (Tezón et al., 1982b, 1985a), but close scrutiny has also revealed differences in rats (Brooks and Tiver, 1984; Zeheb and Orr, 1984; Olson and Hinton, 1985), mice (del Villano et al., 1977), rams (Dacheux and Voglmayr, 1983) and boars (Russell et al., 1984). The differences may be related to the mechanism of insertion of proteins into the membrane (Zeheb and Orr, 1984).

c. Micropuncture

Radioactivity associated with unwashed spermatozoa recovered from the epididymis after injection of labelled amino acids (del Rio et al., 1979) does not necessarily indicate that secreted proteins are bound, but the retention of activity on the sperm surface after washing (Kopečný, 1971; Cuasnicú et al., 1982) is more convincing of a firm attachment.

d. Radioautography

At the light microscopical level radioactive label in the epididymal lumen after injection of tracer amino acids is often assumed to be associated with poteins bound to luminal spermatozoa (Glucksman et al., 1955; Kopečńy, 1971; Kopečńy and Pech, 1977; del Rio, 1979), but silver grains may originate from proteins in fluid surrounding the sperm cells. Similar doubts are shown for radioautographic studies with the electron microscope where individual silver grains located over spermatozoa in the lumen are too large to be designated as originating from a particular structure (Turner, 1979; Flickinger, 1981).

Radioactive material in the epididymal lumen 10 days after injection of labelled fucose (Kopečńy and Pech, 1977) must be present in luminal fluid, since the spermatozoa, which within the testis also incorporate fucose, only reach the cauda 19 days after injection (Kopečńy, 1976). Whether the luminal proteins bind to the sperm is not clear from these studies.

e. Immunological Evidence

In contrast to radioautography, the location of antibodies in histological sections can identify whether coating proteins are on the sperm head or tail. Immunological assessment does require cautionary interpretation, however, as there may be doubt as to the nature of the antigen against which the antibody was raised. For example, there is evidence that some proteins isolated from sperm-free epididymal fluid may have originated from spermatozoa during degeneration or maturation (Bostwick et al., 1980; Hunter and Schellpfeffer, 1981; Brooks and Tiver, 1984). The similar location in the epididymis of proteins obtained from the sperm-free epididymis of androgen-primed pre-pubertal rats (Dravland and Joshi, 1981) and those of the pubertal animal, however, indicate that these considerations may not be too important. Conversely, antigens on washed spermatozoa (Fox et al., 1982; Gaunt, 1982; Gaunt et al., 1983; Moore et al., 1985; Ellis et al. 1985) may include bound, secreted epididymal proteins that require, *e.g.* high ionic strength, to elute them (Rifkin and Olson, 1985) as well as those from an intracellular origin (Jones et al., 1983a).

Data from washed sperm and epididymal fluid in-vitro reveal immunological identity between proteins in epididymal fluid and those on mature spermatozoa from rats (Jones and Brown, 1982; Brown et al., 1983; Brooks and Tiver, 1983, 1984; Brooks, 1983d, 1985; Klinefelter and Hamilton, 1985) and hamsters (Moore, 1981a).

That antibodies raised to specific epididymal proteins cause agglutination of spermatozoa (Moore, 1981a) and that sperm cells can block the interaction of epididymal fluid proteins with their own antibodies (Kohane et al., 1979; Wong and Tsang, 1982, 1983) also suggest that epididymal specific proteins are present on the surface of the spermatozoa.

f. Other Evidence

Indirect evidence for the presence of coating proteins on the surface of sperm cells include the presence, on the surface of epididymal spermatozoa, of proteins that are absent from testicular spermatozoa (Brooks and Higgins, 1980), although unmasking of integral membrane components cannot be excluded. The appearance of epididymal glycoproteins in epithelial cells in the corpus epididymidis of mice only when spermatozoa are absent from the lumen, has been taken to indicate that the spermatozoa normally bind these proteins and thus prevent their uptake by epithelial cells (Abe et al., 1982a,b, 1983a). By a similar reasoning, sperm-binding proteins that accumulate in azoospermic semen and bind to the equatorial segment of sperm from normal patients, are thought to arise from the human epididymis (del Rio et al., 1984; Abresca et al., 1985). The most compelling evidence for the binding of secreted epididymal proteins to sperm cells is the direct demonstration of this uptake. Bovine serum albumin binds to spermatozoa from rabbits (Blank et al., 1976) and rats (Davis and Gergely, 1979) in-vitro, but Brooks and Higgins (1980) found no evidence for the binding of pre-albumins to mature spermatozoa (see Section III.G.9.b).

9. Binding of Proteins to Maturing Spermatozoa

a. Acquisition of Proteins by Maturing Spermatozoa

The first appearance of secreted proteins on the surface of epididymal spermatozoa occurs in (a) the proximal caput of hamsters (H1, H2: Moore, 1980, 1981b), rats (AEG: Lea et al., 1978; White et al., 1982; SEP : Kohane et al., 1980; SP: Faye et al., 1980; Bayard et al., 1981; α-lactalbumin: Byers et al., 1984), mice (SSEA-1: Fox et al., 1982) and rabbits (R1–R3: Moore, 1980, 1981a); (b) the distal caput of mice (SMA-4: Vernon et al., 1982) and man (R_f = 0.3,0.43,1.0: Tezon et al., 1985b) or (c) the corpus of rabbits (ASF: Thomas et al., 1984).

SSEA-1 (Fox et al., 1982) and α-lactalbumin (Byers et al., 1984) are only detectable within cells of the epithelium in the proximal caput and their presence more distally is solely in the lumen, whereas other proteins are found in the epithelium of more distal regions in both principal and clear cells (Kopecny and Pech, 1977; Lea et al., 1978; Kohane et al., 1980; Abe et al., 1982a, 1983a; Thomas et al., 1984).

Many changes occur to spermatozoa during their maturation and in the rat some proteins are absorbed onto their surface from rete testis fluid as well as from epididymal fluid (Brooks and Tiver, 1984). Antibodies to a dimeric acidic glycoprotein (DAG) secreted by Sertoli cells in culture are localised in epididymal cells and on testicular as well as epididymal spermatozoa (Sylvester et al., 1984).

Other proteins in epididymal, but not testicular, fluid bind to epididymal, but not testicular, spermatozoa (Jones and Brown, 1982; Brown et al., 1983; Brooks and Tiver, 1983, 1984). Proteins which are lost from the sperm surface are either found in epididymal fluid or remain in the membrane either masked or internalised (Brooks and Tiver, 1984; Dacheux and Voglmayr, 1983).

Antibodies to one 32kDa protein from cauda epididymal fluid bind more to caput than cauda cells suggesting that more of this protein is present on the immature cell surface (Wong and Tsang, 1982) but another 32kDa protein from epididymal fluid does not bind to caput sperm (Zeheb and Orr, 1984). A different

(16kDa) protein of epididymal origin binds equally well to testicular and cauda epididymal spermatozoa (Brooks, 1983d).

Elution of proteins from the sperm surface with high ionic strength solutions has demonstrated greater amounts of proteins associated with mature spermatozoa from hamsters (González Echeverría et al., 1982), rats (Cameo and Blaquier, 1976; Kohane et al., 1980), guinea pigs (Aonuma et al., 1973) and man (Tezón et al., 1985b).

b. The Ability of Immature Spermatozoa to Bind Proteins

Although the amount of epididymal-specific proteins (SEP) available to rat spermatozoa in the caput exceeds that in the cauda because of low sperm numbers proximally (Kohane et al., 1979), they are not associated with the surface of the immature spermatozoa. The presence of specific proteins within epithelial cells in certain regions of the epididymis where they are not demonstrable on the surface of luminal sperm from rats (AEG: Lea et al., 1978; α-lactalbumin: Byers et al., 1984), rabbits (R1: Moore, 1980), hamsters (H1, H2: Moore, 1980) and mice (SEA-1: Fox et al., 1982; SMA-4: Vernon et al., 1982) may not just indicate a failure to detect such binding, but could indicate a failure of the epithelial cells to release the protein into the lumen or an inability of the immature sperm cell to bind the proteins or. In-vitro studies, however, show that immature spermatozoa are able to bind epididymal proteins.

Proteins foreign to the epididymis (*e.g.* seminal vesicle trypsin inhibitors) bind more to cauda sperm than caput cells (Irwin et al., 1981) but monoclonal antibodies to mature sperm antigens may either bind less (Feuchter et al., 1981; Gaunt, 1982) or more (Gaunt et al., 1983) to immature spermatozoa, depending on the source of the antigen. Ovine testicular spermatozoa, but not mature cauda ones, can take up a 24kDa protein from epididymal fluid in-vitro (Voglmayr et al., 1980) into trypsin-resistant sites (Voglmayr et al., 1982) and corpus sperm are able to bind agglutinins (ram: Dacheux et al. 1983) and decapacitation factors (boar: Hunter et al., 1978) in cauda epididymal fluid.

Experiments employing proteins from the epididymis are complicated by the possibility of prior binding of endogenous proteins before the binding assay. An example of this is the binding of a gold-labelled corpus-specific protein to testicular and caput sperm but not those from the corpus or cauda. Preincubation of testicular sperm with unlabelled corpus protein prevents the subsequent demonstration of labelled protein binding (Courtens et al. 1982). Similarly, the binding of ConA lectin to ovine testicular sperm is prevented by their incubation in epididymal fluid (Fournier-Delpech et al., 1983). This "saturation" of surface binding sites by endogenous proteins may explain why cauda sperm to not take up epididymal proteins (Voglmayr et al., 1980) or FMP (Acott and Hoskins, 1981) when testicular and caput sperm do so.

10. The Nature of the Interaction

a. Peripheral Proteins

i. Surface Charge

The removal of surface proteins by simple washing (Aonuma et al., 1973; Cameo and Blaquier, 1976) does not necessarily indicate that the eluted proteins were tightly bound to the sperm surface, but the increasing electronegativity of maturing spermatozoa (Table 19) and the acidic nature of most epididymal proteins (Table 20) make ionic interactions probable. High ionic strength solutions do elute proteins from the surface of spermatozoa from mice (Oliphant and Brackett, 1973), rabbits (Oliphant, 1976), rats (Wong and Tsang, 1982; Kohane et al., 1980b) and hamsters (González Echeverría et al., 1982). The binding to spermatozoa of a basic protein (histone), however, is not purely electrostatic (Majumder, 1981). Reducing surface charge by urea is effective at removing surface glycoproteins from ejaculated human spermatozoa (Silvestroni et al., 1982, 1983).

ii. Surface Carbohydrates

The work of Fournier-Delpech et al. (1983) showing that (a) ovine testicular spermatozoa, but not mature epididymal spermatozoa, bind the lectin ConA; (b) ConA does bind to mature sperm from castrated rams but not androgen-supplemmented castrates and (c) preincubation of testicular spermatozoa with cauda epididymal fluid from intact rams prevents the binding of ConA, attests that the epididymis in this species secretes an androgen-dependent glycoprotein that binds specifically to carbohydrate moieties on the sperm surface. The changing carbohydrates on the sperm surface (Table 19) and in epididymal fluid (Table 20) provide a wide range of possibilities for interactions. Enzymes in epididymal fluid (Table 21), especially carbohydrases, could act to modify proteins in the fluid and on the surface.

b. Integral Proteins

The high proportions of non-polar amino acids are characteristic of some secreted proteins (Lea et al., 1978; Lea and French, 1981; Jones et al., 1980b; Brown and Jones, 1983) raises the possibility of intercalation of proteins into the membrane leaflets (Bigelow, 1967). While some proteins can only be removed from the sperm by destroying the membrane (Olson and Orgebin-Crist, 1982; Zeheb and Orr, 1984; Hamilton et al., 1986) there is little evidence on the nature of the interaction (Brown et al., 1983; Zeheb and Orr, 1984). Modification of apolar regions of the sperm plasma membrane that occur during maturation (Mercado-Pichardo et al., 1982) would affect the uptake of putative integral membrane proteins from the surrounding fluid, but it is not known if or how these changes are mediated by the epididymis.

Changes in the lipid content of sperm membranes (see Section II.B.9.b) may also facilitate the insertion of hydrophobic proteins and could influence any attachment of proteins *via* phosphatidylinositol (see Low and Kincade, 1985). The prevalence of free sulphydryl groups on a 22kDa protein on the rat sperm surface while an otherwise identical protein free in epididymal fluid contained more disulphide bonds suggested to Brooks (1985) that attachment of luminal fluid proteins to the sperm could be by suphydryl bonds. In this case, changes in the oxidation state of sulphydryl groups (see Section II.A.6.d) could also modify the attachment of proteins bearing surface thiol groups.

11. Site of Binding

Some epididymal proteins are bound to the acrosome in rats (SEP: Kohane et al., 1979, 1980; SP: Bayard et al. 1981; D and E: Brooks and Tiver, 1983) and mice (SMA-4: Vernon et al., 1982) whereas AEG binds to predominantly to the flagellum (Lea et al., 1978) and rabbit ASF binds to both the acrosome and flagellum (Thomas et al., 1984).

12. Role of Proteins in Epididymal Function

a. Non-Specific Effects

i. On the Epididymis

Growth factors in rete testis fluid (Brown et al., 1982a,b), which are synthesised in Sertoli cells (Feig et al., 1980, 1983), and other Sertoli cell products (*e.g.* insulin and transferrin, (see Section III.G.2.b) which maintain cells in culture (Barnes and Sato, 1980) may support the epithelial cells in the proximal epididymis.

ii. Interactions with Sperm Membrane Proteins

Peripheral proteins can restrict the motion of integral (intramembranous particle) proteins (Gingell, 1976; Barber, 1982). Over the rat sperm head the adherence of a "Variable Affinity Glycoproteins" (VAG) with the plasma membrane over the acrosome is associated with an organised disribution of intramembranous particles in the mid-caput (Suzuki and Nagano, 1980). That neither immature sperm, lacking VAG, nor more mature sperm from which VAG has dissociated, display such ordered particle lattices, supports a role for the secreted epididymal protein in modifying the sperm membrane. In the guinea pig, however, there are ordered arrays of intramembranous particles in the plasma membrane over the acrosome of sperm from the corpus that do not have surface proteins (Bearer and Friend, 1980) and there are no ordered particle arrays in the boar sperm head membranes (Suzuki, 1981).

The ability of extraneous proteins to influence the position of intramembranous particles is also determined by the sterol content of the membrane and its temperature, both of which decrease along the length of the epididymis (see Section II.B.3.b and Brooks, 1973), and may together influence local permeability.

iii. Permeability

The permeability of liposomes towards glucose is increased by albumin (Sweet and Zull, 1969), but it is not known if the differences in permeability of immature and mature spermatozoa (see Section II.A.d.ii) reflect the action of proteins. Epididymal proteins from the cauda, but not corpus, reduce the uptake of potassium and sodium ions into immature, but not mature spermatozoa (Zimmermann et al., 1979). That some basic proteins obtained from bovine seminal plasma can reduce the calcium uptake into mature epididymal cells (Rufo et al., 1982, 1984) suggest that in the epididymis there is little restriction to calcium transport into spermatozoa (see Section III.C.5.b).

iv. Protective Agents

Both transferrin (Papkoff, 1966) and caeruloplasmin (Al-Timin and D'Ormandy, 1977) can prevent oxidation of tissues and may have general protective effects on spermatozoa. As well as transporting iron (Aisen and Litowski, 1980), transferrin may protect cells against heavy metal contamination *via* chelation (Emery, 1980).

v. Sperm Survival Factors (SSF)

Epididymal fluid can prevent the loss of motility that occurs on dilution of epididymal contents in the hamster (Morita and Chang, 1971; Morton and Chang, 1973) and which is accompanied by release of intracellular LDH. Proteinaceous "Sperm Survival Factors", which prevent the loss of motility and LDH from spermatozoa, are found in epididymal fluid from hamsters, dogs, and bulls (Morton and Chang, 1973; Morton et al., 1979a). SSF are not specific and their prevention of the loss of motility on dilution can be mimicked by other large molecular weight compounds together with an inhibitors of phosphodiesterase (Morton et al., 1979b). Epididymal fluid from rams and bulls does not prolong the survival of sperm motility in these species and in other species albumin alone can maintain sperm motility (Harrison et al., 1978; Dott et al., 1979; Harrison et al., 1982). The action of an epididymal protein that promotes the motility of ejaculated human spermatozoa (Sheth et al., 1979) and epididymal goat sperm (Chaudhury and Majumder, 1983) may be similar to SSF or, like FMP (see Section II.A.6.c.ii), reduce sticking of cells to the observation chamber (Roy et al., 1985).

vi. Antagglutinins

"Clusterin" in rete testis fluid has agglutinating ability towards many cell types (Fritz et al., 1983; Blaschuk et al., 1983; Blaschko and Fritz, 1984; Tung and Fritz, 1985) but its significance for the agglutination of sperm has not been demonstrated. There are antagglutinins in epididymal fluid from boars and rams that prevent the dilution-induced agglutination of immature sperm cells (Dacheux et al., 1983).

vii. Immunosuppression

Proteins in epididymal fluid from mice (Anderson and Tartar, 1982) and man (Tartar and Alexander, 1984), including uteroglobin (see Section III.B.3.c.ii), are immunosuppressive and antagonise complement-mediated haemolysis in-vitro. While they may play a role in preventing immunological recognition of spermatozoa, there is evidence that epididymal fluid from one species does not prevent immune responses to proteins obtained from epididymal fluid from another species (Esponda and Bedford, 1985).

b. Specific Effects

i. Enzymes

The significance for spermatozoa of the multitude of enzymes in epididymal fluid (Table 21) is not known. Esterases and carbohydrases acting on extracellular proteins and glycoproteins may modify membrane function with consequences for motility or sperm-egg binding (see Sections III.G.12.b.ii and iii).

ii. Sperm Motility

AEG stimulates the motility of immature rat spermatozoa (see Section II.A.6.b) and FMP can convert this flagellation to forward progression (see Section II.A.6.c). In one case of human azoospermia due to obstruction of the cauda epididymidis, seminal fluid ejaculated after epididymovasostomy lacked a protein that was present in normal seminal plasma and detectable in human caput tissue. The ejaculated sperm were immotile, and also lacked from their surfaces the same protein that was present on spermatozoa from normal ejaculates (Wong et al., 1982b). Although the immotile sperm were capable of binding the proteins upon incubation in normal seminal plasma, the motility of these protein-coated spermatozoa was not examined.

Androgen binding-proteins may affect motility by modifying the luminal androgen content (see Section III.B.3.c.ii); others may inhibit motility of spermatozoa *in-situ* by virtue of their viscosity (see Section III.C.4.a.iv).

iii. Sperm-Egg Binding

Epididymal sialoproteins that bind to rat testicular and hamster corpus spermatozoa can promote the binding of sperm to eggs (see Section II.C.6.a.i) but the acquisition by sperm cells of these proteins in the epididymis (Table 20) does not parallel the aquisition of sperm-egg binding (Table 8). The carbohydrases in epididymal fluid, together with galactosyltransferases and α-lactalbumin, may be important for modifying the sperm surface with consequences for sperm-egg binding (see Section II.C.4.a). Compounds preventing the binding of androgens to ABP in-vitro can also inhibit sperm-egg binding after administration to rats (see Section III.B.3.c.ii; Sirett et al., 1984).

iv. Capacitation

Decapacitation activity in epididymal fluid, including ASF (Table 20), should prevent the acrosome reaction occurring in in the epididymis and acrosin inhibitors should limit the damage done to the epididymal epithelium by enzymes leaking from the acrosome of degenerating spermatozoa (see Section II.B.3.a).

iv. Sperm Fertilising Ability

Rats rendered infertile by the antifertility agents α-chlorohydrin and 6-chloro-6-deoxyglucose lack specific proteins found on the surface of sperm from fertile animals (Tsang et al., 1981).

13. Peptides

Ovine epididymal fluid contains polypeptides (Neumark and Schindler, 1967) and the rat cauda, but not corpus, luminal fluid contains a low molecular weight peptide secreted by the cauda in-vivo (Wong et al., 1980d). This peptide can substitute for Ca^{2+} in the Na^{+}-dependent maintenance of forward motility (Wong et al., 1980, 1982a).

The tripeptide glutathione is present in the murine epididymis with higher concentrations in the caput (Kochakian, 1976) and it is present in goat epididymal fluid (Li, 1975). There is less glutathione in testicular sperm that those from the cauda of rats (Chinoy and Sarajeenen, 1980). In the rat epididymis it is androgen-dependent (Robaire and Hales, 1982). Glutathione may be involved in amino acid transport (see Section III.G.3.a) and the presence of GSH-sulphydryltransferase (Hales et al., 1980) may be involved in sulphoconjugation of amino acids. A protective action against membrane thiol oxidation may be the role of GSH reductase and GSH peroxidase which are present in spermatozoa from the goat epididymis (Li, 1975).

An increasing number of peptides, bioactive in other tissues, have been demonstrated in epididymal tissue by immunocytochemical and biochemical methods. They include β-endorphin (Sharp et al., 1980), enkephalin (Rama Sastry et al., 1982), somatostatin 14 and 28 (Pekary et al., 1984), TRH (Pekary et al., 1980, 1983), epidermal growth factor (Elson et al., 1984), retinol binding-proteins (Ong et al., 1982), angiotensin converting enzyme (ACE: Cushman and Cheung, 1971), neuropeptide Y (Adrian et al., 1984) and vasoactive intestinal peptide (VIP: Larsson et al., 1977). VIP has been localised in the interstitium associated with nerves and muscle and the epididymal capsule (Alm et al., 1977, 1980), whereas β-endorphin and ACTH (Tsong et al., 1980, 1982a,b,) and binding proteins for retinol (Kato et al., 1985; Porter et al., 1985) are within the epididymal epithelial cells. Retinoic acid binding protein is found on the surface of spermatozoa (Porter et al., 1985) and angiotensin-converting enzyme is present within the epididymal lumen (Hohlbrugger et al., 1982; Jaiswal et al., 1984, 1985; Strittmatter and Snider, 1984); Vanha-Perttula et al., 1985). ACE is synthesised by the epididymis (Seike et al., 1985) and is androgen-dependent (Jaiswal et al., 1985). The role of the peptides is unclear but they are clearly in appropriate positions to modulate contractions of the capsule and tubules or to modify the capillary bed and thus influence secretions of the epithelium and their interactions with spermatozoa.

There is little immunocytochemical staining of prolactin in epididymal epithelial cells, but tissue sections bind exogenous prolactin (Witorsch, 1978). Prolactin receptors are present in the epididymis of rabbits (Orgebin-Crist and Jiane, 1979) and rats (Aragona and Friesen, 1975) but prolactin does not augment testosterone-stimulated protein synthesis in the rat caput (Jones et al., 1983b). As in the mammary gland, prolactin may have a function in the control of α-lactalbumin synthesis. In-vitro prolactin promotes the uptake of calcium by human epididymal spermatozoa (Reyes et al., 1979).

14. Resorption of Proteins by the Epididymis

Iodinated albumin perfused through the epididymal lumen is taken up into lysosomes (Mayorga and Bertini, 1983), most of which are probably derived from the clear cells (see Section III.A.9) since horse radish peroxidase introduced into the epididymal lumen is taken up into clear cell vacuoles (Moore and Bedford, 1979; Cooper and Yeung, 1980; Yeung and Cooper, 1982). Endogenous epididymal proteins are also taken up by the clear cells: AEG (Lea et al., 1978; White et al., 1982), SEP (Faye et al., 1980; Bayard et al., 1981), DAG (Sylvester et al., 1984) in rats and proteins labelled after injection of labelled amino acids in mice (Kopečńy and Pech, 1977; Kopečńy et al., 1984).

The testicular androgen-binding protein (ABP) is taken up by principal cells of the efferent ducts (Byers et al., 1985) and caput epididymidis (Pelliniemi et al., 1981; Attramadal et al., 1981; Feldman et al., 1981; White et al., 1982; Musto et al., 1983). While ABP is expectedly lost from epithelial cells in the caput of castrated rats and in the ipsilateral epididymis of hemi-castrated rats, it is still visible in the cells after ligation of the efferent ducts (Musto et al., 1983).

The principal cells of the mouse epididymis have also been implicated in the removal of luminal proteins because they accumulate material that stains with periodic acid-Schiff's reagent when spermatozoa are prevented from entering the epididymis (Abe et al., 1982a,b, 1983c, 1984a). These cellular inclusions are not found if the corpus is ligated, suggesting that the proteins are secreted by the caput and bind there to spermatozoa. This observation also implies either that the amount of glycoprotein secreted by the epididymis is related to the number of luminal spermatozoa, or that the uptake of proteins by the epithelium is inhibited by the flow of testicular secretions.

The fluid rather than the sperm appear to be involved in carrying the absorbed material because when the spermatogenic function of the testis is interrupted by X-irradiation, but testicular fluid secretion is maintained as judged from the height of the epithelium in the initial segment, PAS-positive material is still accumulated by principal cells as sperm disappear from the lumen. Conversely, when spermatogenesis restarts and sperm appear in the epididymis, PAS-positive material is lost from the epithelial cells (Abe et al., 1983a). These data suggest not only that the sperm-bound proteins do not enter the epithelial cells but also that the epithelial cells that once contained the proteins are able to digest them.

The possibility of transepithelial transport of proteins has been discussed in Section III.A.9.

15. Summary

A large variety of proteins are synthesised and secreted by the epididymal epithelium. Some of these adhere in a specific way to maturing spermatozoa whereas other are non-specific. There are epididymal proteins that influence the form of sperm motility, increase the ability of the sperm to recognise an egg and act as decapacitation factors that prevent acrosome reactions. An array of enzymes are present in luminal fluid, some of which are secreted by the epididymis, that can modify the sperm surface and may be involved in the interaction of spermatozoa with secreted components or with the zona pellucida.

16. References

Aafjes JH and Vreeburg JTM (1972) Distribution of 5α-dihydrotestosterone in the epididymis of bull and boar, and its concentration in rat epididymis after ligation of efferent testicular ducts, castration and unilateral gonadectomy. J Endocr 53:85–93

Aarons D, Robinson R, Richardson R and Poirier GR (1985) Competition between seminal and exogenous proteinase inhibitor for sites on murine epididymal sperm. Contraception 31:177–184

Abe K, Takano H and Ito T (1982a) Response of the epididymal duct in the corpus epididymis to efferent or epididymal duct ligation in the mouse. J Reprod Fert 64:69–72

Abe K, Takano H and Ito T (1982b) Appearance of peculiar epithelial cells in the epididymal duct of the mouse ligated epididymis. Biol Reprod 26:501–509

Abe K, Takano H and Ito T (1983)Response of the epididymal duct to the temporary depletion of spermatozoa induced by testicular irradiation in mice. Anat Record 207:17–24

Abe K, Takano H and Ito T (1984) Appearance of peculair multivesicular bodies in the principal cells of the epididymal duct after efferent duct cutting in the mouse. Arch Histol Japonicum 47:121–135

Abresca P, Lombardi G, de Rosa M, Quagliozzo L, Guardiola J and Metafora S (1985) Identification and preliminary characterization of a sperm-binding protein in normal human semen. J Reprod Fert 73:71–77

Acott TS and Hoskins DD (1981) Bovine sperm forward motility protein: binding to epididymal spermatozoa. Biol Reprod 24:234–240

Acott TS and Carr DW (1984) Inhibition of bovine spermatozoa by cauda epididymal fluid: II. Interaction of pH and a quiescence factor. Biol Reprod 30:926–935

Adrian TE, Gu J, Allen JM, Tatemoto K, Polak JM and Bloom SR (1984) Neuropeptide Y in the human male genital tract. Life Sci 35:2643–2648

Aisen P and Litowsky I (1980) Iron transport and storage proteins. Ann Rev Biochem 49:357–393

Albert Z, Orlonwski J, Orlowsky M and Szenczuk A (1964) Histochemical and biochemical investigations of gamma-glutamyl transpeptidase in the tissues of man and laboratory rodents. Acta Histochem 18:78–89

Alm P, Alumets J, Hakansen and Sundler F (1977) Peptidergic (vasoactive intestinal peptide) nerves in the genito-urinary tract. Neuroscience 2:751–754

Alm P, Alumets J, Hakansen R, Owman C, Sjoberg NO, Sundler F and Walles B (1980) Origin and distribution of VIP (vasoactive intestinal peptide)-nerves in the genitourinary tract. Cell Tiss Res 205:337–347

Al-Timini DJ and Dormandy TL (1977) The inhibition of lipid autoxidation by human caeruloplasmin. Biochem J 168:283–288

Alumot E, Lensky Y and Schindler H (1971) Separation of proteins in the epididymal fluid of the ram. J Reprod Fert 25:349–353

Amann RP, Killian GJ and Benton AW (1973) Differences in the electrophoretic characteristics of bovine rete testis fluid and plasma from the cauda epididymidis. J Reprod Fert 35:321–330

Amir D and Lavon U (1976) Changes in total nitrogen, lipoproteins and amino acids in epididymal and ejaculated spermatozoa from bulls treated orally with ethylene dibromide. J Reprod Fert 47:73–76

Anderson DJ and Tartar TH (1982) Immunosuppressive effects of mouse seminal plasma components in vivo and in vitro. J Immunol 128:535–539

Aonuma S, Mayumi T, Suzuki K, Noguchi T, Iwai I and Okabe M (1973) Studies on sperm capacitation. I. The relationship between a guinea-pig sperm-coating antigen and a sperm capacitation phenomenon. J Reprod Fert 35:425–432

Aragona C and Friesen HG (1975) Specific prolactin binding sites in the prostate and testis of rats. Endocrinology 97:677–684

Arora R, Dinakar N and Prasad MRN (1975) Biochemical changes in the spermatozoa and luminal contents of different regions of the epididymis of the Rhesus monkey, *Macaca mulatta*. Contraception 11:689–700

Attramadal A, Bardin CW, Gunsalas GL, Musto NA and Hansson V (1981) Immunocytochemical localization of androgen binding protein in rat Sertoli and epididymal cells. Biol Reprod 25:983–988

Back DJ, Glover TD, Shenton JC and Boyd CP (1977) Some effects of cyproterone acetate on the reproductive physiology of the male rat. J Reprod Fert 49:237–243

Barber J (1982) The control of membrane organization by electrostatic forces (review). Bioscience Reports 2:1–13

Barnes D and Sato G (1980) Methods for growth of cultured cells in serum-free medium. Anal Biochem 102:255–270

Bartsch G, Oberholzer M, Holliger O, Weber J, Weber A and Rohr HP (1978) Stereology: a new quantitative method to study epididymal function. Androl 10:31–42

Bayard F, Duguet A, Mazzuca M and Faye JC (1981) Study of a glycoprotein produced by the rat epididymis. In: Reproductive Processes and Contraception. Ed KW McKerns. Plenum Pub Co. pp

Bearer EL and Friend DS (1980) Anionic lipid domains: correlation with functional topography in a mammalian cell membrane. Proc Natn Acad Sci US 77:6601–6605

Bennett G (1970) Migration of glycoprotein from Golgi apparatus to cell coat in the columnar cells of the duodenal epithelium. J Cell Biol 45:668–673

Bennett G and Leblond CP (1970) Formation of cell coat material for the whole surface of columnar cells in the rat small intestine, as visualised by radioautography with L-fucose-^{3}H. J Cell Biol 46:409–416

Bennett G, Leblond CP and Haddad A (1974) Migration of glycoprotein from the Golgi apparatus to the surface of various cell types as shown by radioautography after labelled fucose injection into rats. J Cell Biol 60:258–284

Bernals A, Torres J, Reyes A and Rosado A (1980) Presence and regional distribution of sialyltransferase in the epididymis of the rat. Biol Reprod 23:290–293

Bigelow CC (1967) On the average hydrophobicity of proteins and the relation between it and protein structure. J Theoret Biol 16:187–211

Blank M, Soo L and Britten JS (1976) Adsorption of albumin on rabbit sperm membranes. J Memb Biol 29:401–409

Blaquier JA (1973) An in vivo action of androgens on protein synthesis by epididymal tubules maintained in organ culture. Biochem Biophys Res Commun 52:1177–1183

Blaquier JA (1975) The influence of androgens on protein synthesis by cultured rat epididymal tubules. Acta Endocr 79:403–416

Blaquier JA and Breger D (1974) The in vitro effects of androgens on RNA synthesis by cultured rat epididymal tubules. Endocrine Res Commun 1:247–260

Blaquier JA and Breger J (1975) The in vitro effects of androgens on RNA synthesis by cultured rat epididymal tubules. Endocr Res Commun 1:247–260

Blaquier JA, Cameo MS and Calandra R (1974) Studies on the correlation between androgen binding and activation of rat epididymal tubules in culture. In: Hormone Binding and Target Cell Activation in the Testis. Eds ML Dufau and AR Means, Current Topics in Endocrinology volume 1, Plenum Press, New York. pp 329–341

Blaquier JA, Breger D, Cameo MS and Calandra R (1975) The activation of cultured epididymal tubules by androgens. J Steroid Biochem 6:573–576

Blaschuk OW and Fritz IB (1984) Isoelectric forms of clusterin isolated from ram rete testis fluid and from secretions of primary cultures of ram and rat Sertoli cell-enriched preparations. Can J Biochem Cell Biol 62:456–461

Blaschuk O, Brudzy K and Fritz IB (1983) Purification and characterization of a cell-aggregating factor (Clusterin), the major glycoprotein in ram rete testis fluid. J Biol Chem 258:7714–7720

Bose AR and Kar AB (1968) Distribution of sialic acid in the genital organs of male rhesus monkeys: effect of castration and replacement therapy. Current Sci (India) 37:168–169

Bose AR, Kar AB and Dasgupta PR (1966) Sialic acid in the genital organs of the male rat. Current Sci (India) 35:336–337

Bose TK, Prasad MRN and Moudgal NR (1975) Changing patterns of silaic acid in the spermatozoa and luminal plasma of the epididymis and vas deferens of the hamster, *Mesocricetus auratus* (Warterhouse). Ind J exp Biol 13:8–11

Bostwick EF, Bentley MD, Hunter AG and Hammer R (1980) Identification of a surface glycoprotein on porcine spermatozoa and its alteration during epididymal maturation. Biol Reprod 23:161–169

Brooks DE (1973) Epididymal and testicular temperature in the unrestrained conscious rat. J Reprod Fert 35:157–160

Brooks DE (1977) The androgenic control of the composition of the rat epididymis determined by efferent duct ligation or castration. J Reprod Fert 49:383–385

Brooks DE (1979) Carbohydrate metabolism in the rat epididymis: evidence that glucose is taken up by tissue slices and isolated cells by a process of facilitated transport. Biol Reprod 21:19–26

Brooks DE (1981) Secretion of proteins and glycoproteins by the rat epididymis: regional differences, androgen-dependence, and effects of protease inhibitors, procaine, and tunicamycin. Biol Reprod 25:1099–1117

Brooks DE (1982) Purification of rat epididymal proteins "D" and "E", demonstration of shared immunological determinants, and identification of regional synthesis and secretion. Int J Androl 5:513–524

Brooks DE (1983a) Effect of androgens on protein synthesis and secretion in various regions of the rat epididymis, as analyzed by two-dimensional gel electrophoresis. Mol Cell Endocrinol 29:255–270

Brooks DE (1983b) Influence of incubation conditions, tunicamycin and castration on incorporation of (^{3}H)mannose and (^{3}H)fucose into rat epididymal glycoproteins in vitro. J Reprod Fert 67:97–105

Brooks DE (1983c) Epididymal protein secretion. In: The Sperm Cell. Ed J Andre, Martinus Nijhoff, The Hague. pp 175–178

Brooks DE (1983d) Selective binding of specific rat epididymal secretory proteins to spermatozoa and erythrocytes. Gamete Res 7:367–376

Brooks DE (1984) Protein secretion by the rat epididymis can be selectively modified in vitro by local anaesthetics, glucose deprivation, dinitrophenol, ouabain, and ionophores. J Androl 5:351–360

Brooks DE and Higgins SJ (1980) Characterization and androgen-dependence of proteins associated wih luminal fluid and spermatozoa in the rat epididymis. J Reprod Fert 59:363–375

Brooks DE and Tiver K (1983) Localization of epididymal secretory proteins on rat spermatozoa. J Reprod Fert 69:651–657

Brooks DE and Tiver K (1984) Analysis of surface proteins of rat spermatozoa during epididymal transit and identification of antigens common to spermatozoa, rete testis fluid and cauda epididymal plasma. J Reprod Fert 71:249–257

Brooks DE, Means AR, Wright EJ, Singh SP and Tiver KK (1986) Molecular cloning of the cDNA for two major androgen-dependent secretory proteins of 18.5 kilodaltons synthesized by the rat epididymis. J Biol Chem 26:4965–4961

Brown KD, Blakeley DM, Henville A and Setchell BP (1982a) Rete testis fluid contains a growth factor for cultured fibroblasts. Biochem Biophys Res Commun 105:391–397

Brown KD, Henville A and Setchell BP (1982b) A growth factor in rete testis fluid. Ann New York Acad Sci US 383:428–429

Brown CR, von Glos KI and Jones R (1983) Changes in plasma membrane glycoproteins of rat spermatozoa during maturation in the epididymis. J Cell Biol 96:256–264

Byers SW, Qasba PK and Paulson HL (1984) Immuno cytochemical localization of α-lactalbumin in the male reproductive tract. Biol Reprod 30:171–177

Byers SW, Musto NA and Dym M (1985) Culture of ciliated and nonciliated cells from rat ductuli efferentes. J Androl 6:271–278.

Cameo MS and Blaquier JA (1976) Androgen-controlled specific proteins in rat epididymis. J Endocr 69:47–55

Chapdelaine P, Tremblay RR, Dube RR, St-Yves C and Mailhot J (1981) Origin of maltase and variations in infertile men. Archs Androl 1:61–68

Charreau EH, Baldi A, Blaquier J and Wasserman G (1970) Studies on the mechanism of action of testosterone and dihydrotestosterone uptake and effects on RNA metabolism in castrated rats. Acta Endocr Panam 1:39–48

Chaudhury DP and Majumder GC (1983) Evidence for the reconstitution of motility by epididymal plasma-protein factors(s) in immotile washed spermatozoa from goat cauda epididymis. Andrologia 15:495–497

Chinoy NJ and Sanjeevan AG (1980) Effects of an intra-epididymal and intra-scrotal copper device on rat spermatozoa. Int J Androl 3:719–737

Chulavatnatol M (1982) Motility initiation of quiescent spermatozoa from rat caudal epididymis: effects of pH, viscosity, osmolality and inhibitors. Int J Androl 5:425–436

Courtens JL, Rozinek J and Fournier-Delpech S (1982) Binding of epididymal proteins to the spermatozoa of ram. Andrologia 14:509–514

Cuasnicú PS, González Echeverría F, Piazza A and Blaquier JA (1984) Addition of androgens to cultured hamster epididymis increases zona recognition by immature spermatozoa. J Reprod Fert 70:541–547

Cushman DW and Cheung HS (1971) Concentrations of angiotensin converting enzyme in tissues of the rat. Biochem Biophys Acta 250:261–265

Cooper TG and Waites GMH (1979) Investigation by luminal perfusion of the transfer of compounds into the epididymis of the anaesthetized rat. J Reprod Fert 56:159–164

Cooper TG and Yeung CH (1980) Epithelial structure of the rat cauda epididymidis after luminal perfusion. Int J Androl 3:367–374

Dacheux JL (1980) An in vitro luminal perfusion technique to study epididymal secretion. IRCS Med Sci 8:137

Dacheux JL and Voglmayr JK (1983) Sequence of sperm cell surface differentiation and its relationship to exogenous fluid proteins in the ram epididymis. Biol Reprod 29:1033–1046

Dacheux JL, Paquignon M and Combarnous Y (1983) Head-to-head agglutination of ram and boar epididymal spermatozoa and evidence for an epididymal antagglutinin. J Reprod Fert 67:181–189

Dadoune JP (1981) Structural and metabolic findings on epididymal cells. In: Epididymis and Fertility: Biology and Pathology. Eds C Bollack and A Clavert, S Karger, London. Progr Reprod Biol 8:34–47

D'Agnostino A, Jones R, White R and Parker MG (1980) Androgenic regulation of messenger RNA in rat epididymis. Biochem J 190:505–512

Danzo BJ, Eller BC and Orgebin-Crist MC (1976) Approaches to fertility regulation through an understanding of the molecular control of epididymal function. In: Sperm Action. Progress in Reproductive Biology Volume 1. Eds PO Hermite, M L'Hermite and J Schirren, Karger, Basel. pp 74–88

Danzo BJ, Cooper TG and Orgebin-Crist MC (1977) Androgen binding protein (ABP) in fluid colected from the rete testis and cauda epididymidis of sexually mature and immature rabbits and observations on morphological changes in the epididymis following ligation of the ductuli efferentes. Biol Reprod 17:64–77

Davis BK and Gergely AF (1979) Studies on the mechanism of capacitation: changes in plasma membrane proteins of rat spermatozoa during incubation. Biochem Biophys Res Commun 88:613–618

DeLap LV, Tate SS and Meister A (1977) γ-glutamyl transpeptidase and related enzyme activities in the reproductive system of the mae rat. Life Sci 20:673–680

del Rio AG (1979) Macromolecular secretion into various segments of the guinea pig epididymis. Archs Androl 3:231–237

del Rio AG, DeSanchez LZ and Sirena A (1984) Evaluation of epididymal function through specific protein on spermatozoa. Archs Androl 12:195–196

del Villano BC, Lerner RA, Kennel SJ and Elder JH (1977) Expression of an oncornavirus gene product in the mouse genital tract. In: The Testis in Normal and Infertile Men. Eds P Troen and HR Nankin, Raven Press, New York pp 173–183

Devine KM and Carroll J (1985) Peptide synthesis by the murine epididymis. Biochem Soc Trans 13:506–507

Dostál J and Veselský L (1972) Proteins in the seminal plasma and the accessory sexual gland fluids of the boar. J Reprod Fert 30:255–267

Dott HM, Harrison RAP and Foster GCA (1979) The maintenance of motility and the surface properties of epididymal spermatozoa from bull, rabbit and ram in homolgous seminal and epididymal plasma. J Reprod Fert 55:113–124

Dravland E and Joshi MS (1981) Sperm-coating antigens secreted by the epididymis and seminal vesicle of the rat. Biol Reprod 25:649–658

Durliat M and Vendrely C (1972) Etude de l'action des androgenes sur l'incorporation de cytidine dans les organes du tractus genital de la souris male. 1. Influence de la castration et de l'implantation d'androgene chez les castrats. Comptes Rendus Soc de Biol 166:1578–1585

Emery T (1980) Iron deprivation as a biological defence mechanism. Nature (Lond) 287:776–777

Ellis DH, Hartman T and Moore (1985) Maturation and function of the hamster spermatozoon probed with monoclonal antibodies. J Reprod Immunol 7:299–314

Elson SD, Browne CA and Thorburn GD (1984) Identification of epidermal growth factor-like activity in human male reproductive tissues and fluids. J Clin Endocr 58:589–594

Esponda P and Bedford JM (1985) Epididymal fluid macromolecules do not at as auto- or alloantigens. J Androl 6:359–364

Fain-Maurel MA, Dadoune JP and Alfons MF (1981) High-resolution autoradiography of newly formed proteins in the epididymis after incorporation of tritiated amino acids. Archs Androl 6:249–266

Fain-Maurel MA, Dadoune JP and Jauzein-Leau F (1983) Protein secretion by the principal cells of mouse epididymis as evidenced by in vitro incorporation of tritiated leucine. Reprod Nutr Develop 23:175–182

Faye JC, Duguet L, Mazzuca M and Bayard F (1980) Purification, radioimmunoassay, and immunohistochemical localization of a glycoprotein produced by the rat epididymis. Biol Reprod 23:423–432

Feig LA, Klagsbrun and Bellve AR (1983) Mitogenic poplypeptide of the mammalian seminiferous epithelium: biochemical characterization and partial purification. J Cell Biol 97:1435–1443

Feldman MS, Lea OA, Petrusz P, Tres LL, Kierszenbaum AL and French FS (1981) Androgen-binding protein. Purification from rat epididymis, characterization, and immunocytochemical localization. J Biol Chen 256:5170–5175

Feuchter FA, Vernon RB and Eddy EM (1981) Analysis of the sperm surface with monoclonal antibodies: topographically restricted antigens appearing in the epididymis. Biol Reprod 24:1099–1110

Fleischer B, Fleischer S and Ozawa H (1969) Isolation and characterization of Golgi membranes from bovine liver. J Cell Biol 43:59–79

Flickinger CJ (1979) Synthesis, transport and secretion of protein in the initial segment of the mouse epididymis as studied by electron microsocpe radioautography. Biol Reprod 20:1015–1030

Flickinger CJ (1981) Regional differences in synthesis, intracellular transport, and secretion of protein in the mouse epididymis. Biol Reprod 25:871–883

Flickinger CJ (1983) Synthesis and secretion of glycoprotein by the epididymal epithelium. J Androl 4:157–161

Flickinger CJ (1985) Radioautographic analysis of the secretory pathway for glycoproteins in principal cells of the mouse epididymis exposed to (^{3}H)fucose. Biol Reprod 32:377–389

Flickinger CJ, Wilson KM and Gray HA (1984) The secretory pathway in the mouse epididymis as shown by electron microscope radioautography of principal cells exposed to monensin. Anat Rec 210:435–448

Fournier S (1966) Repartition de l'acide sialique dans l'appareil genital rat Wistar d'adulte normal et castré. Comptes Rendus Soc de Biol 160:1087–1090

Fournier-Delpech S, Bayard F and Boulard C (1973a) Contribution à l'étude de la maturation du sperme. Etude d'une proteine acide de l'épididyme chez le rat. Dependence androgene, relation avec l'acide sialique. CR Soc de Biol 167:1989–1996

Fournier-Delpech S, Danzo BJ and Orgebin-Crist MC (1977) Extraction of concanavalin A affinity material from rat testicular and epididymal spermatozoa. Ann biol anim Bioch Biophys 17:207–213

Fournier-Delpech S, Hamamah S, Colas G and Courot M (1983a) Aquisition of zona binding structures by ram spermatozoa during epididymal passage. In: The Sperm Cell. Ed J Andre, Martinus-Nijhoff, The Hague. pp 103–106.

Fritz IB, Burdyzy K, Setchell BP and Blaschuk D (1983) Ram rete testis fluid contains a protein (Clusterin) which influences cell-cell interactions in vitro. Biol Reprod 28:1173–1188

Fox N, Damjanov I, Knowles BB and Solter D (1982) Teratocarcinoma antigen is secreted by epididymal cells and coupled to maturing sperm. Exp Cell Res 137:485–486

Garberi JC, Kohane AC, Cameo MS and Blaquier JA (1979) Isolation and characterization of specific rat epididymal proteins. Mol Cell Endocr 13:73–82

Gaunt SJ (1982) A 28K-dalton cell surface autoantigen of spermatogenesis: characterization using a monoclonal antibody. Develop Biol 89:92–100

Gaunt SJ, Brown CR and Jones R (1983) Identification of mobile and fixed antogens on the plasma membrane of rat spermatozoa using monoclonal antibodies. Exp Cell Res 144:275–284

Gingell D (1976) Electrostatic control of membrane permeability *via* intramembranous particle aggregation. In: Mammalian Cell Membranes Vol 1 General Concepts. Eds GA Jamieson and DM Robinson, Butterworths, London. pp 198–223

Glenner GG, Folk JE and McMillan PJ (1962) Histochemical demonstration of a gamma-glutamyl transpeptidase-like activity. J Histo Cytochem 10:481–489

Glucksmann A, Howards A and Pele SR (1955) Incoportaitn of ^{35}S-DL-methionine in mouse tissue as indicated by autoradiography. 1. Testis, epididymis and seminal vesicle. J Anat 89:13–18

Goldbarg JA, Friedman OM, Pinera EP, Smith EE, Chatterji R, Stein EH and Rutenberg AM (1960) The colorimetric determination of γ-glutamyltranspeptidase with synthetic substrate. Arch Biochem Biophys 91:61–70

González Echeverría FM, Cuasnicú PS and Blaquier JA (1982) Identification of androgen-dependent glycoproteins in the hamster epididymis and their association with spermatozoa. J Reprod Fert 64:1–7

Guérin JF, Rollet J, Perrin P, Menezo Y, Orgiazzi A and Czyba JC (1981) Enzymes in the seminal plasma from azoospermic men: correlation with the origin of their azoospermia. Fert Steril 36:368–372

Gunn RMC (1936) Fertility in sheep. Artificial production of seminal ejaculation and the characteristics of the spermatozoa contained therein. CSIRO Bulletin No. 94, Commonwealth of Australia

Gupta G, Rajalakshmi M, Prasad MRN and Moudgal NR (1974) Alteration in epididymal function and its relation to maturation of spematozoa. Andrologia 6:35–44

Hales BF, Hachey C and Robaire B (1980) The presence and longitudinal distribution of the glutathione S-transferases in rat epididymis and vas deferens. Biochem J 189:135–142

Hamilton DW (1981a) Future trends in cell biological research on the male reproductive system. In: Goals in Male Reproductive Research. Eds S Boyarsky and K Polakoski, Pergamon Press, Oxford. pp 19–27

Hamilton DW (1981b) Evidence for α-lactalbumin-like activity in reproductive tract fluids of the male rat. Biol Reprod 25:385–392

Hamilton DW, Wenstrom JC and Baker JB (1986) Membrane glycoproteins from spermatozoa: partial characterization of an integral $M_r = \sim 24{,}000$ molecule from rat spermatozoa that is glycosylated during epididymal maturation. Biol Reprod (in press)

Harrison RAP, Dott HM and Foster GC (1978) Effect of ionic strength, serum albumin and other macromolecules on the maintenance of motility and the surface of mammalian spermatozoa in a simple medium. J Reprod Fert 52:65–73

Harrison RAP, Dott HM and Foster GC (1982) Bovine serum albumin, sperm motility, and the "dilution effect". J exp Zool 222:81–88

Higgins SJ, Brooks DE and Fuller FM (1981) Isolation of cells from rat seminal vesicles and epididymis and their use in studying androgen action. Mol Cell Endocr 23:207–223

Hinton BT and Howards SS (1982a) Rat testis and epididymis can transport (^{3}H)3-*O*-methyl-D-glucose, (^{3}H)inositol and (^{3}H)α-aminoisobutyric acid across its epithelia in vivo. Biol Reprod 27:1181–1189

Hoffer AP, Hamilton DW and Fawcett DW (1973) The ultrastructure of the principal cells and intraepithelial leucocytes in the initial segment of the rat epididymis. Anat Rec 175:169–202

Hoffmann DS and Killian GJ (1981) Isolation of epithelial cells from the corpus epididymidis and analysis for glycerylphosphorylcholine, sialic acid and protein. J exp Zool 217:93–102

Hohlbrugger G, Schweisfurth H and Dahlheim H (1982) Angiotensin I converting enzyme in rat testis, epididymis and vas deferens under different conditions. J Reprod Fert 65:97–103

Huang HFS and Johnson AD (1975) Amino acid composition of epididymal plasma of mouse, rat, rabbit and sheep. Comp Biochem Physiol 50B:359–362

Huggenvik J and Griswold MD (1981) Retinol binding protein in rat testicular cells. J Reprod Fert 61:403–408

Hunter AG and Schellpfeffer DA (1981) Concentrations of sperm, protein and a sperm membrane glycoprotein within boar epididymal luminal fluids. J Anim Sci 52:575–579

Hunter RHF, Holtz W and Hermann H (1978) Stabilizing role of epididymal plasma in relation to the capacitation time of boar spermatozoa. Anim Reprod Sci 1:161–166

Irwin M, Nicholson N, Haywood JT and Poirier GR (1983) Immunological localization of a murine seminal vesicle proteinase inhibitor. Biol Reprod 28:1201–1206

Iusem ND, de Larminat MA, Tezon JG, Blaquier JA and Belocopitow E (1984) Androgen dependence of protein *N*-glycosylation in rat epididymis. Endocrinology 114:1448–1453

Jaiswal A, Joshi P, Kumar MV, Panda JN and Singh LN (1984) Angiotensin converting enzyme in the tesis and epididymis of mammals. Andrologia 16:410–416

Jaiswal AK, Panda JN, Vijaykumar M and Joshi P (1985) Androgen dependence of testicular and epididymal angiotensin converting enzyme. Andrologia 17:92–97

Jégou B, Dacheux JL, Garnier DH,Terqui M, Colas G and Courot M (1979) Biochemical and physiological studies of androgen-binding protein in the reproductive tract of the ram. J Reprod Fert 57:311–318

Jones R (1974) Absorption and secretion in the cauda epididymidis of the rabbit and the effects of degenerating spermatozoa on epididymal plasma after castration. J Endocr 63:157–165

Jones R (1977) Effects of testosterone, testosterone metabolites and anti-androgens on the function of the male accessory glands in the rabbit and rat. J Endocr 74:75–78

Jones R (1978) Comparative biochemistry of mammalian epididymal plasma. Comp Biochem Biophys 61B:365–370

Jones R and Brown CR (1982) Association of epididymal secretory proteins showing α-lactalbumin-like activity with the plasma membrane of rat spermatozoa. Biochem J 206:161–164

Jones R and Dott HM (1980) Changes in luminal plasma and disappearance of spermatozoa from the ligated cauda epididymidis of the androgen-deficient rabbit. J Reprod Fert 60:65–72

Jones R and Glover TD (1973) The effects of castration on the composition of rabbit epididymal plasma. J Reprod Fert 34:405–411

Jones R, von Glos KI and Brown CR (1980a) The synthesis of a sperm-coating protein in the initial segment of the rat epididymis is stimulated by factors in testicular fluid. IRCS Med Sci 8:56

Jones R, Brown CR, von Glos KI and Parker MG (1980b) Hormonal regulation of proteins synthesis in the rat epididymis. Characterizatiom of androgen-dependent and testicular fluid-dependent proteins. Biochem J 188:667–676

Jones R, von Glos KI and Brown CR (1981) Characterization of hormonally regulated secretory proteins from the caput epididymidis of the rabbit. Biochem J 196:105–114

Jones R, Fournier-Delpech S and Willadson SA (1982) Identification of androgen-dependent proteins synthesised in vitro by the ram epididymis. Reprod Nutr Develop 22:495–504

Jones R, Brown CR, Cran DG and Gaunt SJ (1983a) Surface and integral antigens of rat spermatozoa distinguished using monoclonal antibodies. Gamete Res 8:255–265

Jones R, Riding PR and Parker MG (1983b) Effects of prolactin on testosterone-induced growth and protein synthesis in rat accessory glands. J Endocr 96:407–416

Jones R, von Glos KI and Brown CR (1983c) Changes in the protein composition of rat spermatozoa during maturation in the epididymis. J Reprod Fert 67:299–308

Kanka J and Kopecny V (1977) An autoradiographic study of macromolecular synthesis in the epitheliumn of the ductus epididymidis in the mouse. I. DNA, RNA and Protein. Biol Reprod 16:421–427

Kato M, Weng KS, Kato K and Goodman DS (1985) Immunohistochemical studies on the localization of cellular retinol-binding protein in rat testis and epididymis. Biol Reprod 32:178–189

Kierszenbaum AL, Lea O, Petrusz P, French FS and Tres LL (1981) Isolation, culture, and immunocytochemical characterization of epididymal epithelial cells from pubertal and adult rats. Proc Natn Acad Sci US 78:1675–1679

Killian GJ and Chapman DA (1980) Glycerylphosphorylcholine, sialic acid and protein in epithelial cells isolated from the rat caput epididymidis by elutriation. Biol Reprod 22: 846–850

Kissinger C, Skinner MK and Griswold MDN (1982) Analysis of Sertoli cell-secreted proteins by two dimensional gel electrophoresis. Biol Reprod 27:233–240

Klinefelter GR and Hamilton DW (1984) Organ culture of rat caput epididymal tubules in a perifusion chamber. J Androl 5:243–258

Klinefelter GR and Hamilton DW (1985) Synthesis and secretion of proteins by perifused caput epididymal tubules, and association of secreted proteins with spermatozoa. Biol Reprod 33:1017–1027

Klinefelter GR, Amann RP and Hammerstedt RH (1982) Culture of principal cells from the rat caput epididymidis. Biol Reprod 26:885–901

Kochakian CD (1975) Free amino acids of sex organs of the mouse: regulation by androgen. Am J Physiol 228:1231–1235

Kochakian CD (1976) Influence of testosterone on the concentration of hypotaurine and taurine in the reproductive tract of the male guinea pig, rat, and mouse. In: Taurine. Eds R Huxtable and A Barbeau, Raven Press, New York. pp 327–334

Kohane AC, Garberi JC, Cameo MS and Blaquier JA (1979) Quantitative determination of specific proteins in rat epididymis. J Steroid Biochem 11:671–674

Kohane AC, Cameo MS, Piñeiro L, Garberi JC and Blaquier JA (1980) Distribution and site of production of specific proteins in the rat epididymis. Biol Reprod 23:181–187

Kohane AC, Piñeiro L and Blaquier JA (1983) Androgen-controlled synthesis of specific proteins in the rat epididymis. Endocrinology 112:1590–1596

Kopečńy V (1971) Epididymal luminal concents labelling after ^{14}C- or ^{3}H-lysine administration in the mouse. Acta Histochem 40:116–122

Kopečńy V (1976) Labelling of acrosomes in mouse spermatozoa by L-(1-^{3}H)fucose. J Reprod Fert 47:403–405

Kopečńy V and Pech V (1977) An autoradiographic study of macromolecular synthesis in the epithelium of the ductus epididymidis in the mouse. II. Incorporation of L-fucose-1-^{3}H. Histochem 50:229–238

Kopečný V, Fléchon JE and Pivko J (1984) Binding of secreted glycoproteins to spermatozoa in the mammalian epididymis: a fine-structure autoradiographic study. Anat Rec 208:197–206
Koskimies AI and Kormano M (1975) Protein in fluids from different segments of the rat epididymis. J Reprod Fert 43:345–348
Laporte P (1970) Acides sialiques epididymaires et maturite sexualle chez le rat. CR Soc de Biol 164:2079–2984
Larsson LI, Fahrenkrug J, Schaffalitzky de Muckadell OB (1977) Occurrence of nerves containing vasoactive intestinal polypeptide immunoreactivity in the male genital tract. Life Sci 21:503–508
Lavon U, Volcani R and Danon D (1971) The proteins of bovine spermatozoa from the caput and cauda epididymidis. J Reprod Fert 24:219–232
Letts PJ, Meistrich ML, Bruce WR and Schacter H (1974) Glycoprotein glycosyl-transferase levels during spermatogenesis in mice. Biochem Biophys Acta 343:192–207
Lea OA, Petrusz P and French FS (1978) Purification and localization of acidic epididymal glycoprotein (AEG): a sperm coating protein secreted by the rat epididymis. Int J Androl Suppl 2:592–607
Lea OA and French FS (1981) Characterization of an acidic glycoprotein secreted by principal cells of the rat epididymis. Biochem Biophys Acta 668:370–376
Li TK (1975) The glutathione and thiol content of mammalian spermatozoa and seminal plasma. Biol Reprod 12:641–646
Lojda Z (1975) The use of hexazonium-p-rosanilin in the histochemical demonstration of peptidases. Histochem 44:323–335
Low MG and Kincade PW (1985) Phosphatidylinositol is the membrane-anchoring domain of the Thy1-glycoprotein. Nature (Lond) 318:62–64
Majumder GC (1981) Histone-mediated agglutination of epididymal spermatozoa and the occurence of histone receptors on the rat sperm surface. J Reprod Fert 62:459–466
Matsuzawa T (1981) Substrate specificity and histochemical distribution of aminopeptidase in rat testis and epididymis. Endocr Japon 28:469–475
Mayorga L and Bertini F (1983) Two populations of hydrolase-containing particles in rat epididymis. J Cell Physiol 117:135–139
Meister A (1973) On the enzymology of amino acid transport. Transport in kidney and probably other tissues is mediated by a cycle of enzyme reactions involving glutathione. Science 180:33–39
Meister A (1974) Glutathione, metabolism and function via the γ-glutamylcycle. Life Sci 15:177–190
Meister A and Tate SS (1972) Glutathione and related γ-glutamylcompounds: biosynthesis and utilization. Ann Rev Biochem 45:559–604
Mercado Pichardo E, Villalobos Ramirez M, Dominguez Aguilar R and Rosado Garcia A (1982) Modifications in the apolar regions of the membrane of rabbit sperm cells during epididymal maturation. Archos Invest Med 13:173–180
Mongkolsirikieat S and Chulavatnatol M (1984) Phosphorylated secretory proteins from rat epididymis and their androgenic control. J Reprod Fert 72:423–428
Moore HDM (1980) Localization of specific glycoproteins secreted by the rabbit and hamster epididymis. Biol Reprod 22:705–718
Moore HDM (1981a) Effects of castration on specific glycoprotein secretions of the epididymis in the rabbit and hamster. J Reprod Fert 61:347–354
Moore HDM (1981b) Glycoprotein secretions of the epididymis in the rabbit and hamster: localization on epididymal spermatozoa and the effect of specific antibodies on fertilization in vivo. J exp Zool 215:77–85
Moore HDM and Bedford JM (1979) Short-term effects of androgen withdrawal on the structure of different epithelial cells in the rat epididymis. Anat Rec 193:293–312
Moore HDM and Pryor JP (1981) The comparative ultrastructure of the epididymis in monkeys and man: a search for a suitable animal model for studying epididymal physiology in Primates. Am J Primatol 1:241–250
Moore HDM, Hartman TD, Brown AC, Smith CA and Ellis DH (1985) Expression of sperm antigens during spermatogenesis and maturation detected with monoclonal antibodies. Expl clin Immunogenet 2:84–96

Morton B and Chang TSK (1973) The effect of fluid from the cauda epididymidis, serum components and caffeine upon the survival of diluted epididymal hamster spermatozoa. J Reprod Fert 35:255–263

Morton BE, Fraser CF and Albagli L (1979a) Studies on factors in hamster caudal epididymal plasma and other sources which inhibit sperm dilution damage. Fert Steril 32:99–106

Morton BE, Fraser CF and Sagadraca R (1979b) Inhibition of sperm dilution damage by purified factors from hamster caudal epididymal plasma and by defined diluents. Fert Steril 32:107–114

Musto N, Gunsalas G, Cheng CY, Tsong SD, Goldstein M, Phillips D and Bardin (1983) Identification of androgen-binding proteins and their localization in the testes and male reproductive tract. In: Recent Advances in Male Reproduction. Molecular Basis and Clinical Implications. Eds R D'Agato, MB Lipsett, P Polosa and HJ van der Molen, Raven Press, New York. pp 37–45

Neutra M and Leblond CP (1966) Radioautographic comparison of the uptake of galactose-H^3 and glucose -H^3 in the Golgi region of various cells secreting glycoproteins or mucopolysaccharides. J Cell Biol 30:137–150

Neumark H and Schindler H (1967) Amino acids, amines and peptides of ram epididymal semen. J Reprod Fert 14:469–471

Nicander L and Malmqvist M (1977) Ultrastructural observations suggesting merocrine secretion in the initial segment of the mammalian epididymis. Cell Tiss Res 184:487–490

Niemi M and Tuohimaa P (1971) The mitogenic activity of testosterone in the accessory glands of the rat in relation to its conversion to dihydrotestosterone. In: Basic Action of Sex Steroid on Target Organs. Eds PO Hubinot, F Leroy and P Galand, Karger, Basel. pp 258–264

Oliphant G (1976) Removal of sperm-bound seminal plasma components as a prerequisite to induction of the rabbit acrosome reaction. Fert Steril 27:28–38

Oliphant G and Brackett BG (1973) Capacitation of mouse spermatozoa in media with elevated ionic strength and reversible decapacitation with epididymal extracts. Fert Steril 24:948–955

Olson GE and Hinton BT (1985) Regional differences in luminal fluid polypeptides of the rat testis and epididymis revealed by two-dimensional gel electrophoresis. J Androl 6:20–34

Olson GE and Orgebin-Crist MC (1982) Sperm surface changes during epididymal maturation. Ann NY Acad Sci 383:372–391

Olson GE, Jonas-Davies J, Hoffman LH and Orgebin-Crist MC (1983) Structural features of cultured epithelial cells from the adult rat epididymis. J Androl 4:347–360

Ong DE, Crow JA and Chytil F (1982) Radioimmunochemical determination of cellular retinol- and cellular retinoic acid-binding proteins in cytosols of rat tissues. J Biol Chem 257:13385–13389

Orgebin-Crist MC and Jiane JD (1979) Properties of prolactin receptor from the rabbit epididymis. Biol Reprod 21:135–139

Orgebin-Crist MC and Ménézo Y (1980) A continuous flow method for organ cultrue of rabbit epididymis: morphology, amino acid utilization, glucose uptake, RNA, and protein synthesis. J Androl 1:289–298

Papkoff H (1966) Glycoproteins with biological activity. In: Glycoproteins and their composition, structure, and function, Section 9. Ed A Gottschalk, Elsevier Pub Co, Amsterdam. pp 532–557

Pekary AE, Meyer NV, Vaillant C and Hershman JM (1980) Thyrotropin-releasing hormone and a homologous peptide in the male rat reproductive system. Biochem Biophys Res Commun 95:993–1000

Pekary AE, Ross R, Kernion JOE and Hershman JM (1983) Human reproductive tissues contain thyrotropin releasing hormone (TRH) and TRH-homologous peptides. J Androl 4:408–414

Pekary AE, Yamada T, Sharp B, Bhasin S, Swerdloff RS and Hershman JM (1984) Somatostatin-14 and -28 in the male rat reproductive-system. Life Sci 34:939–946

Pelliniemi LJ, Dym M, Gunsalas GL, Musto NA, Bardin CW and Fawcett DW (1981) Immunocytochemical localization of androgen-binding protein in the male rat reproductive tract. Endocrinology 108:925–931

Peyre A and Laporte P (1966) Action de la testosterone et de l'oestradiol sur les sialo-proteins de la queue de l'épididyme de rat impubère castré. CR Soc de Biol 160:2178–2180

Poirier GR and Jackson J (1981) Isolation and characterization of two proteinase inhibitors from the male reproductive tract of mice. Gamete Res 4:555–569

Poirier CR and Nicholson N (1984) Distribution of a proteinase inhibitor of epididymal origin in the tissues and secretions of the male reproductive tract of mice. J exp Zool 230:465–472

Porter SB, Ong DE, Chytil F and Orgebin-Crist MC (1985) Localization of cellular retinol-binding protein and cellular retinoic acid binding protein in the rat testis and epididymis. J Androl 6:197–212

Purvis K and Hansson V (1978) Androgens and androgen-binding protein in the rat epididymis. J Reprod Fert 52:59–63

Rajalakshmi M (1985) Hormonal-regulation of specific proteins in the rat epididymis. Archs Androl 14:181–185

Rajalakshmi M and Prasad MRN (1968) Changes in the sialic acid content of the accessory glands of the rat. J Endocr 41:471–476

Rajalakshmi M and Prasad MRN (1969) Changes in sialic acid in the testis and epididymis of the rat during the onset of puberty. J Endocr 44:379–385

Rama Sastry BV, Janson VE, Owens LK and Tayeb, OS (1982) Enkephalin- and substance P-like immunoreactivities of mammalian sperm and accessory sex glands. Biochem Pharmacol 31:3519–3522

Reyes A, Parra A, Chavarria ME, Goicoechea B and Rosado A (1979) Effect of prolactin on the calcium binding and/or transport of ejaculated and epididymal human spermatozoa. Fert Steril 31:669–672

Riar SS, Setty BS and Kar AB (1973) Studies on the physiology and biochemistry of mammalian epididymis. Biochemical composition of epididymis. A comparative study. Fert Steril 24:355–363

Robaire B and Hales BF (1982) Regulation of epididymal glutathione *S*-transferases: effects of orchidectomy and androgen replacement. Biol Reprod 26:559–565

Rohr H, Oberholzer M, Bartsch G and Keller M (1976) Morphometry in experimental pathology: methods, baseline data, and applications. In: International review of Experimental Pathology. Volume 15. Eds GW Richter and MA Epstein, Acad Press, London pp 233–325

Rosenweig JL, Havrankova J, Lesniak MA, Brownstein M and Roth JK (1980) Insulin is ubiquitous in extrahepatic tissues of rats and humans. Proc Natn Acad Sci Us 77:572–576

Roy N, Majumder GC and Chakrabarty CK (1985) Occurrence of specific glycoprotein factors(s) in goat epididymal plasma that prevent adhesion of spermatozoa to glass. Andrologia 17:200–206

Rufo GA, Singh JP, Babcock DF and Lardy HA (1982) Purification and characterization of a calcium transport inhibitor protein from bovine seminal plasma. J Biol Chem 257:4627–4632

Rufo GA, Schoff PK and Lardy HA (1984) Regulation of calcium content in bovine spermatozoa. J Biol Chem 259:2547–2552

Russell LD, Peterson RN, Hunt W and Strack LE (1984) Posttesticular surface modifications and contributions of reproductive tract fluids to the surface polypeptide composition of boar spermatozoa. Biol Reprod 30:959–978

Rutenburg AM, Kim H, Fischbein JW, Hanker JS, Wasserkrug HL and Seligman AM (1969) Histochemical and ultrastructural demonstration of γ-glutamyl transpeptidase activity. J Histo Cyto 17:517–526

Schachter H (1978) Glycoprotein synthesis. In: The glycoconjugates, Volume II. Eds MI Horowitz and W Pigman, Acad Press, New York. pp 87–181

Sedláková E, Dostál J and Matousek J (1968) Serum albumin in pig epididymis. Comp Biochem Physiol 26:143–148

Seike H, Yokoyama M, Iwata H, Ochi K, Takeuchio M, Takada Y, Hiwada K and Kokubu T (1985) Production of angiotensin-converting enzyme in rat epididymis. Andrologia 17:406–410

Setchell BP, Hinks NT, Voglmayr JK and Scott TW (1967) Amino acids in ram testicular fluid and semen and their metabolism by spermatozoa. Biochem J 105:1061–1066

Setty BS, Riar SS and Kar AB (1979) Androgen control of epididymal function in rhesus monkey and rabbit. Fert Steril 28:674–681

Sexton TJ. Amann RP and Flipse RJ (1971) Free amino acids and protein in rete testis fluid, vas deferens plasma, and accessory sex gland fluid and seminal plasma of the conscious bull. J Dairy Sci 54:412–416

Sexton TJ, Amann RP and Flipse RJ (1972) Oxidative metabolism of glutamate by testicular, epididymal, and ejaculated bovine spermatozoa in the presence of rete testis fluid and seminal plasma. J Dairy Sci 55:1005–1009

Sharp B, Pekary AE, Meyer NV and Hersham JM (1980) β-endorphin in male rat reproductive organs. Biochem Biophys Res Commun 95:618–623

Shimazaki J, Yamanaka H, Taguchi I and Shida K (1976) Free amino acids in the caput and the cauda epididymis of adult rats. Endocr Japon 23:149–156

Sheth AR, Gunjikar AN and Shah GV (1981) The presence of progressive motility sustaining factor (PMSF) in human epididymis. Andrologia 13:142–146

Silvestroni L, Satori C, Modesti A and Frajese G (1982) Harmless technique for removal of the cell coat from human spermatozoa. Experientia 38:672–673

Silvestroni L, Sartori C, Modesti A and Frejese G (1983) Advances in molecular andrology: biochemical and ultrastructural studies on human sperm plasma membrane. In: Recent Advances in Male Reproduction. Molecular Basis and Clinical Implications. Eds R D'Agat, MB Lipsett, P Polosa and HJ Van Der Molen, Raven Press, New York. pp 1–11

Sirett DAN, Rolin Jacqemyns CF, Quivy JI and Rousseau GG (1984) Effects of dicyclohexane compounds in vivo on rat epididymal spermatozoa. Proc of 23rd Harden Conference, "Molecular and Cellular Aspects of Reproduction", Wye Colege, University of London, Sept 1984 (Abst)

Skinner MK and Griswold MD (1980) Sertoli cells synthesize and secrete transferrin-like protein. J Biol Chem 255:9523–9525

Skinner MK and Griswold MD (1982) Secretion of testicular transferrin by cultured Sertoli cells is regulated by hormones and retinoids. Biol Reprod 27:211–221

Skinner MK and Griswold MD (1983) Sertoli cells synthesize and secrete a ceruloplasmin-like protein. Biol Reprod 28:1225–1229

Staněk R and Dostál J (1974) Proteins in bull rete testis and cauda epididymis plasma. Comp Biochem Biophys 48B:613–618

Strittmatter SM and Snyder SH (1984) Angiotensin-converting enzyme in the mnale rat reproductive system: autoradiographic visualization with (^{3}H)captopril. Endocrinology 115:2332–2341

Suzuki F (1981) Changes in intramembranous particle distribution in epididymal spermatozoa of the boar. Anat Rec 199:361–376

Suzuki F and Nagano T (1980) Epididymal maturation of rat spermatozoa studied by thin sectioning and freeze fracture. Biol Reprod 22:1219–1231

Sweet C and Zull JE (1969) Activation of glucose diffusion from egg lecithin liquid crystals by serum albumin. Biochem Biophys Acta 173:94–103

Sylvester SR, Skinner MK and Griswold MD (1984) A sulphated glycoprotein synthesized by Sertoli cells and by epididymal cells is a component of the sperm membrane. Biol Reprod 31:1087–1801

Tadolini B, Wilson J, Reddy PRK and Williams-Ashman HG (1977) Characteristics and hormonal control of some glycoprotein glycosyltransferase reactions in male reproductive organs. Advances in Enzyme Regulation, Ed G Weber, Volume 15. pp 319–336

Tarter TH and Alexander NJ (1984) Complement-inhibiting activity of seminal plasma. Am J Reprod Immunol 6:28–32

Tezón JG and Blaqier JA (1981) The organ culture of human epididymal tubules and their response to androgens. Mol Cell Endocr 21:233–242

Tezón JG, Cuasnicú PS, Scorticati C and Blaquier JA (1982a) Development and characterization of a model system for the study of epididymal function in man. In: Physiopathology of Hypophysial Disturbances and Diseases of Reproduction. Eds A DeNicola, JA Blaquier and RJ Soto, Alan Liss Inc. New York. pp 251 –275

Tezón JG, Vazquez M, Piñeiro L, De Larminat MA and Blaquier J (1982b) The effect of androgens in the human epididymis in organ culture. In: Therapy in Andrology. Pharmacological, Surgical and Pathological Aspects. Eds GF Mechini-Fabris, W Pasini and L Martini, Elsevier, New York. pp. 53–60

Tezón JG, Vazquez MH, Piñeiro L, deLarminat MA and Blaquier JG (1985a) Identification of androgen-induced proteins in human epididymis. Biol Reprod 32:584–590

Tezón JG, Ramella E, Cameo MS, Vazquez MH and Blaquier, JG (1985b) Immunocytochemical localization of secretion antigens in the human epididymis and their association with spermatozoa. Biol Reprod 32:591–597

Thomas TS, Reynolds AL and Oliphant G (1984) Evaluation of the site of synthesis of rabbit sperm acrosome stabilizing factor using immunocytochemical and metabolic labeling techniques. Biol Reprod 30:693–705

Tkacz JS and Lampen JO (1975) Tunicamycin inhibition of polyisoprenol *N*-acetylglucosaminyl pyrophosphate formation in calf-liver microsomes. Biochem Biophys Res Commun 65:248–257

Toowicharanont P and Chulavatnatol M (1983a) Measurement of anionic sites of rat epididymal spermatozoa using tritiated polycationized ferritin. J Reprod Fert 69:303–306

Tremblay RR, Chapdelaine P and Mailhot J (1979) α-1,4-glucosidase activity in human semen: variations with number and motility of spermatozoa. Fert Steril 31:592–593

Tremblay RR, Chapdelaine P and Dube JY (1982) Neutral α-1,4-glucosidase in human seminal plasma: molecular forms in varicocele abd after vasectomy. Fert Steril 38:344–348

Tsang AYF, Lee WM and Wong PYD (1981) Effects of antifertility drugs on epididymal protein secretion, aquisition of sperm surface proteins and fertility in male rats. Int J Androl 4:703–712

Tsong SD, Phillips D, Halmi N, Liotta AS, Margioris A, Bardin CW and Krieger DT (1982a) ACTH and β-endorphin-related peptides are present in multiple sites in the reproductive tract of the male rat. Endocrinology 110:2204–2206

Tsong SD, Phillips DM, Halmi N, Krieger D and Bardin CW (1982b) β-endorphin is present in the male reproductive tract of five species. Biol Reprod 27:755–764

Tung PS and Fritz IB (1985) Immunolocalization of clusterin in the ram testis, rete testis, and excurrent ducts. Biol Reprod 33:177–186

Turner TT (1979) On the epididymis and its function. Invest Urol 16:311–321

Turner TT and Giles RD (1982) Sperm motility-inhibiting factror in rat epididymis. Am J Physiol 242:R199–R203

Turner TT and Giles RD (1982b) The effects of cyclic adenine nucleotides, phosphodiesterase inhibitors, and cauda epididymal fluid on the motility of rat epididymal spermatozoa. J Androl 3:134–139

Turner TT and Howards SS (1978) Factors involved in the initiation of sperm motility. Biol Reprod 18:571–578

Turner TT and Reich GW (1985) Cauda epididymal sperm motility: a comparison among 5 species. Biol Reprod 32: 120-128

Turner TT, Plesums JL and Cabot CL (1979) Luminal fluid proteins of the male rat reproductive tract. Biol Reprod 21:883–890

Usselman MC and Cone RA (1983) Rat sperm are mechanically immobilized in the caudal epididymis by "immobilin" a high molecular weight glycoprotein. Biol Reprod 29:1241–1253

Usselman MC, Cone RA and Rossignol P (1985) Rat cauda epididymal fluid is a mucus. J Androl 6:315–320

Vanha-Perttula T, Mather JP, Bardin CW, Moss SB and Bellvé AR (1985) Localization of the Angiotensin-converting enzyme-activity in testis and epididymis. Biol Reprod 33:870–877

Velázquez A, Rosado A and Rosado A (1975) Participation of vitamin A in the maturation of rabbit spermatozoa. Int J Fert 20:151–155

Vendrely C and Durliat M (1968) Etude autoradiographique de l'influence de l'hormone male sur l'incorporation de leucine dans le tractus genital de la souris blanche. Archive d'Anat Histol Embryol 51:735–740

Vermouth NT, Carriazo CS, Ponce RH and Blanco A (1986) Lactate dehydrogenase-X, malate-dehydrogenase and total protein in rat spermatozoa during epididymal transit. Comp Biochem 83B:381–384

Vernon RB, Mueller CH, Herr JC, Feuchter FA and Eddy EM (1982) Epididymal secretion of a mouse sperm surface component recognized by a monoclonal antibody. Biol Reprod 26:523–535

Vreeburg JTM (1975) Distribution of testosterone and 5α-dihydrotestosterone in rat epididymis and their concentrations in efferent duct fluid. J Endocr 67:203–210

Voglmayr JK, Musto NA, Saksena SK, Brown-Woodman PDC, Marley PB and White IG (1977) Characteristics of semen collected from the cauda epididymidis of conscious rams. J Reprod Fert 49:245–251

Voglmayr JK, Fairbanks G, Jackowitz MA and Colella JR (1980) Post-testicular developmental changes in the ram sperm cell surface and their relationship to luminal fluid proteins of the reproductive tract. Biol Reprod 22:655–667

Voglmayr JK, Fairbank G, Vespa DB and Colella JR (1982) Studies on the mechanisms of surface modifications in ram spermatozoa during the final stages of differentiation. Biol Reprod 26:483–500

Wenstrom JC and Hamilton DW (1980) Dolichol concentration and biosynthesis in rat testis and epididymis. Biol Reprod 23:1054–1059

Wenstrom JC and Hamilton DW (1984) Synthesis and secretion of proteins in vitro by isolated epithelial strips from the proximal and distal vas deferens of the rat. Int J Androl 7:215–235

White MG, Huang YS, Tres LL and Kierszenbaum AL (1982) Structural and functional aspects of cultured epididymal epithelial cells isolated from pubertal rats. J Reprod Fert 66:475–484

Wilson RM and Griswold MD (1979) Secreted proteins from rat Sertoli cells. Exp Cell Res 123:127–13

Witorsch RJ (1978) Immunohistochemical studies of prolactin binding in sex accessory organs of the male rat. J Histo Cyto 26:565–580

Wong PYD and Tsang AYF (1982) Studies on the binding of a 32K rat epididymal protein to rat epididymal spermatoza. Biol Reprod 27:1239–1246

Wong PYD and Tsang AYF (1983) Radioimmunoreactivity and receptor binding activity of specific epididymal proteins of the rat and other species. Biochem Biophys Res Commun 114:653–656

Wong PYD, Au CL and Ngai HK (1978) Electrolyte and water transport in rat epididymis. Its possible role in sperm maturation. Int J Androl Suppl 5:608–628

Wong PYD, Au CL and Ngai HK (1979) Some characteristics of salt and water transport in the rat epididymis. In: The Spermatozoon. Eds DW Fawcett and JM Bedford, Urban and Schwarzenberg, Baltimore. pp 57–63

Wong PYD, Tsang AYF, Lee WM and Li CM (1980) Secretion of the rat cauda epididymidis. Archs Androl 5:327–336

Wong PYD, Tsang AYF and Lee WM (1981) Origin of the luminal fluid proteins of the rat epididymis. Int J Androl 4:331–341

Wong PYD, Tsang AYF and Lee WM (1982a) Secretion of macromoleculaes by the rat epididymis. Int J Androl Suppl 5:34–47

Wong PYD, Tsang AYF and Wang C (1982b) Absence of a sperm coating protein after epididymovasostomy. J Androl 3:205–208

Wright WW, Musto NA, Mather JT and Bardin CW (1981) Sertoli cells secrete both testis-specific and serum proteins. Proc Natn Acad Sci US 78:7565–7569

Yeung CH and Cooper TG (1982) Ultrastructure of the perfused rat epididymis: effect of luminal sodium ion concentration. Cell Tiss Res 226:407–425

Yoo BY, Ji I and Ji TH (1982) Changes in protein composition of bovine spermatozoa during epididymal maturation. Can J Zool 60:2970–2972

Zeheb R and Orr GA (1984) Characterization of a maturation-associated glycoprotein on the plasma membrane of rat caudal epididymal sperm. J Biol Chem 259:839–848

Zimmerman KJ, Crabo BG, Moore R, Weisberg S, Diebel FC and Graham EF (1979) Movements of sodium and potassium into epididymal boar spermatozoa. Biol Reprod 21:173–179

Part IV. Conclusions

1. The Importance of the Epididymis in Male Fertility

The numerous actions of the epididymis described in the Parts II and III of this review can be considered in terms of two complementary phases of the interaction of sperm cells with the epididymal environment. From this point of view, the epididymal function of sperm storage is intimately linked with, and necessarily consequent upon, that of sperm maturation. The two phases are, firstly, promoting the ability of sperm to respond appropriately to conditions within the female tract so that they can achieve fertilisation, and secondly, preventing the expression of these propensities within the male tract itself (see Fig. 4).

The former activity includes, for example, the addition of carnitine and AEG to induce the motility of sperm cells, so that forward progression and hyperactivated motility may later occur; the addition of FMP to promote forward progression, so that transport beyond certain regions of the female tract can be achieved; and the addition of glycoproteins, or the supply of enyzmes or cofactors, to modulate the nature of the sperm surface, such that recognition of components of the zona pellucida and vitellus is permitted. In many cases these secretion products are known to be androgen-dependent and the dependence of sperm maturation on circulating androgens was seen in Part I to provide evidence for the involvement of the epididymis in the process of sperm maturation. The resorptive functions of the epididymis are also androgen-dependent and could help to concentrate the secretion products and provide optimal conditions of ionic strength and pH for sperm-secretion interaction. The mechanisms whereby sperm membranes are altered to a state permitting fusion (either with other sperm membranes, as in the acrosome reaction, or with egg membranes as at gamete fusion) are less understood, and may reflect some epithelial-dependent changes in membrane lipids.

The inhibitory influences of the epididymis that act to keep spermatozoa quiescent, but fertile, in the epididymis before ejaculation include the depression of motility (and hence the related energy demand) by secretion of viscous proteins, which physically prevent flagellation, and the maintenance of high potassium ion and low sodium ion concentrations, of which the former is inhibitory and the latter stimulatory to motility expressed in-vitro. The removal of Na^+ from the epididymal lumen is associated with resorption of fluid, and this not only packs the sperm cells together, but concentrates within the lumen secretions from testicular fluid and of epididymal origin that do not permeate the epithelium, thereby raising the osmolality of the fluid and contributing to hyperosmotic dehydration of the cells. This, together with the close packing of spermatozoa acts to reduce further the possibility of sperm motility within the lumen.

Premature capacitation of spermatozoa within the epididymis is prevented by the secretion of decapacitation factors, including sterol sulphates, proteins (*e.g.* ASF) together with potassium and magnesium ions which depress the acrosome reaction in-vitro. Trypsin inhibitors in epididymal fluid would prevent damage to the epithelium by enzymes leaking from degenerating spermatozoa. While the ionic environment of spermatozoa is rapidly altered at ejaculation, permitting the expression of their fertilising potential, a more gradual loss of bound proteins (including FMP?) would permit the exaggerated hyperactivated motility of capacitated sperm to occur only in the vicinity of the eggs after successful negotiation of the cervix or uterotubal junction has been achieved by forward progressing, non-capacitated cells.

While the central role of the epididymis is mainly involved with promoting the access and contact of gametes, epididymal androgens may also limit the extent of disulphide bond formation within sperm chromatin to that compatible with the reducing power of the egg and so influence post-fusion events.

Because the events leading up to fertilisation, summarised in Part I, are interrelated and sequential, interruption of any of the steps should have consequences for fertility. However, as these events all take place in the female tract, the contraceptives derived from basic investigations into the mechanisms of fertilisation, with a view to inhibiting these processes *per se*, are more likely to be of use to women. By recognising that spermatozoa only gain their fertilising potential after their stay in the epididymis, however, a rational approach to the design of a contraceptive for men becomes feasible, with the epididymis deserving consideration as a site of attack of an antifertility agent. Attempting to interfere with its function should have repercussions for the functional competence of its contained spermatozoa; furthermore, a readily reversible antifertility action would stem from a reversible action on the epithelial cells, leading to irreversible damage to spermatozoa, providing testicular function is not impaired. In this regard, further systematic study of the activities of the epididymis are warranted.

2. Summary

The epididymis acts in two ways to prepare spermatozoa for fertilisation; firstly, promoting the ability of spermatozoa to interact successfully with eggs, and secondly, preventing the premature expression of this ability in the male tract. While more is becoming known of the elaboration and characteristics of secretion products of the epididymis, their interaction with maturing spermatozoa and the consequences of the interaction in biochemical terms remain an area for future research. While this is a potentially rewarding area for further studies, until more is known of the basic physiology and biochemistry of the human epididymis there is little hope of understanding male infertility stemming from epididymal dysfunction or designing antifertility agents acting upon this organ.

Appendix I

Table 1. The fertilising ability of spermatozoa from the testis, various parts of the epididymis[a] and vas deferens after in-vivo insemination

Species	Testis	Caput			Corpus			Cauda			Vas	Site[b]	Meth[c]	Ref
		P	M	D	P	M	D	P	M	D				
Rabbit	–		—4—		–	–	–		—98—		–	O, V	1	1
	–	2	–	0	13	–	78	77	–	79	–	O	1	2
	–	–	–	–	–	–	–	–	–	94	–	U	1	3
	–		—0—			—88—			—89—		–	U	1	4
	–		—2—		0	–	57	74	93	95	–	U	1	5
	–		—0—			—0—			—84—		–	V	1	4
	–		—0—		–	–	–	75	–	–	–	V	2	6
	0		—0—			—0—			—50—		100	V	2	7
	0		—17—			—25—		14	–	50	57	V	2	7
	–		—0.5—		–	–	–		—5.0—		–	V	3	8
Boar	–	–	–	–	10	30	53	92	92	100	91	O	1	9
	–	–	–	–	–	–	25	50	–	79	–	U?	2	10
	–		—8—			—26—		83	–	79	–	U?	2	10
	–	–	–	–		—0—		–	50	–	–	U?	2	11
Ram	0	–	–	–	–	–	–	–	–	–	–	C	1	12
	–	–	–	–		—30—			—85—		–	U	1	13
	–		—0—		–	13	88	100	–	80	–	U	2	14
	–		—0—		8	1	54	78	–	80	–	U	2	15
Guinea pig	–	–	–	–	–	–	–	32[d]	–	71	–	U	2	16
	–	–	–	–	–	–	–	33[d]	–	68	–	U	2	17
	–	–	–	–			—25[e]—			91	–	U	2	18
	–		—0[f]—				—43[g]—			–	–	P	2	19
Rat	–		—8—		–	–	–		—93—		–	U	1	21
	–		—0—			—6—		48	–	57	–	U	1	21
	–		—18—			—54—			—60—		–	U	2	22
Hamster	–		—0—			—0—			—77—		–	U	1	23
	–		—0—		–	–	3	15	89	79	–	U	1	24
	–		—0—		0	–	13	71	–	87	–	U	1	25
Mouse	–	–	–	0	24	–	34	52	65	62	–	U	1	26
Bull	0	–	–	–	–	–	–	–	–	84	–	O, U	1	27
Man	–		—13—		21	–	56	–	–	–	–	C	4	28[h]
	–		—11—			—39—			—67—		–	C	4	29[h,i]
	–	0[j]	–	62[k]	–	–	–	–	–	–	–	C	4	30[h]

[a] Regions of epididymis given as quoted or ascertained from diagrams of regions sampled: *P*, proximal; *M*, middle; *D*, distal

[b] Site of insemination: *O*, oviducts; *U*, uterus; *C*, cervix; *V*, vagina; *P*, peritoneum

[c] Method of assessment: *1*, percentage of recovered eggs fertilised; *2*, percentage of inseminated females pregnant; *3*, average littler size; *4*, percentage of previously azoospermic infertile men, with sperm in their ejaculates following epididymovasostomy at various levels in the epididymis, subsequently fathering children
[d] Region termed "proximal cauda", but includes segments III and IV (ref. 31) or 3 and 4 (ref. 32) containing both single cells and cells stacked in rouleaux
[e] Region includes segments 1–4 (ref. 32) but only those with single cells were inseminated
[f] Region termed "proximal" contained caput (as refs. 32, 33) with single cells only
[g] Regions termed "distal" contained corpus and cauda (as refs. 32, 33) with cells in rouleaux
[h] All surgery performed after epididymal blockage
[i] Epididymal regions termed "high", "medium" and "low"
[j] Sperm taken from the first 7 mm of the epididymis
[k] Sperm taken from beyond the first 8 mm of the epididymis

References

1. Bedford JM (1963) Morphological changes in rabbit spermatozoa during passage through the epididymis. J Reprod Fert 5:169–177
2. Bedford JM (1966) Development of the fertilizing ability of spermatozoa in the epididymis of the rabbit. J exp Zool. 163:319–330
3. Lambiase JT and Amann RP (1973) Infertility of rabbit testicular spermatozoa collected in their native environment. Fert Steril 24:65–67
4. Fulka J and Koefoed-Johnson HH (1966) The influence of epididymal passage in rabbits on different spermatozoan characteristics including fertilizing capacity. Ann Rep Ster Res Inst, The Royal Vet and Agri Coll, Copenhagen. pp 213–225
5. Orgebin-Crist MC (1967) Maturation of spermatozoa in the rabbit epididymis: fertilizing ability and embryonic mortality in does inseminated with epididymal spermatozoa. Ann biol Anim Biochem Biophys 7:373–389
6. Igboeli G and Foote RH (1969) Maturation and aging changes in rabbit spermatozoa isolated by ligatures at different levels of the epididymis. Fert Steril 20:506–520
7. Nishikawa Y and Waide Y (1952) Studies on the maturation of spermatozoa. 1. Spermatozoa in the epididymis and their functional changes. Bull Natn Inst Agric Sci. Chiba, Japan, Series G 3:68–81
8. Paufler SK and Foote RH (1968) Morphology, motility and fertility of spermatozoa recovered from different areas of ligated rabbit epididymides. J Reprod Fert 17:125–137
9. Hunter RHF, Holtz W and Henfrey PJ (1976) Epididymal function in the boar in relation to the fertilizing ability of spermatozoa. J Reprod Fert 46:463–466
10. Holtz W and Smidt D (1976) The fertilizing capacity of epididymal spermatozoa in the pig. J Reprod Fert 46:227–229
11. Schellpfeffer DA and Hunter AG (1976) Specific proteins of the male reproductive tract. In: Regulatory Mechanisms of Male Reproductive Physiology. Eds CH Spilman, TJ Lobl and KT Jirton, Excerpta Medica, Amsterdam. pp 115–132
12. Voglmayr JK, White IG and Parks RP (1978) The fertilizing capacity of ram testicular spermatozoa, freshly collected and after storage in cauda epididymal fluid. Theriogenology 10:313–321
13. Fournier-Delpech S, Colas G and Courot M (1981) Observations sur les premiers clivages des oeufs intratubulaires de brebis après fécondation des spermatozoides épididymaires ou ejaculés. Comptes Rendus Acad Sci (Paris) 292:515–517
14. Fournier-Delpech S, Colas G, Courot M and Ortavant R (1977) Observations on the motility and fertilizing ability of ram epididymal spermatozoa. Ann biol anim Biochem Biophys 17:987–990
15. Fournier-Delpech S, Colas G, Courot M, Ortavant R and Brice G (1979) Epididymal sperm maturation in the ram: motility, fertilizing ability and embryonic survival after uterine insemination in the ewe. Ann biol anim Biochem Biophys 19:597–605
16. Young WC and Simeone FA (1930) Development and fate of spermatozoa in the epididymis and vas deferens in the guinea pig. Proc Soc exp Biol Med 27:838–841

17. Young WC (1931) A study of the function of the epididymis. III. Functional changes undergone by spermatozoa during their passage through the epididymis and vas deferens in the guinea pig. J exp Biol 8:151–162
18. Williamson BR, Shepherd BA and Martan J (1980) Fertility of spermatozoa from the excurrent ducts of the guinea pig. J Reprod Fert 59:515–517
19. Shilon M, Paz (Frenkel) G, Homonnai ZT and Schoenbaum M (1978) The effect of caffeine on guinea pig epididymal spermatozoa: motility and fertilizing capacity. Int J Androl 1:416–423
20. Blandau RJ and Rumery R (1964) The relationship of swimming movements of epididymal spermatozoa to their fertilizing capacity. Fert Steril 15:571–579
21. Dyson ALMB and Orgebin-Crist MC (1973) Effect of hypophysectomy, castration and androgen replacement upon the fertilizing ability of rat epididymal spermatozoa. Endocrinology 93:391–402
22. Paz (Frenkel) G, Kaplan R, Yedwab G, Homonnia ZT and Kraicer PF (1978) The effect of caffeine on rat epididymal spermatozoa: motility, metabolism and fertilizing capacity. Int J Androl 1:145–152
23. Cummins JM (1976) Effects of epididymal occlusion on sperm maturation in the hamster. J exp Zool 197:187–190
24. Horan AH and Bedford JM (1972) Development of the fertilizing ability of spermatozoa in the epididymis of the Syrian hamster. J Reprod Fert 30:417–423
25. González Echeverría F, Cuasnicú PS, Piazza A, Piñeiro L and Blaquier JA (1984) Addition of an androgen-free epididymal protein extract increases the ability of immature hamster spermatozoa to fertilize in vivo and in vitro. J Reprod Fert 71:432–437
26. Pavlok A (1974) Development of the penetrating activity of mouse spermatozoa in vivo and in vitro. J Reprod Fert 36:203–205
27. Amann RP and Griel LC (1974) Fertility of bovine spermatozoa from rete testis, cauda epididymidis and ejaculated semen. J Dairy Sci 57:212–219
28. Schmidt SS, Schoysman R and Stewart SH (1976) Surgical approaches to male infertility. In: Human Semen and Fertility Regulation in Men. Ed ESE Hafez, CV Mosby Co., St. Louis. pp 476–493
29. Schoysman R (1981) Epididymal causes of infertility pathogenesis and management. Progr Reprod Biol 8:102–113
30. Bedford JM (1980) Models for investigation of gamete function in the male and female reproductive tracts. In: Animal Models in Human Reproduction. Eds M Serio and L Martini, Raven Press, New York. pp 233–248
31. Fawcett DW and Hollenberg RD (1963) Changes in the acrosome of guinea-pig spermatozoa during passage through the epididymis. Z Zellforsh Mikr Anat 60:276–292
32. Hoffer AP and Greenberg JA (1978) The structure of the epididymis, efferent ductules and ductus deferens of the guinea pig: a light microscopial study. Anat Rec 190:659–678
33. Frenkel G, Peterson RN and Freund M (1973) Changes in the metabolism of guinea pig sperm from different segments of the epididymis. Proc Soc exp Biol Med 143:1231–1236

Table 2. The in-vitro fertilising ability[a] of spermatozoa from the testis, various regions of the epididymis[b] and vas deferens

Species	Testis	Caput			Corpus			Cauda					
		Prox	Mid	Dist	Prox	Mid	Dist	Prox	Mid	Dist	Vas	Con[c]	Ref
Mouse	–	——	0	——	——	6	——	——	83	——	19	CF	1
	–	–	–	8	3	–	3	23	59	67	–	CI	2
	–	–	–	18	50	–	70	65	78	67	–	ZF	2
Rabbit	8[d]	——	10	——	–	11	–	——	79	——	–	CI	3
Golden hamster	–	0	–	——5	——	——	77—	——	–	——90	—	ZFH	4
	–	–	–	–	–	–	44	–	–	86	–	CI	5
Chinese hamster	–	0	–	0	——	63	——	100	–	100	100	CI	6
Monkey	–	0	–	——20	——	——	71—	——	–	——79	—	ZFH	4
Man	–	——	0	——	——	0	——	——	27	——	–	ZFH	7, 8
	–	——	0	——	11	–	15	——	43	——	–	ZFH	9
	–	——	0	——	3	–	11	——	43	——	–	ZFH	10

[a] Expressed as percentage of eggs fertilised

[b] Regions of the epididymis as quoted or ascertained from diagrams of regions sampled: *Prox*, proximal; *Mid*, middle; *Dist*, distal

[c] Conditions: *CF*, cumulus-free; *CI*, cumulus intact; *ZF*, zona-free; *ZFH*, zona-free hamster ova

[d] Only testicular spermatozoa were stimulated with caffeine

References

1. Hoppe PC (1975) Fertilizing ability of mouse sperm from different epididymal regions and after washing and centrifugation. J exp Zool 192:219–222
2. Pavlok A (1974) Development of the penetrating activity of mouse spermatozoa in vivo and in vitro. J Reprod Fert 36:203–205
3. Brackett BG, Hall JL and Oh YK (1978) In-vitro fertilizing ability of testicular, epididymal, and ejaculated rabbit spermatozoa. Fert Steril 29:571–582
4. Moore HDM (1981) An assessment of the fertilizing ability of spermatozoa in the epididymis of the marmoset monkey (Callithrix jacchus). Int J Androl 4:321–330
5. González Echeverría F, Cuasnicú PS, Piazza A, Piñeiro L and Blaquier JA (1984) Addition of an androgen-free epididymal protein extract increases the ability of immature hamster spermatozoa to fertilize in vivo and in vitro. J Reprod Fert 71:432–437
6. Yanagimachi R, Kamiguchi Y, Mikamo K, Suzuki F and Yanagimachi H (1985) Maturation of spermatozoa in the epididymis of the Chinese hamster. Am J Anat 172:317–330
7. Hinrichsen MJ and Blaquier JA (1980) Evidence supporting the existence of sperm maturation in the human epididymis. J Reprod Fert 60:291–294
8. Hinrichsen MJ and Blaquier JA (1982) Aquisition of fertilizing capacity of spermatozoa in the human epididymis. In: Instrumental Insemination. Eds ESE Hafez and K Semm. Martinus-Nijhof, The Hague. pp 63–65
9. Moore HDM, Hartmann TD and Pryor JP (1983) Development of the oocyte-penetrating capacity of spermatozoa in the human epididymis. Int J Androl 6:310–318
10. Sutherland PD, Matson PL, Moore HDM, Goswamy R, Parsons JH, Vaid P and Pryor JP (1985) Clinical evaluation of the heterolgous oocyte penetration (HOP) test. Br J Urol 57:233–236

Table 3. Effects on immature spermatozoa of incubation in epididymal secretions

Species	Spermatozoa	Secretion	Time	Temp	Effect	Ref
Rabbit	Testicular	Cauda fluid	1–4 h	room	No effect on fertilisation	1, 2
	Caput	$PGF_{2\alpha}$	30 m	37°	No effect on fertilisation	3
Boar	Testicular	Corpus fluid	40–160 m	34°	No effect on fertilisation	4
	Testicular	Cauda fluid	15 m	37°	No effect on fertilisation	5
	Caput	Cauda fluid	15 m	–	No promotion of egg binding	6
	Caput	Cauda fluid	4 h	36°	No effect on fertilisation	7
	Caput	Cauda fluid	15 m	37°	Increase in fertilisation in-vivo	5, 8
	Proximal corpus	Cauda fluid	–	–	Reduction in fertilisation in-vivo	9
Ram	Testicular	Cauda fluid	4–11 d	4°	No effect on fertilisation	10
	Testicular	Cauda fluid	14 d	2–4°	Altered metabolism	11
	Testicular	Cauda fluid	1–6 h	34°	Uptake of specific glycoproteins	12, 13
	Testicular	Steroids	10–12 d	4°	No effect on fertilisation	14
Bull	Caput	Cauda fluid	15 m	37°	Promotion of forward motility	15, 16
Rat	Testicular	Specific sialoprotein	30 m	34°	Promotion of egg binding	17
Hamster	Caput	Cauda fluid	5 m	37°	Promotion of forward motility and fertilisation in-vivo	18, 19
	Distal	Specific glycoprotein	5 m	37°	Promotion of egg binding	20
			5 m, 5 h	37°	promotion of fertilisation in-vitro and in-vivo	21

References

1. Cooper TG and Orgebin-Crist MC (1975) The effect of epididymal and testicular fluids on the fertilising capacity of testicular and epididymal spermatozoa. Andrologia 7:85–93
2. Cooper TG and Orgebin-Crist MC (1977) Effect of aging on the fertilizing capacity of testicular spermatozoa from the rabbit. Biol Reprod 16:258–266
3. Sorgen CD and Glass RH (1972) Lack of effect of prostaglandin $F_{2\alpha}$ on the fertilizing ability of rabbit sperm. Prostaglandins 1:229–233

4. Dacheux JL and Paquignon M (1980) Effects of caffeine on ram and boar spermatozoa; influence of their stage of maturation and the medium; initiation of progressive motility of testicular spermatozoa. In: Testicular Development, Structure, and Function. Eds A Steinberger and E Steinberger, Raven Press, New York. pp 513–522.
5. Dacheux JL and Paquignon M (1983) Influence of initiation of forward motility on the fertilizing ability of immature boar spermatozoa in in vivo homologous and in vitro heterologous systems of insemination. In: The Sperm Cell. Ed J André, Martinus-Nijhoff, The Hague. pp 99–102
6. Peterson RN, Russell LD and Hunt WP (1984) Evidence for specific binding of uncapacitated boar spermatozoa to porcine zonae pellucidae in vitro. J exp Zool 231:137–147
7. Holtz W and Smidt D (1976) The fertilizing capacity of epididymal spermatozoa in the pig. J Reprod Fert 46:227–229
8. Dacheux JL and Paquignon M (1981) Physiologie épididymaire et maturation des spermatozoides. Actualités Gynecologiques, Series 12, Masson Press, Paris. pp 201–213
9. Hunter RHF, Holtz W and Hermann H (1978) Stabilizing role of epididymal plasma in relation to the capacitation tiome of boar spermatozoa. Anim Reprod Sci 1:161–166
10. Voglmayr JK, White IG and Parks RP (1978) The fertilizing capacity of ram testicular spermatozoa, freshly collected and after storage in cauda epididymal fluid. Theriogenology 10:313–321
11. Voglmayr JK, Musto NA, Saksena SK, Brown-Woodman PDC, Marley PB and White IG (1977) Characteristics of semen collected from the cauda epididymidis of conscious rams. J Reprod Fert 49:245–251
12. Voglmayr JK, Fairbanks, G, Jackowitz MA and Colella JR (1980) Post-testicular developmental changes in the ram sperm cell surface and their relationship to luminal fluid proteins of the reproductive tract. Biol Reprod 22:655–667
13. Voglmayr JK, Fairbanks G, Vespa DB and Colella JR (1982) Studies on the mechanisms of surface modifications in ram spermatozoa during the final stages of differentiation. Biol Reprod 26:483–500
14. Voglmayr JK and Gandhi JS (1978) Survival of ram testicular spermatozoa in vitro: effects of glucose, glucose metabolites, rete testis fluid-proteins, selected androgens and phospholipids. Theriogenology 9:463–478
15. Acott TS, Katz DF and Hoskins DO (1983) Movement characteristics of bovine epididymal spermatozoa. Effects of forward motility protein and epididymal maturation. Biol Reprod 29:389–399
16. Acott TS and Hoskins DD (1983) Cinematographic analysis of bovine epididymal sperm motility: epididymal maturation and forward motility protein. J Submicroscop Cytol 15:77–82
17. Orgebin-Crist MC and Fournier-Delpech S (1982) Sperm-egg interactions. Evidence for maturational changes during epididymal transit. J Androl 3:429–433
18. Kann ML and Serres C (1980) Development and initiation of sperm motility in the hamster epididymis. Reprod Nutr Develop 20:1739–1749
19. Kann ML and Raynaud F (1982) In vivo fertilization after initiation of sperm motility in the hamster epididymis. Reprod Nutr Develop 22:455–463
20. Cuasnicú PS, González Echeverría F, Piazza, Piñeiro L and Blaquier JA (1984) Epididymal proteins that mimic the androgenic effect on zona pellucida recognition by immature hamster spermatozoa. J Reprod Fert 71:427–431
21. González Echeverría F, Cuasnicú PS, Piazza A, Piñeiro L and Blaquier JA (1984) Addition of an androgen-free epididymal protein extract increases the ability of immature hamster spermatozoa to fertilize in vivo and in vitro. J Reprod Fert 71:432–437

Table 4. Objective measurements of motility[a] of spermatozoa released from various regions of the epididymis[b] and vas deferens

Species	Initial segment	Caput flexure	Caput		Corpus			Cauda			Vas	Meth[c]	Ref
			Prox	Dist	Prox	Mid	Dist	Prox	Mid	Dist			
Ram	2	4	13	46	54	63	72	–	77	84	92	LLS	1, 2
Boar	3	5	10	33	60	74	75	–	81	82	77	LLS	1, 2
Bull	5	6	19	23	36	56	47	–	61	84	83	LLS	1
Goat	6	8	22	48	40	54	73	–	77	83	83	LLS	1
Horse	–	9	17	27	55	68	77	–	84	83	90	LLS	1
Rat	–	18	–	53	–	45	52	52	–	63	54	ACF	3
	–	–	ca 9	ca 9		ca 12			ca 18		–	CMA	4

[a] Values taken after 10 min incubation in-vitro for LLS and ACF; mm travelled in 30 min for CMA; *ca*, circa: values taken from graphs

[b] Regions of epididymis as quoted or ascertained from diagrams of the regions sampled: *Prox*, proximal; *Mid*, middle; *Dist*, distal

[c] Method: *LLS*, Laser Light Scattering (see ref. 5); *ACF*, Area Change Frequency (see ref. 6); *CMA*, Capillary Migration Assay (see ref. 4)

References

1. Paquignon M, Dacheux JL, Jeulin C and Fauquenot AM (1983) Laser light scattering study of spermatozoa motility of domestic animals. In: The Sperm Cell. Ed J André, Martinus-Nijhoff, The Hague. pp 332–335
2. Dacheux JL, Paquignon M and Combarnous Y (1983) Head-to-head agglutination of ram and boar epididymal spermatozoa and evidence for an epididymal antagglutinin. J Reprod Fert 67:181–189
3. Hinton BT, Dott HM and Setchell BP (1979) Measurement of the motility of rat spermatozoa collected by micropuncture from the testis and from different regions along the epididymis. J Reprod Fert 55:167–172
4. Turner TT and Giles RD (1981) The effects of carnitine, glycerylphosphorylcholine, caffeine, and egg yolk on the motility of rat epididymal spermatozoa. Gamete Res 4:283–295
5. Jouannet P (1979) Measurement of human sperm motility based on an optical Doppler effect. In: The Spermatozoon. Eds DW Fawcett and JM Bedford, Urban and Schwarzenberg, Baltimore. pp 427–430
6. Dott HM and Foster GCA (1979) The estimation of sperm motility in semen, on a membrane slide, by measuring the area change frequency with an image analysing computer. J Reprod Fert 55:161–166

Table 5. Osmotic pressure of fluids in the lumen of various regions of the epididymis [a]

Species	Units	Caput	Corpus	Cauda		Ref
				Prox	Dist	
Hamster	mOsmol/kg [b]	417.0	408.4	385.5	339.7	1
		ca 390	ca 420	–	ca 325	2
		ca 376	ca 390	–	ca 335	2
Rat	mOsmol/kg	315	340	–	335	3
Ram	T_f	0.61	–		0.64	4
	mOsmol/kg [c]	330	–		340	

[a] Regions of the epididymis as quoted or ascertained from diagrams of regions sampled: *Prox*, proximal; *Dist*, distal

[b] *ca*, circa: values taken from graphs

[c] Calculated from ref. 4

References

1. Johnson AL and Howards SS (1976) Hyperosmolarity in intratubular fluids from hamster testis and epididymis: a micropuncture study. Science 195:492–493
2. D'Addario DA, Turner TT and Howards SS (1980) Effect of vasectomy on the osmolarity of hamster testicular and epididymal intraluminal fluid. J Androl 1:167–170
3. Levine N and Marsh DJ (1971) Micropuncture studies of the electrochemical aspects of fluid and electrolyte transport in individual seminiferous tubules, the epididymis and vas deferens. J Physiol 213:557–570
4. White IG (1978) Accessory sex organs and fluids of the male reproductive tract. In: Animal Models for Research on Contraception and Fertility. Ed NJ Alexander, Harper and Row, Hagerstown. pp 105–123

Table 6. Mitochondrial function[a] in spermatozoa from various regions of the rat epididymis

Buffer[b]	Caput		Corpus		Proximal cauda		Distal cauda	
	aRTF	aCEP	aRTF	aCEP	aRTF	aCEP	aRTF	aCEP
Endogenous	16.0 2.5	29.0 8.0	11.3	21.9 2.4	12.3 2.2	19.5 2.3	9.5 1.5	20.1 2.4
αGP[c]	40.1 9.6	45.1 7.7	58.2	57.5 12.6	33.2 7.2	27.6 3.6	35.3 6.8	39.0 8.1
Oligo[d]	14.4 5.0	4.3 1.3	0.3	3.8 1.3	11.7 3.7	3.4 1.5	12.8 5.5	5.5 1.9
Sperm conc $\times 10^6$/ml	15.6 0.8	14.7 2.2	19.4	16.2 1.3	22.8 3.3	19.1 1.6	32.5 3.7	18.8 1.4
No. exps	5	5	2	6	5	6	9	7

[a] Oxygen uptake: nmol-$O_2/10^8$ cells/min; mean $\pm$ SEM
[b] *aRTF*, artificial Rete Testis Fluid; 14 mM-KCl, 130 mM-NaCl, albumin 1 mg/ml, 33.3 mM-MOPS/13 mM-NaOH, pH 6.9; *aCEP*, artificial Cauda Epididymal Plasma: 50 mM-KCl, 7 mM-NaCl, 88 mM-cholineCl, albumin 1 mg/ml, 33.3 mMMOPS/13 mM-NaOH, pH 6.9
[c] 4 mM-rac-glycerol 1-phosphate
[d] Oligomycin, inhibitor of mitochondrial ATPase 2.3 μM

Reference

Cooper TG, unpublished observations. Sperm were collected by microcannulation of the duct, diluted in aRTF or aCEP and washed by centrifugation (500 × for 5 min then 1200 × g for 10 min) through Ficoll (7.5% w/v) made up in aRTF or aCEP respectively before respiratory measurements were made on 0.5 ml aliquants

Table 7. Effect of capacitation time in-vivo on the fertilising capacity of immature rabbit spermatozoa

Source of sperm	No. male donors	No. does inseminated	Site[a]	Time[b]	No. ova fertilised / No. ova recovered	No. ova with sperm[c]
Rete testis[d]	1	3	O	24	0/24	0
	1	3	O	14	0/41	0
	1	2	O	4	0/4	2 (1)
	1	1	U	4	0/4	0
Caput epididymidis	1	2	U	4	0/16	0
	1	2	U	4	0/9	0
	1	2	U	4	0/13	0

[a] *O*, Oviduct, 8.2–13.5 × 10^6 sperm inseminated; *U*, Uterus, 3.8–4.6 × 10^6 sperm inseminated
[b] Hours in female tract before ovulation
[c] No. of sperm per egg in parenthesis
[d] Rete testis fluid collected 2 days after ligation of the efferent ducts

Reference

Cooper TG and Orgebin-Crist MC, unpublished observations

Table 8. Binding to eggs in-vivo and in-vitro by spermatozoa from the testis and various regions of the epididymis[a]

	Testis	Caput			Corpus				Cauda			Method[b]	Eggs[c]	Ref
		Prox	Mid	Dist	Prox	Mid 1	Mid 2	Dist	Prox	Mid	Dist			
(a) % Eggs with surface sperm[d]														
Boar														
	–	0	0	7	40	82	100	100	100	100	100	1	HZF	1
Man														
	–		100			100				100		1	HZF	1
Ram[e]														
	0		50			100				100		1	RCF	4
	0	0	–	–	–	31		100	–	–	100	1	RCF	5
Hamster														
	–		0		39	–	–	83		100		1	HCF	6
Rat[f]														
	0	–	–	–	–	–	–	–	–	–	100	1	RCF	7
	–	ca 4	–	–	ca 30	ca 45		–	–	–	ca 50	1	RCI	8
(b) *Number of sperm/egg*														
Boar														
	–	0	0	1–10	1–100	1–100	1–>100	1–>100	1–>100	1–>100	1–>100	1	HZF	1
	–	–	–	–	12	–	–	53	–	–	–	2[g]	PCI	9
	–		<10			50				96		1[h]	PCF	10
Man														
	–		0–1		5–10	–	–	5–10		10–20		1	HZF	11
	–		1–10			1–10				1–10		1	HZF	2, 3
Monkey														
	–	–	–	1.0				1.2		–	1.4	1	HZF	12

Hamster														
	–	–	–	1.0				1.8		–	2.4	1	HZF	12
	–		0		1.95	–	–	> 20		> 20		1	HCI	6
Mouse														
	–		ca 1				ca 5			ca 60		1	MCF	13
	–		ca 1				ca 5			ca 35		1	MZF	13
Ram														
	0	0	–	–	–		5–25		5–25	> 25		1	RCF	5

[a] Regions of epididymis given as quoted or ascertained from diagrams of regions sampled: *Prox*, proximal; *Mid*, middle; *Dist*, distal
[b] Method: *1*, in-vitro; *2*, in-vivo
[c] Eggs: *CI*, cumulus intact; *CF*, cumulus-free; *ZF*, zona-free; *H*, hamster; *R*, rat; *P*, pig; *M*, mouse
[d] *ca*, circa: approximate values taken from graphs
[e] % eggs with > 10 sperm/egg
[f] % eggs with 1 sperm/egg at 4.5 h incubation
[g] At 4–4.5 h post-insemination
[h] At 30 min, caput and corpus: 2.10^6/ml; cauda: 5.10^6/ml

References

1. Dacheux JL and Paquignon M (1983) Influence of initiation of forward motility on the fertilizing ability of immature boar spermatozoa in in vivo homolgous and in vitro heterolgous systems of insemination. In: The Sperm Cell. Ed J André, Martinus-Nijhoff, The Hague. pp 99–102
2. Hinrichsen MJ and Blaquier JA (1980) Evidence supporting the existence of sperm maturation in the human epididymis. J Reprod Fert 60:291–294
3. Hinrichsen MJ and Blaquier JA (1982) Aquisition of fertilizing capacity of spermatozoa in the human epididymis. In: Instrumental Insemination. Eds ESE Hafez and K Semm. Martinus-Nijhoff, The Hague. pp 63–65
4. Fournier-Delpech S, Hamamah S, Colas G and Courot M (1983) Aquisition of zona-binding structures by ram spermatozoa during epididymal passage. In: The Sperm Cell. Ed J André, Martinus-Nijhoff, The Hague. pp 103–106
5. Fournier-Delpech S, Courtens JL, Pisselet CL, DeLaleu B and Courot M (1982) Aquisition of zona-binding by ram spermatozoa during epididymal passage, as revealed by interaction with rat oocytes. Gamete Res 5:403–408
6. Cuasnicú PS, González Echeverría F, Piazza A and Blaquier JA (1984) Addition of androgens to cultured hamster epididymis increases zona recognition by immature spermatozoa. J Reprod Fert 70:541–547
7. Orgebin-Crist MC and Fournier-Delpech S (1982) Sperm-egg interactions. Evidence for maturational changes during epididymal transit. J Androl 3:429–433
8. Fournier-Delpech S, Hamamah S, Tananis-Anthony C, Courot M and Orgebin-Crist MC (1984) Hormonal regulation of zona-binding ability and fertilizing ability of rat epididymal spermatozoa. Gamete Res 9:21–30
9. Hunter RHF, Holtz W and Henfrey PJ (1976) Epididymal function in the boar in relation to the fertilizing ability of spermatozoa. J Reprod Fert 46:463–466
10. Peterson RN, Russell LD and Hunt WP (1984) Evidence for specific binding of uncapacitated boar spermatozoa to porcine zonae pellucidae in vitro. J exp Zool 231:137–147
11. Moore HDM, Hartmann TD and Pryor JP (1983) Development of the oocyte-penetrating capacity of spermatozoa in the human epididymis. Int J Androl 6:310–318
12. Moore HDM (1981) An assessment of the fertilizing ability of spermatozoa in the epididymis of the marmoset monkey (*Callithrix jacchus*). Int J Androl 4:321–330
13. Saling PM (1982) Development of the ability to bind to zonae pellucidae during epididymal maturation: reversible immobilization of mouse spermatozoa by lanthanum. Biol Reprod 26:429–436

Table 9. The percentage of freely motile spermatozoa from various regions of the mouse epididymis undergoing hyperactivated motility[a] in-vitro

Incubation time (h)	Distal corpus (n = 5)	Proximal cauda (n = 6)	Distal cauda (n = 10)
0	29.0 ± 5.2	18.3 ± 6.7	4.5 ± 2.1
1	35.6 ± 4.0	15.5 ± 8.5	12.3 ± 4.6
2	19.4 ± 10.6	25.8 ± 8.1	13.7 ± 4.9
3	46.0 ± 5.0	36.7 ± 9.2	17.3 ± 3.6
4	55.8 ± 5.5	51.7 ± 3.8	31.3 ± 3.9
5	47.8 ± 8.6	45.3 ± 13.0	37.5 ± 2.0
6	46.6 ± 13.9	63.8 ± 11.9	40.8 ± 5.4

[a] Hyperactivated motility was detected by stroboscopic illumination during incubation in low ionic strength medium containing 2 mM caffeine and 32 mg/ml bovine serum albumin (see Cooper and Woolley, 1982). Cooper TG, unpublished observations

Reference

Cooper TG and Woolley DM (1982) Stroboscopic illumination for the assessment of hyperactivated motility of mouse spermatozoa. J exp Zool 223:35–42

Table 10. Permeability of the epithelium in various regions of the epididymis[a]

Region	Compound	% Plasma (% Bath)	Time (h)	Method[b]	Ref
Rat					
Caput	^{3}H-inositol	450	3	MP	1
	^{3}H-carnitine	120	12	MP	2[c]
		40	12	MP	2[d]
	^{3}H-water	104	2	MP	3
	^{14}C-urea	97	2	MP	3
	^{3}H-α-aminoisobutyrate	75	3	MP	1
	^{3}H-3-*O*-methyl glucose	35	3	MP	1, 4
	^{3}H-L-glucose	5	3	MP	1
	^{3}H-inulin	3	2	MP	3
		1	1	MP	5
	^{125}I-bovine serum albumin	2	2	MP	3
	^{3}H-polyethylene glycol	1	2	MP	3
Corpus	^{3}H-carnitine	80	4	LP	6
		50	3	LP/V	7
	^{3}H-inositol	18–23	3	LP/V	8
	^{3}H-D-glucose	19	3	LP/V	8
	^{3}H-inulin	3	3	LP/V	8
Cauda	^{3}H-water	86–96	1–3	LP	9
	^{14}C-urea	60–70	2	MP	10
	^{3}H-carnitine	50	3	LP/V	7
		40	4	LP	6[c]
		20	12	MP	2
		20	4	LP	6[d]
	^{3}H-dehydroepiandrosterone	40–60	1–2	LP	11[e]
	^{3}H-androstenedione	36–61	1–2	LP	11
	^{3}H-dihydrotestosterone	30–48	1–2	LP	11
	^{3}H-testosterone	41	1–2	LP	12[e,f]
	^{3}H-progesterone	38	1–2	LP	11
	^{3}H-pregnenolone	33	1–2	LP	11
	^{3}H-inositol	61	3	LP	27
		30	3	MP	1
		16–29	4	LP/V	8
		26	1–3	LP	13
	^{3}H-D-glucose	17–20	3	LP/V	8
		1–7	1–3	LP	9
	^{3}H-3-*O*-methylglucose	1–7	1–3	LP	9
	^{14}C-cyproterone acetate	14	1–2	LP	11
	glycerol	10	1–3	LP	14
	^{3}H-cholesterol	0	–	LP	11
	^{14}C-dextran	0	–	LP	9
	^{3}H-inulin	4–6	3	LP/V	8
		3	1	MP	5, 26
		0	1–3	LP	9
	Immunoglobulin (IgG)	0	2	LP	15

Table 10 (continued)

Region	Compound	% Plasma (% Bath)	Time (h)	Method[b]	Ref
Hamster					
Caput	^{3}H-inulin	8	1.5	MP	16
Corpus	^{3}H-inulin	6.5	1.5	MP	16
Cauda	^{14}C-urea	95	1–3	MP	17, 18
		90	1–3	MP	19
	^{3}H-water	90	1–2	MP	19
		88	1–2	MP	18
	^{3}H-progesterone	33	2	MP	20, 21
	^{3}H-testosterone	31	2	MP	20, 21
	^{3}H-dihydrotestosterone	20	2	MP	20, 21
	^{3}H-3-*O*-methyl-D-glucose	7	2–4	MP	22, 25
	^{3}H-inulin	3.9	1.5	MP	16
		< 1	3	MP	10, 19
Rabbit					
Cauda	Immunoglobulin (IgG)	0.4	6 days	MP	23, 24

[a] Concentration of radioactivity in the lumen expressed as a percentage of that in blood plasma (MP, LP) or in the bath (LP/V)
[b] *MP*, micropuncture; *LP*, luminal perfusion; *LP/V*, luminal perfusion in-vitro
[c] Proximal region
[d] Distal region
[e] BSA (38 mg/ml) in lumen
[f] Slow perfusion rate (ca. 1 µl/min)

References

1. Hinton BT and Howards SS (1982) Rat testis and epididymis can transport (^{3}H)3-*O*-methylglucose, (^{3}H)inositol and (^{3}H)α-aminoisobutyric acid across its epithelia in vivo. Biol Reprod 27:1181–1189
2. Hinton BT and Setchell BP (1980) Concentration and uptake of carnitine in the rat epididymis. A micropuncture study. In: Carnitine Biosynthesis, Metabolism, and Function. Eds RA Frenkel and JD McGarry, Acad Press, New York. pp 237–250
3. Hinton BT and Howards SS (1981) Permeability characteristics of the epithelium in the rat caput epididymidis. J Reprod Fert 63:95–99
4. Hinton BT, Hernandez H and Howards SS (1983) The antifertility agents α-chlorohydrin, 5-thio-D-glucose, and 6-chloro-6-deoxy-D-glucose interfere with sugar transport across the epithelium of the rat caput epididymidis. J Androl 4:216–221
5. Hoffer AP and Hinton BT (1984) Morphological evidence for a blood-epididymis barrier and the effects of gossypol on its integrity. Biol Reprod 30:991–1004
6. Yeung CH, Cooper TG and Waites GMH (1980) Carnitine transport into the perfused epididymis of the rat: regional differences, stereospecificity, stimulation by choline and effects of other luminal compounds. Biol Reprod 23:294–304
7. Cooper TG, Gudermann TW, Yeung CH (1986) Characteristics of the transport of carnitine into the cauda epididymidis of the rat as ascertained by luminal perfusion in-vivo. Int J Androl (to be published)
8. Cooper TG, Yeung CH, Lui W and Yang CZ (1985) Luminal secretion of myo-inositol by the rat epididymis perfused in vitro. J Reprod Fert 74:135–144
9. Cooper TG and Waites GMH (1979) Investigation by luminal perfusion of the transfer of compounds into the epididymis of the anaesthetized rat. J Reprod Fert 56:159–164
10. Turner TT, D'Addario DA and Howards SS (1979) Effects of vasectomy on the blood-testis barrier of the hamster. J Reprod Fert 55:323–328
11. Cooper TG (1980) The general importance of proteins and other factors in the transfer of steroids into the rat epididymis. Int J Androl 3:333–348
12. Cooper TG and Waites GMH (1979) Factors affecting the entry of testosterone into the lumen of the cauda epididymidis of the anaesthetized rat. J Reprod Fert 56:165–174

13. Cooper TG (1982) Secretion of inositol and glucose by the perfused rat cauda epididymidis. J Reprod Fert 64:373–379
14. Cooper TG and Brooks DE (1981) Entry of glycerol into the rat epididymis and its utilization by epididymal spermatozoa. J Reprod Fert 61:163–169
15. Wong PYD, Tsang AYF and Lau HK (1983) Restricted entry of an anti-rat epididymal protein IgG into the rat epididymis. Int J Androl 6:275–282
16. Turner TT, D'Addario DA and Howards SS (1981) The blood epididymal barrier to (^{3}H)-inulin in intact and vasectomized hamsters. Invest Urol 19:89–91
17. Turner TT and Howards SS (1978) The effects of vasectomy in the movement of ^{14}C-urea in the seminiferous tubule of the hamster. Invest Urol 16:237–239
18. Turner TT, Hartmann PK and Howards SS (1979) Urea in the seminiferous tubule: evidence for active transport. Biol Reprod 20:511–515
19. Howards SS, Jessee SJ and Johnson AL (1976) Micropuncture studies of the blood seminiferous tubule barrier. Biol Reprod 14:264–269
20. Turner TT, Cochran RC and Howards SS (1981) Transfer of steroids across the hamster blood testis and blood epididymal barriers. Biol Reprod 25:342–348
21. Turner TT (1979) On the epididymis and its function. Invest Urol 16:311–321
22. Turner TT, D'Addario DA and Howards SS (1983) The transepithelial movement of ^{3}H-3-*O*-methyl-D-glucose in the hamster seminiferous and cauda epididymal tubules. Fert Steril 40:530–535
23. Bedford JM, Weininger RB, Fisher S and Rifkin J (1982) Specific IgG levels appearing in the cauda epididymidis of male rabbits. Int J Androl Suppl 5:48–52
24. Weininger RB, Fisher S, Rifkin J and Bedford JM (1982) Experimental studies on the passage of specific IgG to the lumen of the rabbit epididymis. J Reprod Fert 66:251–258
25. Turner TT, D'Addario DA and Howards SS (1980) (^{3}H)3-*O*-methyl-D-glucose transport from blood into the lumen of the seminiferous and epididymal tubules in intact and vasectomized hamsters. J Reprod Fert 60:285–289
26. Turner TT, Giles RG and Howards SS (1981) Effect of oestradiol valerate on the rat blood-testis and blood epididymal barriers to (^{3}H)inulin. J Reprod Fert 63:355–358
27. Sujarit S, Chaturpanich G, Pholpramool C (1985) Evidence for blood myo-inositol as a source of the epididymal secretion in the perfused cauda epididymidis of the rat. Andrologia 17: 321–326

Table 11. Concentrations of steroids in the lumen of various regions of the epididymis[a]

<table>
<tr><th rowspan="2">Species</th><th rowspan="2">Steroid[b]</th><th rowspan="2">Units</th><th rowspan="2">Caput</th><th colspan="2">Corpus</th><th colspan="2">Cauda</th><th rowspan="2">Ref</th></tr>
<tr><th>Prox</th><th>Dist</th><th>Prox</th><th>Dist</th></tr>
<tr><td>Stallion</td><td>T</td><td>ng/ml</td><td>200</td><td>178</td><td>138</td><td>128</td><td>90</td><td>1</td></tr>
<tr><td>Hamster</td><td>T</td><td>ng/ml</td><td>62.3</td><td>329.1</td><td>131.9</td><td>39.3</td><td>19.6</td><td>2</td></tr>
<tr><td>Ram</td><td>T</td><td>ng/ml</td><td>24.8</td><td>–</td><td>–</td><td colspan="2">10.8</td><td>3</td></tr>
<tr><td></td><td>DHEA</td><td>ng/ml</td><td>33.8</td><td>–</td><td>–</td><td colspan="2">25.8</td><td>3</td></tr>
<tr><td>Rat</td><td>T</td><td>ng/ml</td><td colspan="2">5</td><td colspan="3">2</td><td>4[c]</td></tr>
<tr><td></td><td>T</td><td>ng/ml</td><td colspan="2">5.4</td><td>–</td><td colspan="2">7.7</td><td>5</td></tr>
<tr><td></td><td>T</td><td>ng/mg prot</td><td>0.13</td><td>–</td><td>–</td><td colspan="2">0.06</td><td>6[d]</td></tr>
<tr><td></td><td>DHT</td><td>ng/ml</td><td colspan="2">40</td><td colspan="3">5</td><td>4[c]</td></tr>
<tr><td></td><td>DHT</td><td>ng/ml</td><td colspan="2">58.7</td><td>–</td><td colspan="2">4.4</td><td>5</td></tr>
<tr><td></td><td>DHT</td><td>ng/mg prot</td><td>0.51</td><td>–</td><td>–</td><td colspan="2">0.18</td><td>6[d]</td></tr>
<tr><td>Man</td><td>T</td><td>ng/ml</td><td>–</td><td>–</td><td>–</td><td colspan="2">0.08</td><td>7</td></tr>
</table>

[a] Regions of the epididymis given as quoted or ascertained from diagrams of regions sampled: *Prox*, proximal; *Dist*, distal
[b] Abbrevations: *T*, testosterone; *DHEA*, dehydroepiandrosterone; *DHT*, 5α-dihydrotestosterone
[c] Epididymis divided into "proximal" and "distal" portions
[d] Both fluid and spermatozoa extracted for assay

References

1. Ganjam VK, Whelan HA and Brackett BG (1980) Steroid and protein components within the stallion epididymis. J Androl 1:84 (abst)
2. Whelan HA, Ganjam VK and Brackett BG (1980) Experimental alteration of golden hamster epididymal contents. Biol Reprod Suppl 1:41 A (abst)
3. White IG and Hudson B (1968) The testosterone and dehydroepiandrosterone concentrations in fluids of the mammalian male reproductive tract. J Endocr 41:291–292
4. Vreeburg JTM (1975) Distribution of testosterone and 5-α dihydrotestosterone in rat epididymis and their concentrations in efferent duct fluid. J Endocr 67:203–210
5. Turner TT, Jones CE, Howards SS, Ewing LL, Zegeye B and Gunsalas GL (1984) On the androgen microenvironment of maturing spermatozoa. Endocrinology 115:1925–1932
6. Pujol A, Bayard F, Louvet JP and Boulard C (1976) Testosterone and dihydrotestosterone concentrations in plasma, epididymal tissues, and seminal fluid of adult rats. Endocrinology 98:111–113
7. Adamopoulos DA, Lawrence DM, Vassilopoulos R, Kontogeorgos L, Ntounis A and Swyer GIM (1979) Testosterone concentration in different functional compartments of the male reproductive tract. Arch Androl 3:13–18

Table 12. Concentrations (mM) of metal ions in luminal fluid and spermatozoa from the testis and various regions of the epididymis[a]

		Tissue	Testis	Caput	Corpus	Cauda			
						Prox	Dist		
Bull	Na^+	Sperm	–	75	–	–	21.5	20.9	
		Fluid	–	89.3	–	–	36.0	37.8	
	K^+	Sperm	–	75	–	–	48.8	36.2	
		Fluid	–	36.2	–	–	26.3	21.5	
	Ca^{2+}	Sperm	–	6	–	–	3.15	3.98	
		Fluid	–	0.95	–	–	1.28	2.30	
	Mg^{2+}	Sperm	–	7.4	–	–	7.4	6.9	
		Fluid	–	3.15	–	–	1.32	1.30	
	Refs			1			1	2	
Ram	Na^+	Sperm	69	–	–	–	21	21.3	
		Fluid	128	–	–	–	40	40	
	K^+	Sperm	57	–	49.8	44.3	64	63.3	44.5
		Fluid	13	–	49.8	20.0	24	24.1	22.5
	Ca^{2+}	Sperm	5.1	–	–	–	3.6	1.8	
		Fluid	2.6	–	–	–	0.95	0.95	
	Mg^{2+}	Sperm	9.7	–	–	–	8.45	8.5	
		Fluid	0.7	–	–	–	1.3	1.3	
	Refs		3		4	4	3	2	4
Boar	Na^+	Sperm	–	–	44.3	–	39.3		
		Fluid	–	–	64.2	–	51.0		
	K^+	Sperm	–	–	64.1	–	78.1		
		Fluid	–	–	46.4	–	54.2		
	Refs				5		5		

[a] Regions of epididymis as quoted or ascertained from diagrams of regions sampled: *Prox*, proximal; *Dist*, distal

References

1. Drevius LO (1972) The permeability of bull spermatozoa to water, polyhydric alcohols and univalent anions and the effects of the anions upon the kinetic activity of spermatozoa and sperm models. J Reprod Fert 28:41–54
2. Quinn PJ, White IG and Wirrick BR (1965) Studies on the distribution of the major cations in semen and male accessory secretions. J Reprod Fert 10:379–388
3. Setchell B, Scott TW, Voglmayr JK and Waites GMH (1969) Characteristics of testicular spermatozoa and the fluids which transport them into the epididymis. Biol Reprod Suppl 1:40–66
4. Crabo BG and Hunter AG (1975) Sperm maturation and epididymal function. In: Control of Male Fertility. Eds JJ Sciarra, C Markland and JJ Speidel, Harper and Row, London. pp 2–23
5. Zimmerman KJ, Crabo BG, Moore R, Weisberg S, Deibel FC and Graham EF (1979) Movements of sodium and postassium into epididymal boar spermatozoa. Biol Reprod 21:173–179

Table 13. Concentrations of mono- and di-valent ions (mM) in luminal fluid from various regions of the epididymis [a,b]

	Caput			Corpus			Cauda			Refs
	Prox	Mid	Dist	Prox	Mid	Dist	Prox	Mid	Dist	
Sodium										
Bull	113	121	77	–	78	–	–	39	18	6, 12
	–	102	–	–	71	–	–	36	–	6, 10, 11, 17, 18
	–	97	–	–	48	–	–	33	–	6, 10, 11, 12
	–	79	–	–	–	–	–	31	–	9, 12
	–	57	–	–	–	–	–	–	–	10, 17
Rat	–	120	–	–	105	–	26	–	74	1, 3, 4
	–	112	–	–	94	–	–	–	55	2, 13, 18
	–	106	–	–	66	–	–	–	40	3, 18, 19
	–	104	–	–	62	–	–	–	40	1, 13, 18
	–	103	–	–	58	–	–	–	37	2, 13, 18
	–	102	–	–	29	–	–	–	23	4, 20
	–	96	–	–	–	–	–	–	20	2, 18
	–	58	–	–	–	–	–	–	15	4
Boar	76	74	68	–	70	–	39	–	51	6, 7, 17
	–	–	ca 80	–	64	–	ca 35	–	51	7, 14, 17
	–	–	–	–	–	–	25	–	ca 30	14
	–	–	–	–	–	–	24	–	–	6
	–	–	–	–	–	–	23	–	–	16
Ram	–	90	–	–	–	–	–	57	–	8, 10, 17
	–	–	–	–	–	–	–	14	–	19
Hamster	–	28	–	–	31	–	14	–	6	5
Man	–	144	–	113	–	88	–	–	110	9
Potassium										
Bull	–	89	–	–	53	–	–	–	57	11, 17
	–	50	–	–	55	–	–	–	45	10
	–	49	–	–	43	–	–	–	28	6, 12
	–	49	–	–	41	–	–	–	28	6, 11, 12
	–	45	–	–	–	–	–	–	28	10

Table 13 (continued)

	Caput			Corpus			Cauda			Refs
	Prox	Mid	Dist	Prox	Mid	Dist	Prox	Mid	Dist	
Potassium										
Bull	–	44	–	–	–	–	–	–	27	12
	–	44	–	–	–	–	–	–	–	10
	–	30	–	–	–	–	–	–	–	12
	–	28	–	–	–	–	–	–	–	6
Rat	–	28	–	–	41	–	40	–	55	2, 18, 20
	–	25	–	–	40	–	–	–	52	1, 20
	–	23	–	–	38	–	–	–	46	4, 13
	–	21	–	–	37	–	–	–	45	1, 2, 19
	–	21	–	–	31	–	–	–	42	4, 18
	–	21	–	–	30	–	–	–	40	13, 20
	–	21	–	–	–	–	–	–	33	18, 19
	–	16	–	–	–	–	–	–	33	2, 18
Boar	47	36	ca 45	–	40	–	ca 40	–	54	7, 14, 17
	–	–	39	–	39	–	38	–	38	7, 11, 17
	–	–	–	–	–	–	34	–	ca 35	7, 14
	–	–	–	–	–	–	32	–	33	7
	–	–	–	–	–	–	29	–	–	15
Ram	–	101	–	–	–	–	–	–	80	17
	–	49	–	–	–	–	–	–	39	8, 10
Hamster	–	39	–	–	26	–	21	–	24	5
Man	–	33	–	18	–	30	–	–	21	9
	–	–	–	–	–	–	–	–	18	9
Chloride										
Bull	–	93	–	–	85	–	–	–	66	11, 17
	–	92	–	–	85	–	–	–	65	10, 11
	–	81	–	–	78	–	–	–	65	6, 10
	–	58	–	–	–	–	–	–	–	6
Rat	–	36	–	–	45	–	–	–	54	1, 18
	–	31	–	–	44	–	–	–	33	2, 18
	–	26	–	–	39	–	–	–	30	1, 13, 18
	–	25	–	–	34	–	–	–	27	13, 18
	–	24	–	–	24	–	–	–	24	2, 3, 13, 19
	–	17	–	–	–	–	–	–	24	2, 3, 19
Boar	–	44	66	–	39	–	13	–	15	6, 11, 14
	–	–	–	–	39	–	26	–	10	6, 14
	–	–	–	–	34	–	–	–	9	6
	–	–	–	–	–	–	–	–	9	6
Ram	–	63	–	–	–	–	–	–	34	8, 17
	–	62	–	–	–	–	–	–	34	8, 10
	–	62	–	–	–	–	–	–	–	10
Calcium										
Bull	–	1.7	–	–	1.4	–	–	–	1.3	6, 18
	–	1.5	–	–	0.4	–	–	–	1.0	6, 10, 11
	–	1.3	–	–	–	–	–	–	0.4	6, 17

Table 13 (continued)

	Caput			Corpus			Cauda			Refs
	Prox	Mid	Dist	Prox	Mid	Dist	Prox	Mid	Dist	
Calcium										
Bull	–	0.4	–	–	–	–	–	–	0.35	10, 11, 17
	–	0.35	–	–	–	–	–	–	–	10, 11
Rat	–	1.39	–	–	1.2	–	–	–	0.88	18
	–	0.93	–	–	0.51	–	–	–	0.25	3, 13, 19
	–	0.85	–	–	0.5	–	–	–	0.2	1, 13
	–	0.8	–	–	0.28	–	–	–	0.16	1, 18
	–	0.54	–	–	–	–	–	–	0.03	18, 19
Boar	–	2.2	–	–	1.07	–	–	–	2.9	6
	–	–	–	–	0.85	–	–	–	1.5	6, 10
	–	–	–	–	–	–	–	–	1.35	6
	–	–	–	–	–	–	–	–	0.9	6
Ram	–	0.8	–	–	–	–	–	–	0.5	8
	–	0.8	–	–	–	–	–	–	0.95	8, 10, 17
	–	–	–	–	–	–	–	–	0.5	8, 10
Magnesium										
Bull	–	3.2	–	–	2.86	–	–	–	1.35	10, 11, 17
	–	3.15	–	–	2.8	–	–	–	1.3	10, 11, 17
Rat	–	2.78	–	–	4.08	–	–	–	1.38	18
	–	2.57	–	–	2.8	–	–	–	1	1, 18
	–	2.21	–	–	2.61	–	–	–	0.9	3, 13, 19
	–	1.97	–	–	2	–	–	–	0.62	13, 18
	–	1.5	–	–	–	–	–	–	0.24	1, 19
Ram	–	1.5	–	–	–	–	–	–	0.6	8, 10, 17
	–	0.1	–	–	–	–	–	–	–	8, 17
Protons (pH)										
Rat	–	6.5	–	–	–	–	–	–	6.85	2
	–	6.64	–	–	–	–	–	–	–	21
Bull	–	6.7	–	–	6.6	–	–	–	6.8	11

[a] Regions of the epididymis given as quoted or ascertained from diagrams of regions sampled: *Prox*, proximal; *Mid*, middle; *Dist*, distal

[b] *ca*, circa: values taken from graphs

References

1. Howards SS, Johnson A and Jessee S (1975) Micropuncture and microanalytical studies of the rat testis and epididymis. Fert Steril 26:13–19
2. Levine N and Marsh DJ (1971) Micropuncture studies of the electrochemical aspects of fluid and electrolyte transport in individual seminiferous tubules, the epididymis and vas deferens. J Physiol 213:557–570
3. Jenkins AD, Lechene CP and Howards SS (1978) The effect of estrogen on the concentration of seven elements in the intraluminal fluids of the seminiferous tubules, rete testis, and epididymides. Fert Steril 30:732–733
4. Turner TT, Hartmann PK and Howards SS (1977) In vivo sodium, potassium, and sperm concentrations in the rat epididymis. Fert Steril 28:191–194

5. Jessee SJ and Howards SS (1976) A survey of sperm, potassium and sodium concentrations in the tubular fluid of the hamster epididymis. Biol Reprod 15:626–631
6. Crabo B (1965) Studies on the composition of epididymal content in bulls and boars. Acta Vet Scand Vol 6, Suppl 5
7. Zimmerman KJ, Crabo BG, Moore R, Weisberg S, Deibel FC and Graham EF (1979) Movements of sodium and potassium into epididymal boar spermatozoa. Biol Reprod 21:173–179
8. Scott TW, Wales RG, Wallace KC and White IG (1963) Composition of ram epididymal and testicular fluid and the biosynthesis of glycerylphosphorylcholine by the rabbit epididymis. J Reprod Fert 6:49–59
9. Crabo BG and Hunter AG (1975) Sperm maturation and epididymal function. In: Control of Male Fertility. Eds JJ Sciarra, C Markland and JJ Speidel, Harper and Row, London. pp 2–23
10. White IG (1973) Biochemical aspects of spermatozoa and their environment in the male reproductive tract. J Reprod Fert Suppl 18:225–235
11. Wales RG, Wallace JC and White IG (1966) Composition of bull epididymal and testicular fluid. J Reprod Fert 12:139–144
12. Crabo B and Gustafsson B (1964) Distribution of sodium and potassium and its relation to sperm concentration in the epididymal plasma of the bull. J Reprod Fert 7:337–345
13. Jenkins AD, Lechene CP and Howards SS (1980) Concentrations of seven elements in the intraluminal fluids of the rat seminiferous tubules, rete testis, and epididymis. Biol Reprod 23:981–987
14. Crabo BG, Zimmerman KJ, Hunter AG, Graham RF and Moore R (1979) Effect of α-chlorohydrin on epididymal sperm and epididymal plasma in swine. Archs Androl 3:79–87
15. Einarsson S and Gustafsson B (1973) A case of epididymal dysfunction in boar. Andrologia 5:273–279
16. Hirao K and Kishida T (1980) Segmental difference in sperm concentration and some chemical compositions of boar epididymal fluid. Jap J Anim Reprod 26:200–206
17. White IG (1978) Accessory sex organs and fluids of the male reproductive tract. In: Animal Models for Research on Contraception and Fertility. Ed NJ Alexander, Harper and Row, Hagerstown. pp 105–123
18. Jenkins AD, Lechene CP and Howards SS (1983a) Effect of spironolactone on the elemental composition of the intraluminal fluids of the seminiferous tubules, rete testis and epididymis of rat. J Urol 129:851–854
19. Jenkins AD, Lechene CP and Howards SS (1983b) The effect of estrogen administration in vivo on the elemental composition of intraluminal fluids of the seminiferous tubules, rete testis and epididymis of the rat. J Androl 4:272–275
20. Turner TT (1984) Resorption versus secretion in the rat epididymis. J Reprod Fert 72:509–514
21. Levine N and Kelly H (1978) Measurement of pH in the rat epididymis in vivo. J Reprod Fert 52:333–335

Table 14. Concentrations and packed cell volumes (PCV) of spermatozoa in luminal fluid from various regions of the epididymis[a]

Species	Units[b]	Caput			Corpus			Cauda			Refs
		Prox	Mid	Dist	Prox	Mid	Dist	Prox	Mid	Dist	
Rat	A	–	0.75	–	–	1.24	–	–	1.99	–	1
	A	–	0.80	–	–	1.32	–	–	2.03	–	1
	A	–	0.75	–	–	–	–	–	1.75	–	2
	A	–	0.75	–	–	–	–	–	1.95	–	3
	A	–	0.70	–	–	1.4	–	1.5	–	1.9	4
	B	–	38.8	–	–	60.8	–	–	–	54.8	5
	B	21.0	–	29.8	–	33.6	–	37.7	–	57.8	6
	B	23.1	–	–	–	29.8	–	–	–	38.7	7
Bull	B	11.1	42.4	57.8	–	–	43.5	–	–	54.4	8
	B	14.2	52.3	63.2	–	–	39.5	–	–	47.6	8
	B	ca 70	–	–	–	–	–	ca 30	–	ca 30	9
	B	11.2	–	43.6	–	64.2	–	41.0	–	48.9	10
	A	3.4	3.6	3.8	3.7	3.9	1.9	4.9	5.7	–	11
Boar	A	0.21	1.70	2.57	3.53	2.96	2.58	2.70	2.72	2.23	12
	B	13.4	–	47.7	78.5	49.1	39.2	39.0	34.6	27.6	8
Human	B[c]		—— 61 ——		–	–	–		—— 33 ——		13
	B[d]	–	–	–	1	–	30		—— 50 ——		13
Hamster	A	–	1.96	–	–	1.75	–	1.9	–	1.87	14

[a] Regions of epididymis given as quoted or ascertained from diagrams of regions sampled: *Prox*, proximal; *Mid*, middle; *Dist*, distal
[b] $A = \times 10^9$/ml; $B =$ PCV (%)
[c] Young man treated with oestrogens
[d] 71-year old man

References

1. Turner TT and Cesarini DM (1983) The ability of the rat epididymis to concentrate spermatozoa. Responsiveness to aldosterone. J Androl 4:197–202
2. Turner TT (1984) Resorption versus secretion in the rat epididymis. J Reprod Fert 72:509–514
3. Turner TT, Jones CE, Howards SS, Ewing LL, Zegeye B and Gunsalas GL (1984) On the androgen microenvironment of maturing spermatozoa. Endocrinology 115:1925–1932
4. Turner TT, Hartmann PK and Howards SS (1977) In vivo sodium and potassium, and sperm concentrations in the rat epididymis. Fert Steril 28:191–194
5. Levine N and Marsh DJ (1971) Micropuncture studies of the electrochemical aspects of fluid and electrolyte transport in individual seminiferous tubules, the epididymis and vas deferens. J Physiol 213:557–570
6. Hinton BT and Setchell BP (1980) Concentration and uptake of carnitine in the rat epididymis. A micropuncture study. In: Carnitine Biosynthesis, Metabolism, and Function. Eds RA Frenkel and JD McGarry, Acad Press, New York. pp 237–250
7. Pholpramool C, White RW and Setchell BP (1982) Influence of androgens on inositol secretion and sperm transport in the epididymis of rats. J Reprod Fert 66:547–553
8. Crabo BG (1965) Studies on the composition of epididymal content in bulls and boars. Acta Vet Scand Vol 6, Suppl 5
9. Crabo BG, Zimmerman KJ, Hunter AG, Graham RF and Moore R (1979) Effect of α-chlorohydrin on epididymal plasma in swine. Archs Androl 3:79–87
10. Crabo B and Gustafsson B (1964) Distribution of sodium and potassium and its relation to sperm concentration in the epididymal plasma of the bull. J Reprod Fert 7:337–345

11. Pholpramool C, Zupp JL and Setchell BP (1985) Motility of undiluted bull epididymal spermatozoa collected by micropuncture. J Reprod Fert 75:413–420
12. Hunter AG and Schellpfeffer DA (1981) Concentrations of sperm, protein and a sperm membrane glycoprotein within boar epididymal luminal fluids. J Anim Sci 52:575–579
13. Crabo BG and Hunter AG (1975) Sperm maturation and epididymal function. In: Control of Male Fertility. Eds JJ Sciarra, C Markland and JJ Speidel, Harper and Row, Hagerstown. pp 2–23
14. Iessee SJ and Howards SS (1976) A survey of sperm, potassium and sodium concentrations in the tubular fluid of the hamster epididymis. Biol Reprod 15:626–631

Table 15. Concentrations of glycerophosphocholine (mM) in luminal fluid from the testis and various regions of the epididymis[a]

Species	RTF[a]	Caput			Corpus			Cauda			Refs
		Prox	Mid	Dist	Prox	Mid	Dist	Prox	Mid	Dist	
Ram	3.5		18.1		–	–	–		76.0		1
	2.8		15.0		–	–	–		66.9		1
	2.9	–	–	–	–	–	–		76.0		2
	–		7.2		–	–	–		53.6		3
Bull	4.0		9–40			42.5			63.2		4
	–		8.6			19.4			17.4		5
	0.3–0.7	–	–	–	–	–	–		17.4		2
Boar	1.9	–	5.4	–	37.2	57.1	47.7	62–68	–	62.5	4
	15	–	–	53	–	–	–		105		6
	2.1	–	–	–	–	–	–	38.4	–	–	7
Rat	<0.1	40.1	–	48.0	–	51.7	–	48.0	–	41.4	8

[a] Regions of epididymis given as quoted or ascertained from diagrams of regions sampled: *Prox*, proximal; *Dist*, distal
[b] Rete Testis Fluid

References

1. Scott TW, Wales RG, Wallace KC and White IG (1963) Composition of ram epididymal and testicular fluid and the biosynthesis of glycerylphosphorylcholine by the rabbit epididymis. J Reprod Fert 6:49–59
2. White IG (1973) Biochemical aspects of spermatozoa and their environment in the male reproductive tract. J Reprod Fert Suppl 18:225–235
3. Quinn PJ and White IG (1967) Phospholipid and cholesterol content of epididymal and ejaculated spermatozoa and seminal plasma in relation to cold shock. Aust J Biol Sci 20:1205–1215
4. Crabo B (1965) Studies on the composition of epididymal content in bulls and boars. Acta Vet Scand Vol 6, Suppl 5
5. Wales RG, Wallace JC and White IG (1966) Composition of bull epididymal and testicular fluid. J Reprod Fert 12:139–144
6. Crabo BG, Zimmerman KJ, Hunter AG, Graham EF and Moore R (1979) Effect of α-chlorohydrin on epididymal sperm and epididymal plasma in swine. Archs Androl 3:79–87
7. Einarsson S and Gustafsson B (1973) A case of epididymal dysfunction in boar. Andrologia 5:273–279
8. Hinton BT and Setchell BP (1980) Concentrations of glycerophospholine, phosphocholine and free inorganic phosphate in the luminal fluid of the rat testis and epididymis. J Reprod Fert 58:401–406

Table 16. Concentrations of carnitine in luminal fluid and spermatozoa from the testis and various regions of the epididymis[a]

Species	RTF[b]	IS[c]	Caput			Corpus			Cauda			Refs
			Prox	Mid	Dist	Prox	Mid	Dist	Prox	Mid	Dist	
(a) Fluid (mM)[d]												
Ram	0.03	–	–	–	–	–	–	11.2	10.6	–	14.5	1
	ca 0.1	ca 0.1	0.1	0.2	0.8	1.0	1.5	3.5	4.5	–	8.5	2
Hamster	<1	–	–	–	–	–	–	2.36	8.74	–	10.07	1
	–	–		0.019			0.015			0.357		3
Rat	<1	–	<1–3	–	19.1	–	22.2	30.8	51.7	–	53.7	1
Rabbit	–	–	–	–	–	–	–	–	14.5	–	19.0	1
Boar	0.01	–	–	–	5.84	–	4.06	15.6	17.9	–	16.1	1
Bull	–	–	–	–	–	–	–	–	–	–	2.0	3
(b) Spermatozoa (nmol/10^{8} cells)												
Hamster	–	–		40.7			61.5			135		4
	–	–	–	–	–	–	–	41.6	91.6	172	169	4
Rabbit	–	–	5.2	–	6.5	4.9	–	5.4	12.4	–	13.1	5
	–	–	–	–	–	3.1, 3.0	·	4.8	–	–	–	6
Bull	–	–	10	–	7		28		49	–	100	4
Ram[e]	–	–	–	10	19	38		30	43	–	60	7

[a] Regions of epididymis given as quoted or ascertained from diagrams of regions sampled: *Prox*, proximal; *Mid*, middle; *Dist*, distal
[b] Rete Testis Fluid
[c] Initial Segment of the epididymis
[d] *ca*, circa: values taken from graphs
[e] Values given are μmol/10^{8} cells

References

1. Hinton BT, Snoswell AM and Setchell BP (1979) The concentration of carnitine in the luminal fluid of the testis and epididymis of the rat and some other mammals. J Reprod Fert 56:105–111
2. Besançon J, Dacheux JL, Paquin R and Tremblay RR (1985) Major contribution of epididymis to α-glucosidase content of ram seminal plasma. Biol Reprod 33:296–301
3. Casillas ER, Villalobos R and Gonzáles R (1984) Distribution of carnitine and acylcarnitine in the hamster epididymis and in epididymal spermatozoa during maturation. J Reprod Fert 72:197–201
4. Casillas ER (1973) Accumulation of carnitine by bovine spermatozoa during maturation in the epididymis. J Biol Chem 248:8227–8232
5. Casillas ER and Chaiyungpan S (1979) The distribution of carnitine and acetylcarnitine in the rabbit epididymis and the carnitine content of rabbit spermatozoa during maturation. J Reprod Fert 56:439–444
6. Casillas ER and Chaiyungpan S (1982) Carnitine content of rabbit epididymal spermatozoa in organ culture. J Reprod Fert 65:247–251
7. Inskeep PB and Hammerstedt RH (1982) Changes in metabolism of ram sperm associated with epididymal transit or induced by endogenous carnitine. Biol Reprod 27:735–743

Table 17. Concentration of myo-inositol (mM) in luminal fluid from various regions of the epididymis[a]

Species	Caput		Corpus		Cauda		Refs
	Prox	Dist	Prox	Dist	Prox	Dist	
Ram	9.3	–	6.2, 6.0	–	5.1	5.4, 0.9	1
Boar	–	–	–	–	1.2	1.0	1
Rabbit	35.4	–	–	1.4	<0.5	<0.5	1
Rat	11.8–13.2	–	9.2–10.1	–	22.8–28.4	–	2
	8.7	6.9	5.0	5.6	29.5	31.4	1
Hamster	49.2	43.8	43.1	47.4	58.7	90.2	1
Baboon	3.5	–	5.9	–	1.3	–	1
Monkey	25.1	–	18.3	–	17.0	17.4	1

[a] Regions of epididymis given as quoted or ascertained from diagrams of regions sampled: *Prox*, proximal; *Mid*, middle; *Dist*, distal

References

1. Hinton BT, White RW and Setchell BP (1980) Concentrations of myo-inositol in the luminal fluid of the mammalian testis and epididymis. J Reprod Fert 58:395–399
2. Pholpramool C, White RW and Setchell BP (1982) Influence of androgens on inositol secretion and sperm transport in the epididymis of rats. J Reprod Fert 66:547–553

Table 18. Concentrations of proteins (mg/ml) in luminal fluid from various regions of the epididymis[a]

Species	Caput			Corpus			Cauda			Refs
	Prox	Mid	Dist	Prox	Mid	Dist	Prox	Mid	Dist	
Rat		—32.8—			—34.6—			—36.8—		1
		—25.6—		–	–	–		—27.3—		2
		—22.4—		–	–	–		—28.0—		3
Hamster		—107.6—		—153.7—		115.9	83.6	–	56.4	4
Boar	3.0	19.9	32.0	64.6	50.0	41.4	31.6	35.3	28.8	5, 6
	75.1	–	–	–	–	–	45.8	–	44.1	7
Stallion		—108—		114	–	51	55	–	50	8

[a] Regions of epididymis given as quoted or ascertained from diagrams of regions sampled: *Prox*, proximal; *Mid*, middle; *Dist*, distal

References

1. Wong PYD, Tsang AYF and Lee WM (1981) Origin of the luminal fluid proteins of the rat epididymis. Int J Androl 4:331–341
2. Turner TT, Plesums JL and Cabot CL (1979) Luminal fluid proteins of the male rat reproductive tract. Biol Reprod 21:883–890
3. Turner TT (1984) Resorption versus secretion in the rat epididymis. J Reprod Fert 72:509–514
4. Whelan HA, Ganjam VK and Brackett BG (1980) Experimental alteration of golden hamster epididymal contents. Biol Reprod Suppl 1:41 A (abst)

5. Schellpfeffer DA and Hunter AG (1976) Specific proteins of the male reproductive tract. In: Regulatory Mechanisms of Male Reproductive Physiology, Eds CH Spilman, TJ Lobl and KT Jirton, Excerpta Medica, Amsterdam. pp 115–132
6. Hunter AG and Schellpfeffer DA (1981) Concentrations of sperm, protein and a sperm membrane glycoprotein within boar epididymal luminal fluids. J Anim Sci 52:575–579
7. Hirao K and Kishida T (1980) Segmental difference in sperm concentration and some chemical compositions of boar epididymal fluid. Jap J Anim Reprod 26:200–206
8. Whelan HA and Brackett BG (1979) Epididymal luminal fluid proteins. Biol Reprod Suppl 1: 21 A (abst)

Table 19. Evidence for changes in the surface of spermatozoa on maturation in the epididymis

Surface charge (see Hammerstedt, 1979)	
by electrophoresis	
rat, hamster, rabbit	Moore, 1979
rabbit	Bedford, 1963
ram	Hammerstedt, 1979; Hammerstedt et al., 1979, 1982
by charged bead binding	
rat	Chulavatnatol et al., 1978; Eksittikul and Chulavatnatol, 1980
by colloidal iron binding	
rabbit, bull hamster, man	Bedford et al., 1973
ram	Courtens and Fournier-Delpech, 1979
rabbit	Flechon, 1973
by cationic ferritin binding	
rat	Toowicharanont and Chulavatnatol, 1983a
Agglutination	
by virus	
rabbit	Buthala et al., 1971
by antiserum	
rabbit	Bedford, 1965
by antibodies	
rabbit, hamster	Moore, 1981
by dilution	
ram, boar	Dacheux et al., 1983
Antigenicity	
rat	Lea et al., 1978; Faye et al., 1980; Bayard et al., 1981; Gaunt, 1982; Gaunt et al., 1983; Jones et al., 1985
mouse	Feuchter et al., 1981; Vernon et al., 1982; Moore et al., 1985; Eddy et al., 1985
bull	Barker and Amann, 1969, 1970, 1971; Killian and Amann, 1973
boar	Bostwick et al., 1980
hamster	Ellis et al., 1985

Table 19 (continued)

Carbohydrates	
by lectin binding (see Nicolson and Yanagimachi, 1979)	
rat	Baccetti et al., 1978; Lewin et al., 1979; Olson and Danzo, 1981; Arya and Vanha-Perttula, 1984, 1985a,b; Vanha-Perttula and Arya, 1985
rabbit	Gordon et al., 1974, 1975; Nicolson et al., 1977
ram	Courtens and Fournier-Delpech 1979; Fournier-Delpech and Courot, 1980; Hammerstedt et al., 1982
mouse	Bedford and Cooper, 1978; Watanabe et al., 1981; Lee and Damjanov, 1984
bull	Baccetti et al., 1978; Arya and Vanha-Perttula, 1985c
monkey	Fain-Maurel et al., 1983, 1984
by sialic acid measurement	
rat	Toowicharanont and Chulavatnatol, 1983b
ram	Holt, 1980
Lipids (see Hammerstedt, 1979)	
ram	Hammerstedt, 1979; Hammerstedt et al., 1979, 1982; Parks and Hammerstedt, 1985
boar	Nikolopoulou et al., 1985
Proteins	
ATPase	
rat	Chulavatnatol and Yindepit, 1976
Complement activation	
rabbit, hamster	Witkin et al., 1983
Immunoglobulin binding	
rabbit	Suarez et al., 1981
Characterisation of surface proteins	
ram	Vierula and Rajaniemi, 1982
rabbit	Nicolson et al., 1979; Nicolson and Yanagimachi, 1979
rat	Olson and Danzo, 1981; Vierula and Rajaniemi, 1982
chimpanzee	Young et al., 1985
man	Dacheux et al., 1984
Characterisation of surface glycoproteins	
ram	Fournier-Delpech and Courot, 1980
rat	Fournier-Delpech et al., 1977; Olson and Hamilton, 1978; Olson and Danzo, 1981; Olson and Orgebin-Crist, 1982; Toowicharanont and Chulavatnatol, 1983c; Brooks and Tiver, 1983a,b, 1984; Zeheb and Orr, 1984; Brooks, 1985; Rifkin and Olson, 1985
man	Dacheux et al., 1984

References

Arya M and Vanha-Perttula T (1984) Distribution of lectin binding in rat testis and epididymis. Andrologia 16:495–508

Arya M and Vanha-Perttula T (1985a) Lectin staining of rat testis and epididymis: effect of cyproterone acetate and testosterone. Andrologia 17:301–310

Arya M and Vanha-Perttula T (1985b) Effect of castration on lectin staining in rat epididymis. Andrologia 17:327–337
Arya M and Vanha-Perttula T (1985c) Lectin-binding pattern of bull testis and epididymis. J Androl 6:230–242
Baccetti, B, Bigliardi E and Burrini AG (1978) The cell surface during mammalian spermatogenesis. Develop Biol 63:187–196
Barker LDS and Amann RP (1969) Sperm antigens and their localization within the bovine epididymal epithelium. J Reprod Fert 18:155–156
Barker LDS and Amann RP (1970) Epididymal physiology. I. Specificity of antisera against bull spermatozoa and reproductive fluids. J Reprod Fert 22:441–452
Barker LDS and Amann RP (1971) Epididymal physiology. II. Immunofluorescent analyses of epithelial secretion and absorption and of bovine sperm maturation. J Reprod Fert 26:319–332
Bayard F, Duguet A, Mazzuca M and Faye JC (1981) Study of a glycoprotein produced by the rat epididymis. In: Reproductive Processes and Contraception. Ed KW McKerns. Plenum Pub Co. pp 393–405
Bedford JM (1963) Morphological changes in rabbit spermatozoa during passage through the epididymis. J Reprod Fert 5:169–177
Bedford JM (1965) Changes in fine structure of the rabbit sperm head during passage through the epididymis. J Anat 99:891–906
Bedford JM and Cooper GW (1978) Membrane fusion events in the fertilization of vertebrate eggs. In: Membrane Fusion. Eds G Poste and GL Nicolson. Volume 5, Cell Surface Reviews. Elsevier/North Holland, Amsterdam. pp 65–125
Bedford JM, Calvin H and Cooper GW (1973) The maturation of spermatozoa in the human epididymis. J Reprod Fert Suppl 18:199–213
Bostwick EF, Bentley MD, Hunter AG and Hammer R (1980) Identification of a surface glycoprotein on porcine spermatozoa and its alteration during epididymal maturation. Biol Reprod 23:161–169
Brooks, DE (1985) Characterization of a 22 kDa protein with widespread tissue distribution but which is uniquely present in secretions of the testis and epididymis and on the surface of spermatozoa. Biochem Biophys Acta 841:59–70
Brooks DE and Tiver K (1983a) Selective binding of specific rat epididymal secretory proteins to spermatozoa and erythrocytes. Gamete Res 7:367–376
Brooks DE and Tiver K (1983b) Localization of epididymal secretory proteins on rat spermatozoa. J Reprod Fert 69:651–657
Brooks DE and Tiver K (1984) Analysis of surface proteins of rat spermatozoa during epididymal transit and identification of antigens common to spermatozoa, rete testis fluid and cauda epididymal plasma. J Reprod Fert 71:249–257
Buthala DA, Ericsson RJ and Chubb GT (1971) Interaction of Sendai virus and rabbit sperm: transmission and scanning microscopy. Biol Reprod 5:325–329
Chulavatnatol M and Yindepit S (1976) Changes in surface ATPase of rat spermatozoa in transit from the caput to the cauda epididymidis. J Reprod Fert 48:91–97
Chulavatnatol M, Eksittikul T and Toowicharanont P (1978) Control of epididymal sperm motility and approach to male fertility regulation. Int J Androl Suppl 2:629–638
Courtens JL and Fournier-Delpech S (1979) Modification of the plasma membrane of epididymal ram spermatozoa during maturation and incubation in utero. J Ultrstruct Res 68:136–148
Dacheux JL, Paquignon M and Combarnous Y (1983) Head-to-head agglutination of ram and boar epididymal spermatozoa and evidence for an epididymal antagglutinin. J Reprod Fert 67:181–189
Dacheux JL, Chevrier C, Desmee K, Lanneau M, Jouannet P, Zephir D and Lanson Y (1984) Evolution de la motilite et des proteines de surface des spermatozoides humaines lors de leur passage dans l'épididyme. J Urologie (Paris) 90:582–583
Eddy EM, Vernon RB, Muller CH, Hahnel AC and Fenderson BA (1985) Immunodissection of sperm surface modifications during epididymal maturation. Am J Anat 174:225–237
Eksittikul T and Chulavatnatol M (1980) Binding of spermatozoa to positively changed beads as an inexpensive method to isolate sperm heads and tails. Int J Androl 3:643–653

Ellis DH, Hartman T and Moore HDM (1985) Maturation and function of the hamster spermatozoon probed with monoclonal antibodies. J Reprod Immuol 7:299–314

Fain-Maurel MA, Dadoune JP and Reger JF (1983) Surface changes in monkey spermatozoa during epididymal maturation and after ejaculation. In: The Sperm Cell. Ed J André, Martinus-Nijhoff, The Hague. pp 159–162.

Fain-Maurel MA, Dadoune JP and Reger JF (1984) A cytochemical study on surface charges and lectin binding sites in epididymal and ejaculated spermatozoa of *Macaca fascicularis*. Anat Rec 208:375–382

Faye JC, Duguet L, Mazzuca M and Bayard F (1980) Purification, radioimmunoassay, and immunohistochemical localization of a glycoprotein produced by the rat epididymis. Biol Reprod 23:423–432

Feuchter FA, Vernon RB and Eddy EM (1981) Analysis of the sperm surface with monoclonal antibodies: topographically restricted antigens appearing in the epididymis. Biol Reprod 24:1099–1110

Fléchon JE (1973) Ultrastructural and cytochemical modifications of rabbit spermatozoa during epididymal transit. In: The Biology of Spermatozoa. Eds ESE Hafez and CG Thibault, S Karger, Basel, Switzerland, pp 36–45

Fléchon JE (1981) Ultrastructural and cytochemical analysis of the plasma membrane of mammalian sperm during epididymal maturation. In: The Epididymis, Pathology and Biology. Eds C Bollack and A Clavert, S Karger, Munich. Progress in Reproductive Biology 8:90–99

Fournier-Delpech S and Courot M (1980) Glycoproteins of ram sperm plasma membrane. Relationship of protein having affinity for ConA to epididymal maturation. Biochem Biophys Res Comm 96:756–761

Fournier-Delpech S, Danzo BJ and Orgebin-Crist MC (1977) Extraction of concanavalin A affinity material from rat testicular and epididymal spermatozoa. Ann biol Anim Bioch Biophys 17:207–213

Gaunt SJ (1982) A 28 K-dalton cell surface autoantigen of spermatogenesis: characterization using a monoclonal antibody. Develop Biol 89:92–100

Gaunt SJ, Brown CR and Jones R (1983) Identification of mobile and fixed antigens on the plasma membrane of rat spermatozoa using monoclonal antibodies. Exp Cell Res 144:275–284

Gordon M, Dandekar PV and Bartoszewicz W (1974) Ultrastructural localization of surface receptors for concanavalin A on rabbit spermatozoa. J Reprod Fert 36:211–214

Gordon M, Dandekar PV and Bartoszewicz W (1975) The surface coat of epididymal, ejaculated and capacitated sperm. J Ultrastruct Res 50:199–207

Hammerstedt RH (1979) Characterization of sperm surfaces using physical techniques. In: The Spermatozoon. Eds DW Fawcett and JM Bedford, Urban and Schwarzenberg, Baltimore. pp 205–216

Hammerstedt RH, Keith AD, Hay S, DeLuca N and Amann RP (1979) Changes in ram sperm membranes during epididymal transit. Archs Biochem Biophys 196:7–12

Hammerstedt RH, Hay SR and Amann RP (1982) Modification of ram sperm membranes during epididymal transit. Biol Reprod 27:745–754

Holt WV (1980) Surface-bound sialic acid on ram and bull spermatozoa: deposition during epididymal transit and stability during washing. Biol Reprod 23:847–857

Jones R, Brown CR, von Glos KI and Gaunt SJ (1985) Development of a maturation antigen on the plasma membrane of rat spermatozoa in the epididymis and its fate during fertilization. Exp Cell Res 156:31–44

Killian GK and Amann RP (1973) Immunoloelectrophoretic characterization of fluid and sperm entering and leaving the bovine epididymis. Biol Reprod 9:489–499

Lea OA, Petrusz P and French FS (1978) Purification and localization of acidic epididymal glycoprotein (AEG): a sperm coating protein secreted by the rat epididymis. Int J Androl Suppl 2:592–607

Lee MC and Damjanov I (1984) Anatomic distribution of lectin-binding sites in mouse testis and epididymis. Differentiation 27:74–81

Lewin LM, Weissenberg R, Sobel JS, Marcus Z and Nebel L (1979) Differences in concanavalin A-FITC binding to rat spermatozoa during epididymal maturation and capacitation. Archs Androl 2:279–281

Moore HDM (1979) The net surface charge of mammalian spermatozoa as determined by isoelectric focussing. Changes following sperm maturation, ejaculation, incubation in the female tract, and after enzyme treatment. Int J Androl 2:449–462

Moore HDM (1981) Glycoprotein secretions of the epididymis in the rabbit and hamster: localization on epididymal spermatozoa and the effect of specific antibodies on fertilization in vivo. J exp Zool 215:77–85

Moore HDM, Hartman TD, Brown AC, Smith CA and Ellis DH (1985) Expression of sperm antigens during spermatogenesis and maturation detected with monoclonal antibodies. Exp clin Immunogenet 2:84–96

Nicolson GJ and Yanagimachi R (1979) Cell surface changes associated with the epididymal maturation of mammalian spermatozoa. In: The Spermatozoon. Maturation, Motility, Surface Properties and Comparative Aspects. Eds DW Fawcett and JM Bedford, Urban and Schwarzenberg, Baltimore. pp 187–194

Nicolson GL, Usui N, Yanagimachi R, Yanagimachi H and Smith JR (1977) Lectin-binding sites on the plasma membranes of rabbit spermatozoa. Changes in surface receptors during epididymal maturation and after ejaculation. J Cell Biol 74:950–962

Nikolopoulou M, Soucek DA and Vary JC (1985) Changes in the lipid content of boar sperm plasma membranes during epididymal maturation. Biochem Biophys Acta 815:486–498

Olson GEW and Danzo BJ (1981) Surface changes in rat spermatozoa during epididymal transit. Biol Reprod 24:431–443

Olson GE and Hamilton DW (1978) Characterization of the surface glycoproteins of rat spermatozoa. Biol Reprod 19:26–35

Olson GE and Orgebin-Crist MC (1982) Sperm surface changes during epididymal maturation. Ann NY Acad Sci 383:372–391

Parkes JE and Hammerstedt RH (1985) Developmental changes occurring in the lipids of ram epididymal spermatozoa plasma membrane. Biol Reprod 32:653–668

Rifkin JED and Olson GE (1985) Characterization of maturation-dependent extrinsic proteins of the rat sperm surface. J Cell Biol 100:1582–1591

Suarez SS, Hinton BT and Oliphant G (1981) Binding of a marker for immunoglobulins to the surface of rabbit testicular, epididymal and ejaculated spermatozoa. Biol Reprod 25:1091–1097

Toowicharanont P and Chulavatnatol M (1983a) Measurement of anionic sites of rat epididymal spermatozoa using tritiated polycationized ferritin. J Reprod Fert 69:303–306

Toowicharanont P and Chulavatnatol M (1983b) Direct assay of bound sialic acids on rat spermatozoa from the caput and cauda epididymidis. J Reprod Fert67:275–280

Toowicharanont P and Chulavatnatol M (1983c) Characterization of sialoglycoproteins of rat epididymal fluid and spermatozoa by periodate-tritiated borohydride. J Reprod Fert 67:113–141

Vanha-Perttula T and Arya M (1985) Lectin staining of rat testis and epididymis after ligation of excurrent ducts at different levels. Biol Reprod 33:477–485

Vernon RB, Mueller CH, Herr JC, Feuchter FA and Eddy EM (1982) Epididymal secretion of a mouse sperm surface component recognized by a monoclonal antibody. Biol Reprod 26:523–535

Vierula M and Rajaniemi H (1982) Epididymal maturation of the surface protein structure of mammalian spermatozoa. Medical Biology 60: 323–327

Watanabe M, Muramatsu T, Shirane H and Ugai K (1981) Discrete distribution of binding sites for dolichos biflorus agglutinin (DBA) and for peanut agglutinin (PNA) in mouse organ tissues. J Histo Cyto 29:774–790

Witkin SS, Richards JM and Bedford JM (1983) Influence of epididymal maturation on the capacity of hamster and rabbit spermatozoa for complement activation. J Reprod Fert 69:517–521

Yanagimachi R, Noda YD, Fujimoto M and Nicolson GL (1972) The distribution of negative surface changes on mammalian spermatozoa. Am J Anat 135:497–520

Young LG, Hinton BT and Gould KG (1985) Surface changes in chimpanzee sperm during epididymal transit. Biol Reprod 32:399–412

Zeheb R and Orr GA (1984) Characterization of a maturation-associated glycoprotein on the plasma membrane of rat caudal epididymal sperm. J Biol Chem 259:839–848.

Table 20. Characteristics of major luminal fluid proteins secreted by various regions of the epididymis

Protein	MW (kDa)	pI	Carbohydrate*	Refs
Initial Segment				
Rat				
B[a]	16, 18.5	5.9	PAS + ve	1
C[b]	23.1, 22.4	5.56, 5.79, 5.38	binds ConA	1
B and C[a]	18.5			2
–	16	5.7	NAcGlc	1
–	19	–	–	3
I	23, 23.5	5.4	11.1%	3, 4
DAG	70	–	–	5
–	80	–	–	3, 4
Rabbit				
–	90	–	–	6
Ram				
–	23	–	–	7
Caput				
Rat				
B[a]		see above		
C[b]		see above		
D[c]	27, 30	4.4, 4.6	PAS + ve	1, 8
–[d]	37	–	binds ConA	1
II[a]	18.5, 21	5.9	13.2%	4, 9
III[b]	19, 23	5.7	7.6%	4, 9
IV[c]	32	4.2	9.6%	4, 9
–	80	–	–	4, 9
–	34.4	–	–	10
AEG[c]	31.7, 33.0	4.7	7.5%, mannose	11
Sialoprotein	37.5	4.7	low %, binds ConA	12, 13
C[b]	23.1, 22.4	5.38, 5.56, 5.79	PAS + ve	10, 14, 15
D[c]	34.4, 37	5.13	binds ConA	10, 14, 15
E[d]	37	4.95	NAcGlc	10, 14, 15
DAG	27, 35	4.2	–	5
Rabbit				
–	54	–	orcinol + ve	6
–	22.5	–	binds ConA	6
α	20, 22	4.65	NANA	6
β	20, 22	4.25	–	6
γ	20, 22	4.15	–	6
R1	58	3.45	binds ConA	16, 17
R2	–	4.15	–	16, 17
R3	–	4.65	–	16, 17
Hamster				
H1	–	3.4	binds ConA	16, 17
H2	–	4.15	–	16, 17
Ram				
–	24	–	–	7
–	200	–	–	7

Table 20 (continued)

Protein	MW (kDa)	pI	Carbohydrate *	Refs
Man				
–	38	5.8	binds ConA	18
–	21	6.2	–	18
–	29	6.8	–	18
–	69	5.1	–	18
Corpus				
Rat				
D[c]		as above		8
E[d]		as above		8
Rabbit				
ASF	360	–	–	19
–	22.5	–	–	6
Ram				
–	24	–	–	7
PES[e]	64	–	–	7
–	200	–	–	7
Cauda				
Rat				
D[c]		as above		1
E[d]	32, 38	4.15, 4.4, 4.35	PAS + ve	1, 8
–	47	7.9	binds ConA	4, 9
HIS-100	100	–	binds ConA	20
HIS-50	50	–	binds ConA	20
Rabbit				
–	22.5	–	–	6
ASF		as above		19
Ram				
–	24	–	–	7
PES[e]	64	–	–	7
In cauda fluid				
Rat				
–	22	–	orcinol – ve	21
–	26–28	–	acidic	22
REP	32	–	binds ConA	23
–	32	3.8	–	24
Sialoglycoprotein	21	–	–	25
–	66	–	–	25
PES[e]	–	–	PAS + ve, NANA	26
Immobilin	> 1000		56%, no NANA	27
Bull				
FMP[f]	37.5	–	–	28

* *ConA*, Concanavalin A; *PAS*, Periodic Acid Schiff reagent; *NAcGlc*, N-acetylglucosamine; *NANA*, N-acetylneuraminic acid

[a–e] Proteins with the same superscripts are judged by Brooks and Higgins (1980), Brooks (1981b, 1982) and Brooks et al. (1986) to be the same proteins

References

1. Brooks DE and Higgins SJ (1980) Characterization and androgen-dependence of proteins associated with luminal fluid and spermatozoa in the rat epididymis. J Reprod Fert 59:363–375
2. Brooks DE, Means AR, Wright EJ, Singh SP and Tiver KK (1986) Molecular cloning of the cDNA for two major androgen-dependent secretory proteins of 18.5 kilodaltons synthesized by the rat epididymis. J Biol Chem 26:4956–4961
3. Jones R, von Glos KI and Brown CR (1980a) The synthesis of a sperm-coating protein in the initial segment of the rat epididymis is stimulated by factors in testicular fluid. IRCS Med Sci 8:56
4. Jones R, Brown CR, von Glos KI and Parker MG (1980b) Hormonal regulation of proteins synthesis in the rat epididymis. Characterization of androgen-dependent and testicular fluid-dependent proteins. Biochem J 188:667–676
5. Sylvester SR, Skinner MK and Griswold MD (1984) A sulphated glycoprotein synthesized by Sertoli cells and by epididymal cells is a component of the sperm membrane. Biol Reprod 31:1087–1801
6. Jones R, von Glos KI and Brown CR (1981) Characterization of hormonally regulated secretory proteins from the caput epididymidis of the rabbit. Biochem J 196:105–114
7. Jones R, Fournier-Delpech S and Willadson SA (1982) Identification of androgen-dependent proteins synthesised in vitro by the ram epididymis. Reprod Nutr Develop 22:495–504
8. Brooks DE (1982) Purification of rat epididymal proteins "D" and "E", demonstration of shared immunological determinants, and identification of regional synthesis and secretion. Int J Androl 5:513–524
9. Brown CR, von Glos KI and Jones R (1983) Changes in plasma membrane glycoproteins of rat spermatozoa during maturation in the epididymis. J Cell Biol 96:256–264
10. Cameo MS and Blaquier JA (1976) Androgen-controlled specific proteins in rat epididymis. J Endocr 69:47–55
11. Lea OA and French FS (1981) Characterization of an acidic glycoprotein secreted by principal cells of the rat epididymis. Biochem Biophys Acta 668:370–376
12. Faye JC, Duguet L, Mazzuca M and Bayard F (1980) Purification, radioimmunoassay, and immunohistochemical localization of a glycoprotein produced by the rat epididymis. Biol Reprod 23:423–432
13. Bayard F, Duguet A, Mazzuca M and Faye JC (1981) Study of a glycoprotein produced by the rat epididymis. In: Reproductive Processes and Contraception. Ed KW McKerns. Plenum Pub Co. pp 393–405
14. Garberi JC, Kohane AC, Cameo MS and Blaquier JA (1979) Isolation and characterization of specific rat epididymal proteins. Mol Cell Endocr 13:73–82
15. Garberi JC, Fontana JD and Blaquier JA (1982) Carbohydrate composition of specific rat epididymal protein. Int J Androl 5:619–626
16. Moore HDM (1980) Localization of specific glycoproteins secreted by the rabbit and hamster. Biol Reprod 22:705–718
17. Moore HDM (1981) Glycoprotein secretions of the epididymis in the rabbit and hamster: localization on epididymal spermatozoa and the effect of specific antibodies on fertilization in vivo. J exp Zool 215:77–85
18. Tezón JG, Vazquez MH, Piñeiro L, de Larminat MA and Blaquier JG (1985) Identification of androgen-induced proteins in human epididymis. Biol Reprod 32:584–590
19. Thomas TS, Reynolds AL and Oliphant G (1984) Evaluation of the site of synthesis of rabbit sperm acrosome stabilizing factor using immunocytochemical and metabolic labeling techniques. Biol Reprod 30:693–705
20. Rifkin JM and Olson GE (1985) Characterization of maturation-dependent extrinsic proteins of the rat sperm surface. J Cell Biol 100:1582–1591
21. Brooks DE (1985) Characterization of a 22 kDa protein with widespread tissue distribution but which is uniquely present in secretions of the testis and epididymis and on the surface of spermatozoa. Biochem Biophys Acta 841:59–70
22. Olson GE and Orgebin-Crist MC (1982) Sperm surface changes during epididymal maturation. Ann NY Acad Sci 383:372–391

23. Wong PYD and Tsang AYF (1982) Studies on the binding of a 32 K rat epididymal protein to rat epididymal spermatozoa. Biol Reprod 27:1239–1246
24. Zeheb R and Orr GA (1984) Characterization of a maturation-associated glycoprotein on the plasma membrane of rat caudal epididymal sperm. J Biol Chem 259:839–848
25. Toowicharanont P and Chulavatnatol M (1983) Characterization of sialoglycoproteins of rat epididymal fluid and spermatozoa by periodate-tritiated borohydride. J Reprod Fert 67:113–141
26. Fournier-Delpech S, Bayard F and Boulard C (1973) Isolement, extraction, characterization d'une sialoproteine du sperme épididymaire du rat par electrophorèse sur polyacrylamide. Comptes Rendus Soc de Biol 167:543–546
27. Usselman MC, Cone RA and Rossignol P (1985) Rat cauda epididymal fluid is a mucus. J Androl 6:315–320
28. Acott TS and Hoskins DD (1978) Bovine forward motility protein. Partial purification and characterization. J Biol Chem 253:6744–6750

Table 21. The variety of enzymes in luminal fluid from the tail of the epididymis. References in parentheses

Dehydrogenases		
LDH[a]		
rat (1, 2, 11)	hamster (1, 2, 11)	guinea pig (1, 11)
boar (1, 2, 8)	ram (2, 3, 9)	bull (2, 3, 9)
stallion (1)	rabbit (1, 2, 6)	dog (1)
monkey (1)		
GDH[b]		
ram (9)	boar (8)	
Transaminases		
GPT[c]		
ram (14)	boar (13)	bull (23)
GOT[d]		
rat (10, 11)	hamster (11)	guinea pig (11)
boar (8, 12, 13)	ram (9, 14)	bull (23)
rabbit (6)		
Esterases		
Non-specific		
boar (7)	bull (5)	
Non-specific phosphoesterases		
rat (18)	hamster (18)	boar (18)
ram (18)	bull (18)	stallion (18)
rabbit (18)	dog (18)	
Alkaline phosphatase		
rat (1, 2, 10, 11)	hamster (1, 2, 11)	guinea pig (1, 11)
boar (1, 2, 7, 8, 12, 13)	ram (1, 2, 9)	bull (2)
stallion (1)	rabbit (1, 2, 6)	dog (1)
monkey (1)		
Acid phosphatase		
rat (1, 2, 10, 11, 24, 28)	hamster (1, 2, 11)	guinea pig (1, 11)
boar (1, 2, 7, 8)	ram (1, 2, 9)	bull (2, 5)
stallion (1)	rabbit (1, 2, 6)	dog (1)
monkey (1)		

Table 21 (continued)

ATPase		
rat (26a, b)		
Arylsulphatase		
rat (24)		
Angiotensin converting enzyme		
rat (27, 32, 33, 34)		
Nucleolytic enzymes		
Cyclic nucleotide phosphodiesterase		
rat (1, 18)	hamster (1, 18, 20)	guineapig (18)
boar (1, 18)	ram (1, 18, 19)	bull (18)
stallion (1, 18)	rabbit (1, 18)	dog (1, 18)
monkey (1)		
5′-Nucleotidase		
hamster (1, 18)	guinea pig (1)	boar(1, 18)
ram (1, 18)	bull (18)	stallion (18)
rabbit (1, 18)	dog (1, 18)	monkey (1)
Ribonuclease		
rat (18)	hamster (18)	boar (18)
ram (18)	bull (18)	stallion (18)
rabbit (18)	dog (18)	
Acid deoxyribonuclease		
rat (1, 34)	hamster (1)	guinea pig (1)
boar (1)	ram (1)	bull (17)
stallion (1)	rabbit (1, 15, 16)	dog (1)
monkey (1)		
Reverse transcriptase		
mouse (25)		
Carbohydrases		
Hyaluronidase		
rat (1)	hamster (1)	guinea pig (1)
boar (1)	ram (1)	stallion (1)
rabbit (1)	dog (1)	monkey (1)
Alpha-mannosidase		
rat (1)	hamster (1)	guinea pig (1)
boar (1, 2, 3)	ram (1)	stallion (1, 2, 3)
rabbit (1, 6)	dog (1)	monkey (1)
Beta-mannosidase		
stallion (3)		
Alpha-glucosidase		
stallion (3)	bull (30)	man (29)
ram (31)		
Beta-glucosidase		
stallion (3)		
Beta-glucuronidase		
rat (24)	bull (5)	stallion (3)

Table 21 (continued)

Alpha- and beta-galactosidase		
stallion (3)		
N-acetylhexosaminidase		
rat (1, 2)	hamster (1, 2)	guinea pig (1, 2)
boar (1, 2, 3, 4)	bull (2, 5)	stallion (1, 2, 3)
rabbit (1, 2, 6)	dog (1)	monkey (1)
UDP-galactose-N-acetylglucosamine galactosyltransferase		
rat (21)	mouse (22)	

[a] Lactate dehydrogenase
[b] Glutamate dehydrogenase
[c] Glutamate:pyruvate transaminase
[d] Glutamate:oxaloacetate transaminase

References

1. Jones R (1978) Comparative biochemistry of mammalian epididymal plasma. Comp Biochem Biophys 61 B:365–370
2. Jones R and Glover TD (1975) Interrelationships between spermatozoa, the epididymis and epididymal plasma. In: The Biology of the Male Gamete. Eds JG Duckett and PM Racey, Suppl No 1 Biol J Linnaen Soc 7:367–384
3. Conchie J and Mann T (1957) Glycosidases in mammalian sperm and seminal plasma. Nature (Lond) 179:1190–1191
4. Bamberg E, Dorner F and Stockl (1975) Purification of N-acetyl-hexosaminidase from boar epididymis by affinity chromatography. Experientia 31:516–518
5. Amann RP, Killian GJ and Benton AW (1973) Differences in the electrophoretic characteristics of bovine rete testis fluid and plasma from the cauda epididymidis. J Reprod Fert 35:321–330
6. Jones R and Glover TD (1973) The collection and composition of epididymal plasma from the cauda epididymis of the rabbit. J Reprod Fert 34:395–403
7. Dostál J and Veselský L (1972) Proteins in the seminal plasma and the accessory sexual gland fluids of the boar. J Reprod Fert 30:255–267
8. Bowen RE, Graham EF and Crabo B (1973) A technique for collecting epididymal spermatozoa and fluids from the boar virtually free from accessory sex gland secretions. J Reprod Fert 33:319–322
9. Murdoch RN and White IG (1968) Studies on the distribution and source of enzymes in mammalian semen. Aust J Biol Sci 21:483–490
10. Back DJ, Shenton JC and Glover TD (1974) The composition of epididymal plasma from the cauda epididymidis of the rat. J Reprod Fert 40:212–214
11. Back DJ and Shenton JC (1975) A comparison of the composition of epididymal plasma from the cauda epididymidis of the rat, hamster and guinea-pig. Experentia 31:464–465
12. Einarsson S (1971) Studies on the composition of epididymal content and semen in the boar. Acta Vet Scand Suppl 36
13. Einarsson S, Crabo B and Ekman L (1970) A comparative study on the chemical composition of plasma from the cauda epididymidis, semen fractions, and whole semen in boars. Acta Vet Scand 11:156–180
14. Alumot E and Schindler H (1965) Transaminases in the epididymal fluid of the ram. J Reprod Fert 10:261–262
15. Jones R (1974) Absorption and secretion in the cauda epididymidis of the rabbit and the effects of degenerating spermatozoa on epididymal plasma after castration. J Endocr 63:157–165

16. Piñeiro GJ and Roussel JD (1973) On the presence of deoxyribonuclease (DNAse) in the rabbit epididymis and its possible significance in the removal of spermatozoa. Fert Steril 24:145–149
17. Waldschmidt M and Karg H (1973) Review of metabolic and enzymic studies of spermatozoa from the ampullae ductus deferentiae of bulls. Adv Biosci 10:169–180
18. Menella MRF and Jones R (1977) The activity of some nucleolytic enzymes in semen and in secretions of the male reproductive tract. Andrologia 9:15–22
19. Tash JS (1976) Investigations on adenosine 3′,5′-monophosphate phosphodiesterase in ram semen and initial characterization of a sperm-specific isoenzyme. J Reprod Fert 47:63–72
20. Morton B, Harrigan-Lum J, Albagli L and Joos T (1974) The activation of motility in quiescent hamster sperm from the epididymis by calcium and cyclic nucleotides. Biochem Biophys Res Commun 56:372–379
21. Hamilton DW (1980) UDP-galactose:N-acetylglucosamine galactosyltransferase in fluids from rat rete testis and epididymis. Biol Reprod 23:377–385
22. Shur BD and Bennett (1979) A specific defect in galactosyl-transferase regulation on sperm bearing mutant alleles of the T/t locus. Develop Biol 71:243–259
23. Stallcup OT, Brown CJ and Cole WT (1968) Lactic dehydrogenase and transaminases in bovine epididymal fluid. J Reprod Fert 15:317–319
24. Mayorga LS and Bertini F (1982) Effect of androgens on the activity of acid hydrolases in rat epididymis. Int J Androl 5:345–352
25. Kiessling AA (1984) Evidence that reverse transcriptase is a component of murine epididymal fluid. Proc Soc exp Biol Med 176:175–182
26a. Majumder GC and Biswas R (1979) Evidence for the occurence of an ecto-(adenosine triphosphatase) in rat epididymal spermatozoa. Biochem J 183:737–743
26b. Majumder GC (1981) Enzymic characteristics of ecto-adenosine triphosphatase in rat epididymal intact spermatozoa. Biochem J 195:103–110
27. Hohlbrugger G, Schweisfurth H and Dahlheim H (1982) Angiotensin I converting enzyme in rat testis, epididymis, and vas deferens under different conditions. J Reprod Fert 65:92–103
28. Mayorga LS and Bertini F (1985) The origin of some acid hydrolases of the fluid of the rat cauda epididymidis. J Androl 6:243–245
29. Guerin JF, Rollet J, Perrin P, Ménézo Y, Orgiazzi A and Czyba JC (1981) Enzymes in the seminal plasma from azoospermic men: correlation with the origin of their azoospermia. Fert Steril 36:368–372
30. Jauhiainen AJ and Vanha-Perttula T (1985) Acid and neutral α-glucosidase in the reproductive organs and seminal plasma of the bull. J Reprod Fert 74:669–680
31. Paquin R, Chapdelaine P, Dube JY and Tremblay RR (1984) Similar biochemical properties of human seminal plasma and epididymal α-1,4-glucosidase. J Androl 5:277–282
32. Strittmatter SM, Thiele EA, Souza EB and Snyder SH (1985) Angiotensin-converting enzyme in the testis and epididymis: differential development and pituitary regulation of isozymes. Endocrinology 117:1374–1379
33. Strittmatter SM and Snyder SH (1984) Angiotensin-converting enzyme in the male rat reproductive system: autoradiographic visualization with (^{3}H) captopril. Endocrinology 115:2332–2341
34. Vanha-Perttula T, Mather JP, Bardin CW, Moss SB, Bellvé AR (1985) Localization of the angiotensin-converting enzyme activity in testis and epididymis. Biol Reprod 33:870–877

Appendix II

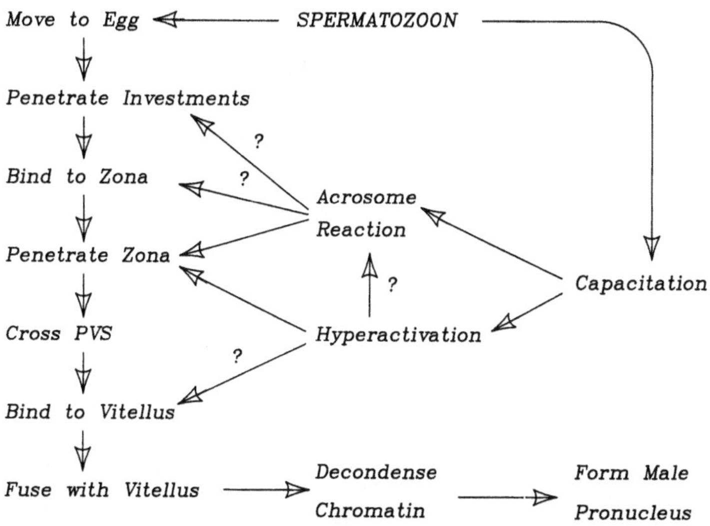

Fig. 1. Prerequisites for fertilisation: Sequential events undergone by the fertilising spermatozoon. Not only do spermatozoa have to reach the eggs, they have to penetrate the egg's investments before fusion with the egg membrane permits donation of the male gamete's hereditary material. Penetration of the investments requires the sperm to undergo the acrosome reaction and hyperactivation, both of which only capacitated sperm can achieve

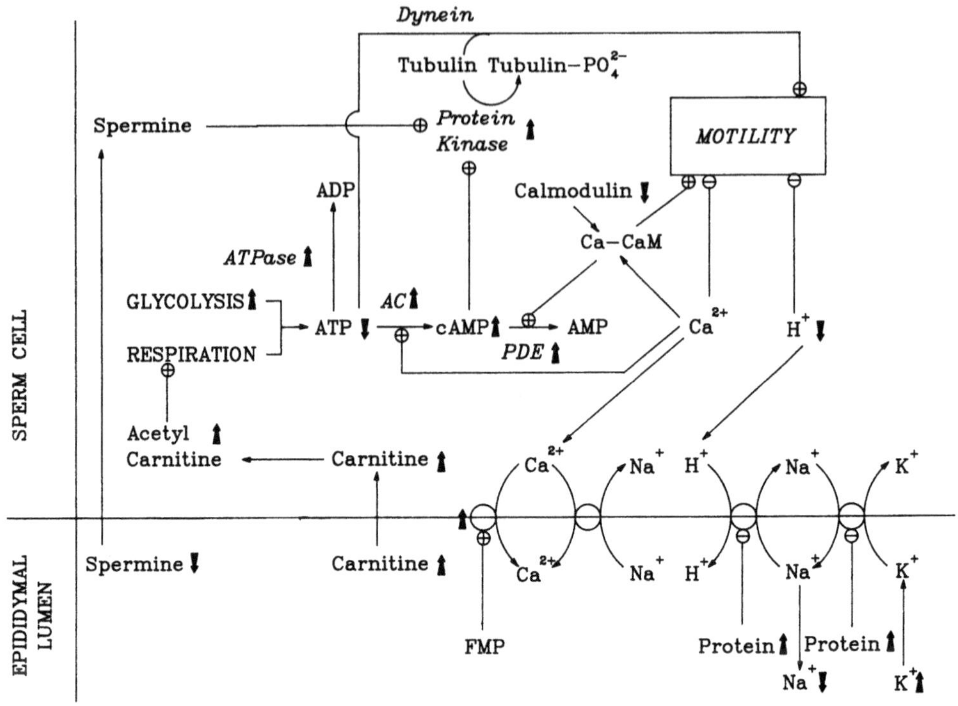

Fig. 2. Summary of possible development of maturation of motility in the epididymis. The potential for sperm motility increases because the activators of motility (cAMP, protein kinase, pH) increase in concentration, activity or value and the inhibitors (Ca^{2+}, calmodulin) decrease. Increased glycolysis, respiration stimulated by secreted carnitine and decreased ATPase activity provide increased ATP for motility or production of cAMP. PDE inhibitors in epididymal fluid promote cAMP production and its target, protein kinase, increases in activity. At the same time, plasma membrane permeability to Ca^{2+} decreases, possibly due to the increased Ca^{2+}-ATPase activity and stimulation by FMP, thus lowering intracellular Ca^{2+}; calmodulin levels also decrease as the sperm mature. The potential for motility is not realised in the epididymal canal because intracellular pH is sub-optimal owing to the acidification of the lumen, the resorption of the requisite counter ion Na^+ and the secretion of proteins that can lower the entry of Na^+ and K^+ into immature cells. Directions of arrows indicate whether factors increase (↑) or decrease (↓) on maturation. Plus or minus represent stimulation (⊕) or inhibition (⊖) of the process indicated

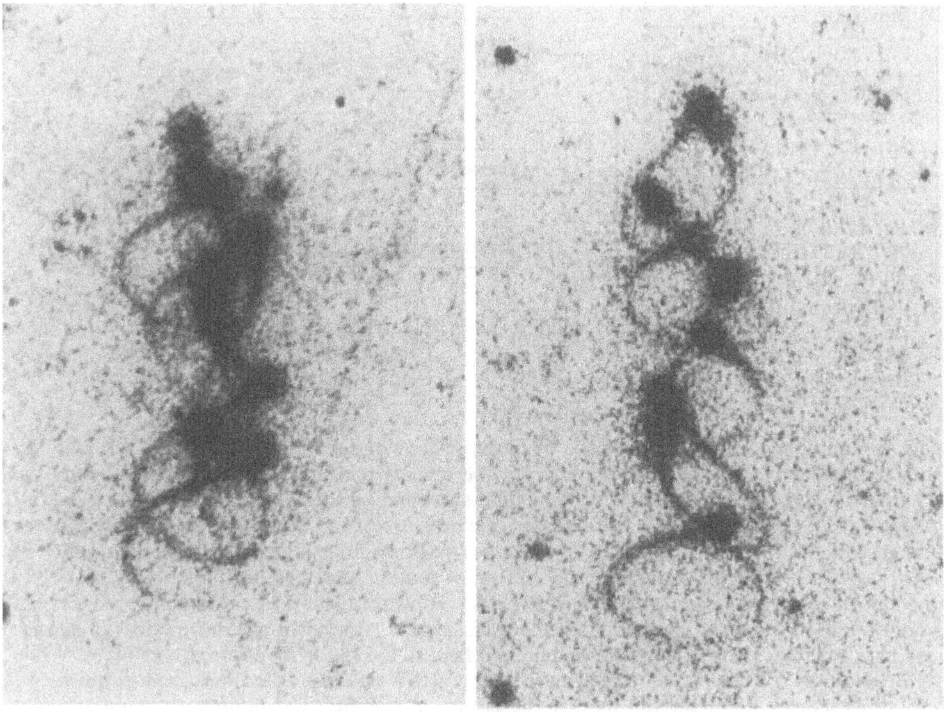

Fig. 3. Human spermatozoa undergoing hyperactivated motility. Illuminating spermatozoa with a stroboscope at 10 Hz during continuous exposure of photographic film for one second produces multiple exposure images of moving spermatozoa. The large bend angles and displacements of the sperm head from the axis of progression are characteristic of spermatozoa that are hyperactivated. × 2000

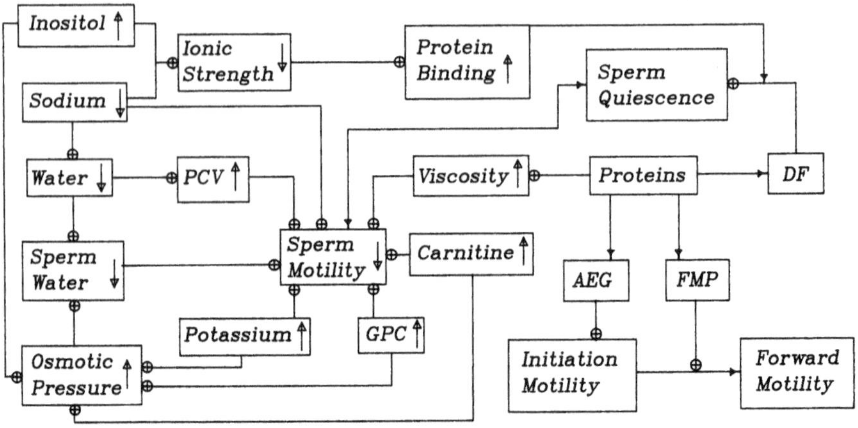

Fig. 4. Consequences for spermatozoa of the resorptive and secretory activities of the epididymal epithelium. The removal of water from the epididymal lumen concentrates not only the spermatozoa but other components of the fluid surrounding them. Together with the secretory activity of the epididymis this raises the osmotic pressure of the fluid and acts to dehydrate the sperm cells. Cellular dehydration and the close packing limits the expression of motility of spermatozoa in the epididymal lumen. Removal of sodium ions from the lumen not only drives the movement of water but together with the secretion of uncharged compounds, reduces the ionic strength of the fluid surrounding the spermatozoa which will prevent the loss of sperm-coating proteins secreted by the epithelium and held to the sperm by electrostatic forces. While decapacitation factors prevent capacitation within the male tract and keep sperm quiescent, other factors can stimulate and modify the nature of motility. Carnitine both stimulates and depresses sperm motility depending on concentrations while low sodium ion concentrations and high concentrations of potassium and glycerophosphocholine (GPC) act in concert to reduce sperm motility in-vitro. Directions of arrows indicate high (↑) or low (↓) concentrations or qualities in the epididymis or the direction of the effect expected on the process indicated. Plus or minus represent stimulation (⊕) or inhibition (⊖) of the quality in the indicated box

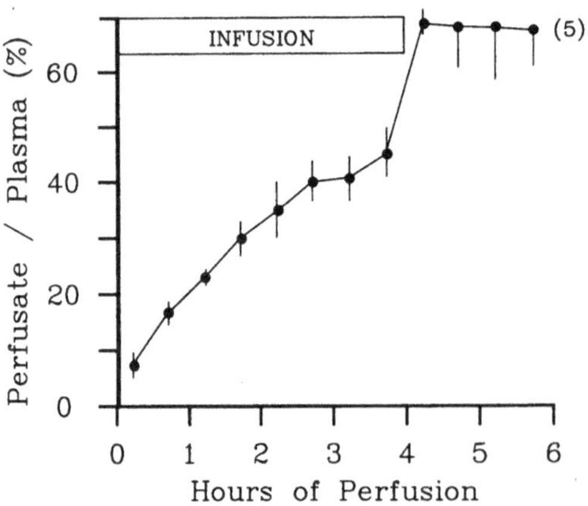

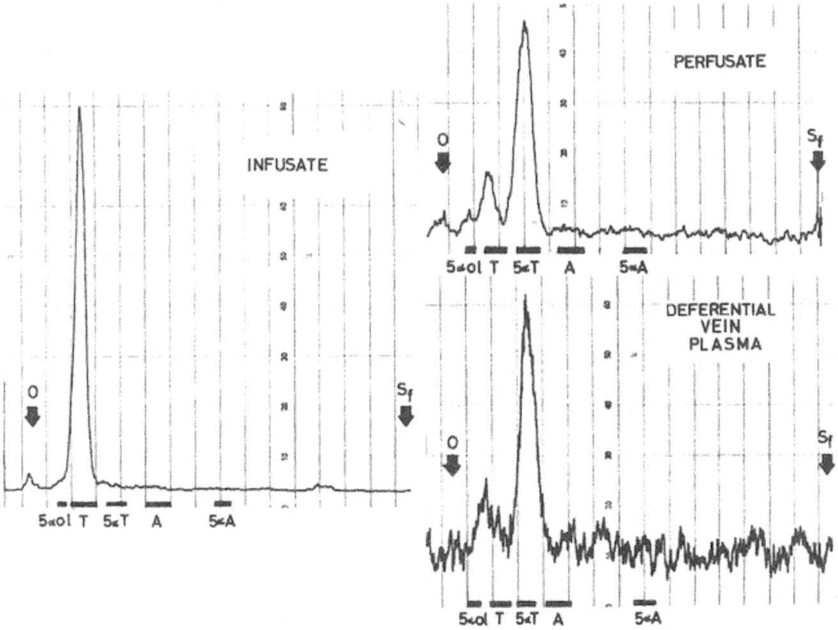

Fig. 6. Secretion of dihydrotestosterone by the rat distal cauda epididymidis perfused in-vivo. Radioactivity scans of thin-layer chromatographic separation of ^{3}H-steroids developed for 15 cm from origin "*O*" to the solvent front "S_f" with authentic unlabelled steroids: 5α-androstan-3α,17β-diol (*5αol*); testosterone (*T*); dihydrotesteosterone (*5αT*); androstenedione (*A*) and 5α-androstanedione (*5αA*). Ether extracts applied to the plates were from: "Infusate" – fluid containing (^{3}H)testosterone perfused through the lumen of the distal cauda epididymidis at 1 μl/min; "Perfusate" – pooled fluid recovered after passage through the epididymal lumen; "Deferential Vein Plasma" – from blood sampled towards the end of the perfusion from the veins draining this region. While testosterone was perfused through the lumen, dihydrotestosterone was the main steroid emerging from the other end and was also present in the venous effluent (Cooper TG, unpublished observations)

Fig. 5. Transport of ^{3}H-lysine to the lumen of the distal cauda epididymidis perfused in-vivo. ^{3}H-lysine in 0.9% saline was infused into the jugular vein of 3 rats during perfusion through the lumen of both distal cauda epididymides with solutions similar in composition to epididymal fluid. Radioactivity in 30-min collections of the outflowing perfusate was expressed as a percentage of that in blood plasma sampled mid-way through each collection period (*ordinate*) during (*horizonatal bar beneath axis*) and after a 4-h infusion period (*abscissa*). Radioactivity (98% of which was not precipitated with trichloracetic acid) rose steadily throughout the infusion and neared a plateau by three hours approaching 50% of the level in blood plasma. When blood levels of radioactivity fell after the end of the infusion, the ratio rose because there was a slower decline in the luminal activity (Cooper TG, unpublished observations)

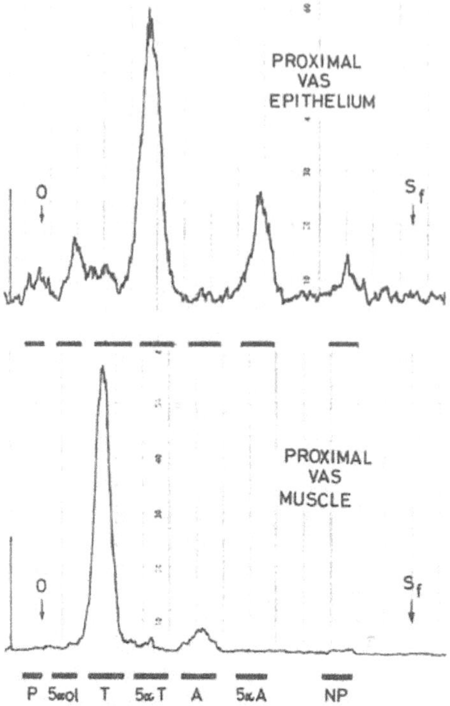

Fig. 7. Conversion of ^{3}H-testosterone to ^{3}H-dihydrotestosterone by epithelium stripped from the vas deferens of the rat. Radio-traces of thin-layer chromatograms. Ether-soluble activity in the medium from 3 h incubations of ^{3}H-testosterone with epithelium (*upper trace*) or muscle (*lower trace*) from the proximal vas deferens were run from the origin (*O*) to solvent front (S_f). The positions of unidentified polar (*P*) and non-polar (*NP*) steroids and standards 5α-androstane-3α,17β-diol (*5αol*), testosterone (*T*), 5α-dihydrotestosterone (*5αT*), androstenedione (*A*), 5α-androstanedione (*5αA*) are given beneath each trace (*bars*). Only the epithelial strips were capable of the 5α-reduction of testosterone, with the oxidation observed in muscle incubations also occurring in tissue free-controls (Cooper TG, unpublished observations)

Fig. 8. A summary of the possible interactions of epididymal secretion products with spermatozoa and the consequences for the fertilising ability of spermatozoa. A whole spectrum of compounds elaborated or accumulated by the epididymis interact with spermatozoa in different ways. Some coating proteins may be required for sperm-egg contact while others may have to be lost before this can occur. Smaller molecules may modify the sperm surface or promote the binding of larger molecules. The motility patterns required just after ejaculation differ from those required around the time of fertilisation so the sperm must be "programmed" by the epididymis, with delayed loss of decapacitation factors. Overall, the epididymis appears to promote the fertilising ability of immature spermatozoa and then take steps to prevent untimely expression of this capacity by adding inhibitory components that are lost later. Plus or minus represent stimulation (⊕) or inhibition (⊖) of the process indicated

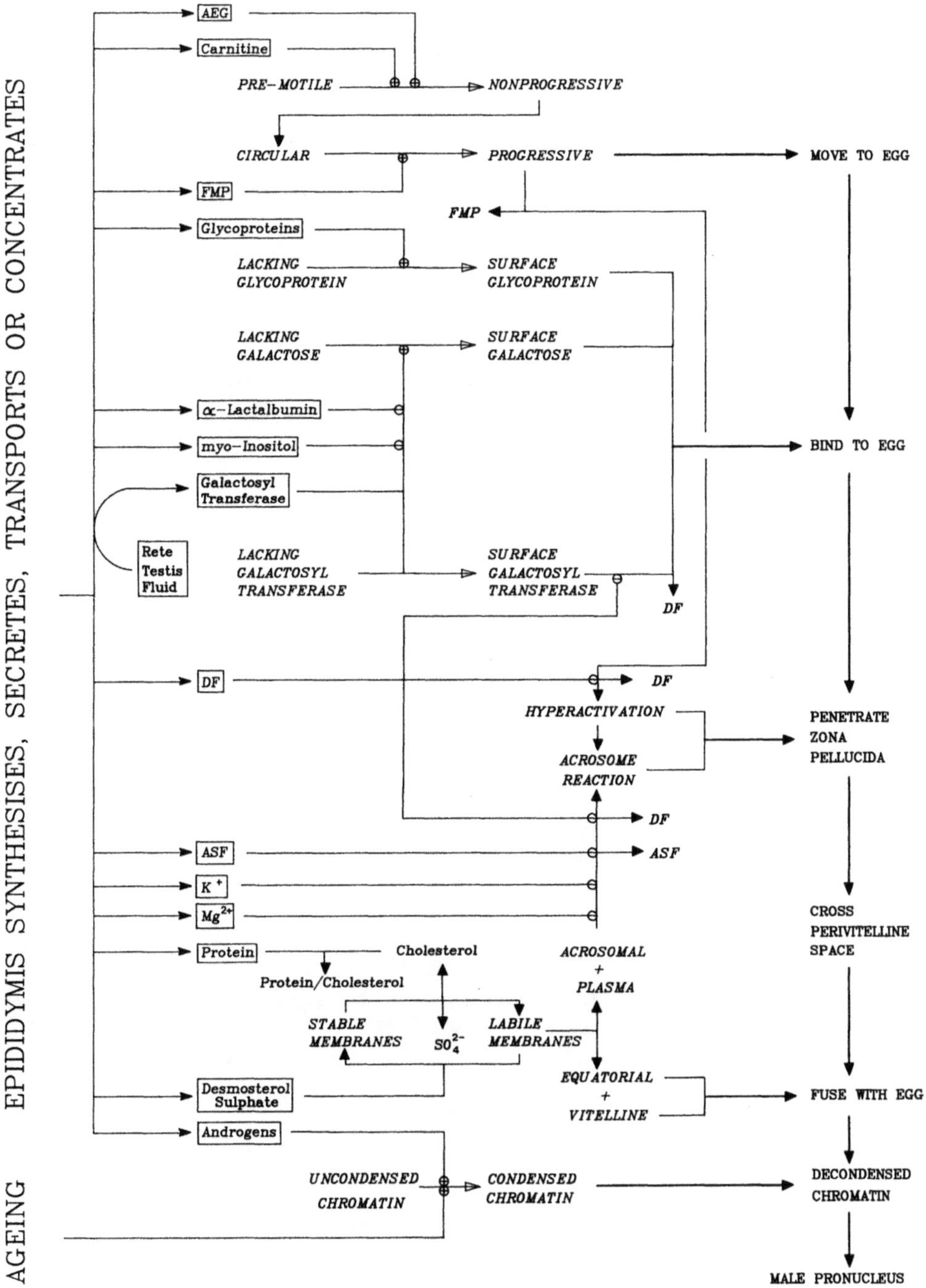
EPIDIDYMIS SYNTHESISES, SECRETES, TRANSPORTS OR CONCENTRATES
AGEING
AEG
Carnitine
PRE-MOTILE
NONPROGRESSIVE
CIRCULAR
PROGRESSIVE
MOVE TO EGG
FMP
FMP
Glycoproteins
LACKING GLYCOPROTEIN
SURFACE GLYCOPROTEIN
LACKING GALACTOSE
SURFACE GALACTOSE
α-Lactalbumin
myo-Inositol
BIND TO EGG
Galactosyl Transferase
Rete Testis Fluid
LACKING GALACTOSYL TRANSFERASE
SURFACE GALACTOSYL TRANSFERASE
DF
DF
DF
HYPERACTIVATION
ACROSOME REACTION
PENETRATE ZONA PELLUCIDA
DF
ASF
ASF
K$^+$
Mg^{2+}
CROSS PERIVITELLINE SPACE
Protein
Cholesterol
Protein/Cholesterol
ACROSOMAL + PLASMA
STABLE MEMBRANES
SO$_4^{2-}$
LABILE MEMBRANES
EQUATORIAL + VITELLINE
Desmosterol Sulphate
FUSE WITH EGG
Androgens
UNCONDENSED CHROMATIN
CONDENSED CHROMATIN
DECONDENSED CHROMATIN
MALE PRONUCLEUS

Subject Index

GPSR Compliance
The European Union's (EU) General Product Safety Regulation (GPSR) is a set of rules that requires consumer products to be safe and our obligations to ensure this.

If you have any concerns about our products, you can contact us on

ProductSafety@springernature.com

In case Publisher is established outside the EU, the EU authorized representative is:

Springer Nature Customer Service Center GmbH
Europaplatz 3
69115 Heidelberg, Germany

www.ingramcontent.com/pod-product-compliance
Lightning Source LLC
Chambersburg PA
CBHW061202240926
51846CB00006B/1092

* 9 7 8 3 6 4 2 7 1 4 7 3 3 *